D1062626

MASTERING IRON

MASTERING IRON

ANNE KELLY KNOWLES

Cartography by Chester Harvey

The Struggle to Modernize an American Industry, 1800–1868

The University of Chicago Press
Chicago and London

Anne Kelly Knowles is a historical geographer who teaches at Middlebury College, where she has been a member of the Department of Geography since 2002. She is the author of *Calvinists Incorporated: Welsh Immigrants on Ohio's Industrial Frontier,* also published by the University of Chicago Press, and the editor of *Placing History: How Maps, Spatial Data, and GIS Are Changing Historical Scholarship.*

The University of Chicago Press, Chicago 60637
The University of Chicago Press, Ltd., London

Printed in China

21 20 19 18 17 16 15 14 13   1 2 3 4 5

ISBN-13: 978-0-226-44859-6 (cloth)
ISBN-10: 0-226-44859-2 (cloth)
ISBN-13: 978-0-226-44861-9 (e-book)
ISBN-10: 0-226-44861-4 (e-book)

Library of Congress Cataloging-in-Publication Data

Knowles, Anne Kelly.
Mastering iron: the struggle to modernize an American industry, 1800–1868 / Anne Kelly Knowles; cartography by Chester Harvey.
pages; cm.
ISBN 978-0-226-44859-6 (cloth: alk. paper)—ISBN 0-226-44859-2 (cloth: alk. paper)—ISBN 978-0-226-44861-9 (e-book)—ISBN 0-226-44861-4 (e-book) 1. Iron industry and trade—United States—History. I. Title.
HD9515.K56 2013
338.4'76691097309034—dc23   2012006826

∞ This paper meets the requirements of ANSI/NISO Z39.48-1992 (Permanence of Paper).

In memory of my father,

Richard Hughes Morris (1919–2008)

CONTENTS

INTRODUCTION

Iron in America

Iron is the most plentiful element on earth. Harder and more durable than copper and much less costly than bronze, iron was for thousands of years the primary metal of agriculture, conquest, and industry throughout the world.[1] It is remarkably adaptable and useful: molten iron can be cast into almost any form, and well-made iron tools last for generations. The volcanic heat required to smelt iron from ore, and the craft skills required to shape hot iron into wares and weapons, historically bequeathed special status to ironworkers. In many societies, iron has been a potent symbol of individual and national strength, power, and prestige.[2]

In modern times, the apogee of iron's importance came in the late eighteenth and early nineteenth centuries. The rise of factory production and the construction of railroads in Europe and North America required an enormous increase in iron production. Historians have acknowledged the central significance of the iron industry to Great Britain's economy during this period. By 1800 Britain had become the world's dominant low-cost, high-volume producer of iron, a position it retained until nearly the end of the nineteenth century.[3] In British historical scholarship, the iron industry is the classic case that demonstrates the tremendous productivity gains achieved in the Industrial Revolution by substituting mineral fuel (coal) for organic fuel (wood and charcoal).[4] In United States historiography, however, the iron industry plays only a minor role, when it is mentioned at all, in accounts of America's dramatic change from a nation of farmers at the beginning of the nineteenth century to an industrial powerhouse by the end of the Civil War. We know far less about this crucial period of the industry's expansion and change than we do about the rise of the corporate steel industry in the late nineteenth and early twentieth centuries.

The scholarly literature on antebellum iron falls mainly into two types of studies. Histories of individual ironworks give a good sense of how iron companies operated and the nature of labor, particularly at small, family-owned works.[5] Economic studies focus on the significance of resource endowments for ironmaking and how price competition influenced the uneven regional and international development of the industry before the Civil War. Economic and labor histories have tended to treat the antebellum iron industry as the prelude to steel rather than as significant in its own right. In most cases they consider a single region, typically Pennsylvania, Virginia, or the Mid-Atlantic region, areas that also draw the most attention in works of national scope.[6] While these regions did account for most of the iron produced during the antebellum period, taking them as the main story has given us a rather skewed, incomplete understanding of the industry. No one has compared development across all the country's iron regions, nor has any study taken a comprehensive approach to the many factors that influenced the industry's development—namely, technological change, labor, management, capital, transportation, and the physical and social contexts in which this heaviest of heavy industries was embedded.

Why hasn't the iron industry been a central part of the narrative of early nineteenth-century American industrialization? That narrative has for some decades mainly told the stories of factory-based industries, particularly those producing textiles, railroads, machine tools, and small arms.[7] Each has come to represent powerful themes in the historiography of antebellum America. Textile mills are the prime example of the substitution of machines for labor, a step in industrialization that was particularly important in the United States because of the relatively high cost of American labor. Textile company owners exemplify the shift of mercantile wealth to domestic manufactures. The many improvements that American mechanics made to British-designed mill machinery fit the image of antebellum America as a nation of mechanics and inventors. Labor also plays a prominent role in the history of the textile industry, whether as a newly feminized workforce in corporate New England factories or as the more mixed labor working for Philadelphia proprietors. Railroads have long fascinated historians because of their heavy capital requirements, the immense wealth some railroad companies generated, and their political clout. They transformed transportation, commerce, and settlement and symbolized the exuberant potential for unifying a vast nation. The machine tool and small arms industries, while less prominent, stand out as examples of American ingenuity and mechanical acumen, of solving practical problems to increase industrial efficiency.

The iron industry differed in many ways from these well-known industries. Iron was the last major industry to be mechanized; by 1870 the process had scarcely begun. Technological innovations that increased productivity did not replace labor—quite the contrary; they created new divisions of labor that increased the number of skilled positions and required more workers. In contrast to the workforce at many textile mills, all ironworkers were male. The demand for skill in the iron industry put a premium on experienced immigrant labor, while the nature of ironmaking technologies created bottlenecks that gave men in certain skilled occupations exceptional control over production. These circumstances contributed to the late development of unions in the American iron industry; the first ironworkers' union was organized in 1858. In relation to other heavy industries, the greatest difference in ironmaking was its sensitivity to the quality of resources, particularly the chemical content of coal and iron ore, and the cost of transporting raw materials and finished goods. Geography and geology influenced the American iron industry's development in fundamental ways, from where ironworks were located to what they produced, how long they survived, what technology they employed, and the size and characteristics of their workforces.

Recognizing the differences between iron and other leading American industries makes the story of industrialization more complicated. There was no one dominant American model of ironmaking before or during the Civil War, just as there was no one model for textile manufacturing, as Philip Scranton showed in his study of textile firms in greater Philadelphia.[8] Between 1800 and 1868, US iron companies used a remarkable array of methods for smelting and refining iron; yet to a greater degree than in perhaps any other major industry, most of those methods were developed outside the United States. Until well after the Civil War, British iron companies and British engineers and artisans developed most of the major improvements in productivity and efficiency for large-scale iron production. When American firms tried to replicate British methods for making iron on a large scale—the main focus of this book—they encountered daunting problems that were, at bottom, geographical or geological—problems that British producers generally did not face because in Britain metallurgical coal, rich iron ore, and limestone deposits were close to one another and relatively near to navigable rivers, coasts, and centers of population. In every European country that tried to compete with Britain, developing a modern iron industry was a prolonged undertaking that produced many more hybrids than replicas, and sometimes as many failures as successes.[9] The same was true in most iron regions of the United States.

Another obstacle preventing iron's inclusion in our national narrative of industrialization may be the industry's great variety of forms, for its multifarious objects and landscapes resist symbolization. In *America as Second Creation*, historian David E. Nye observed, "For those who arrived after Columbus, neither ancient sacred places nor local stories of origin were possible. Instead, the new Americans constructed stories of self-creation in which mastery of particular technologies played a central role." Nye goes on to explain that these stories fixed on iconic objects—the ax, water-powered mills, canals, and railroads—each of which was associated with a characteristic landscape and phase of American economic development. The ax, for example, was the main tool for clearing forests and building log cabins for self-sufficient agriculture on the frontier. Mills anchored villages and towns that housed the country's first industrial working class and created the fortunes of its first industrial capitalist elite.[10]

In each of the objects Nye discusses, iron was the common ingredient, hiding in plain sight. Tench Coxe, an early proponent of American manufacturing, declared in an 1814 report to the US Senate, "Not a building for man, for cattle, nor for the safe keeping of produce or merchandise—not a plough, a mill, a loom, a wheel, a spindle, a carding machine, a fire arm, a sword, a wagon, or a ship, can be provided, without the manufactures of the iron branch."[11] Steel-edged iron axes and scythes chopped firewood and harvested the grain grown on prairies opened to agriculture by the force of hardened iron plowshares. Stirrups, bits, and horseshoes were made of iron, as were the metal rims of wagon wheels. All manner of hand tools for digging in the ground had iron heads, from the spades and hoes that slaves used to ditch and tend rice fields in South Carolina to the picks and shovels Irish laborers wielded to build the Erie Canal. Most hard-wearing parts in antebellum machines were made of iron or steel—from the wire teeth in Eli Whitney's cotton gin and the gears, spindles, and fastenings on power looms to the firing mechanisms of rifles and pistols—as were the precision tools and files used to fit the interchangeable parts together.[12] Iron was essential to the devices that revolutionized American transportation and warfare in the nineteenth century, including steam engines, suspension bridges hung from wire cables, heavy cannons, and the iron cladding on Civil War battleships.[13] A plethora of domestic wares were made of iron, from cast-iron pots, pans, and cookstoves to the nails that held balloon frame houses together. Iron was part of the consumer revolution as well, as a material that permitted low-cost, large-scale production of consumer goods. In 1825, for example, Boston piano maker Alpheus Babcock patented a cast-iron piano frame that was rugged enough to survive long-distance shipment to parlors across the country. In Worcester, Massachusetts, wire manufacturer Ichabod Washburn produced finely drawn piano wire to string American-made instruments, as well as tempered steel wire for hoop skirts, a cheap alternative to whalebone.[14]

In short, iron was ubiquitous in American life from about 1800 through the Civil War, an era that could be called America's iron age. As the list above suggests, however, iron was present in too many things for its symbolic importance to cohere in one cultural talisman. It was also made in too many kinds of places for the industry to lodge in a single iconic landscape. Ironworks ranged from single-hearth bloomeries about the size of a blacksmith's shop to complex industrial sites that covered acres of ground. By the turn of the twentieth century, steel plants in Pittsburgh, Bethlehem, Johnstown, and South Chicago had become icons of heavy industry. Their tall, cylindrical blast furnaces, vast rolling mills, and palls of smoke symbolized the nation's economic muscle as well as the working class and class conflict. No such places encapsulated the meanings of the previous era of iron production. In this regard, the antebellum iron industry confirms Walter Licht's interpretation that the nineteenth century was characterized more by the great diversity of its industrial processes, products, communities, and social relations than by any single kind of production system.[15]

Mastering Iron focuses on the sectors of the iron industry that formed the foundation of iron production and heavy manufacturing: smelting at blast furnaces and the large-scale refining done at rolling mills, forges, and foundries, many of which produced both finished products (such as sheet iron, rails, and nails) and bar iron, rods, and other semifinished goods. This sectoral focus emerged from my initial research question: Why did it take most of the nineteenth century for the United States to match the British iron industry's scale of production? Knowing that coal-fired furnaces and the combined refining methods of puddling and rolling had catapulted the British industry's dramatic growth at the end of the eighteenth century, I set out to discover how those technologies were transferred to the United States. My hope of tracing that transfer through the lives of individual immigrant ironworkers fizzled because I found little record of workers' movements. As other scholars have noted, iron artisans were among the most peripatetic of skilled industrial workers.[16] In the meantime, however, in J. Peter Lesley's directory of the American iron industry, *The Iron Manufacturer's Guide* (1859), I found a source that held the keys to understanding the industry's technological change and geographical expansion. The changes Lesley documented convinced me that the period he covered well, from 1800 to 1858, merited close examination. Case studies of individual iron companies led me to extend the period of study to 1868, in order to include the role of the iron industry in the Civil War, the creation of the first American ironworkers' unions, and the arrival of the Bessemer converter, another imported technology that signaled a fundamental restructuring and rescaling of iron and steel production in the United States.

One purpose of this book is to fill a significant gap in our understanding of American industrialization. Another is to explain why some US iron companies swiftly implemented British technologies while others struggled or failed in the attempt, or never ventured to try large-scale production. I discovered the range of technologies employed at antebellum ironworks by mapping the data in Lesley's *Guide*, using the methods of historical geographic information systems (HGIS) (see the appendix). I could not have determined the regional patterns of technology and production, or the temporal dynamics of the iron industry, without the exploratory mapping that GIS makes possible. Visual evidence and visual argument are also important parts of my method. The maps, fine art, and historical photographs I have included are not illustrations; they are sources of evidence and modes of representation that are essential for many of my arguments. Taken together, my use of GIS and visual analysis make this book a significant contribution to the emerging fields of historical GIS, spatial history, and the digital humanities.[17] At the same time, this book is deeply anchored in archival research, which was necessary to understand the ideas, individuals, and historical circumstances that propelled and inhibited modernization.

While piecing together the course of technology transfer is a central concern of my research, I take a rather different approach than have most historians of technology. The transfer of technology always involves displacement, as people, ideas, machines, and ways of working are moved from their places of origin to new places. As a geographer, I am particularly interested in the places that anchor each end of the transfer. How different were those places? How well did the individuals who were attempting to effect the transfer of a given technology understand those differences? To what extent could the physical and social conditions in the new location be changed to fit the requirements of the technology's original context? If conditions could not be altered much, could the technology be changed without losing its efficacy?

These questions give a new twist to Bruno Latour's elegant formulation of scientific change as a process of accumulating information from remote places in centers of calculation.[18] In the case of Americans' efforts to modernize the domestic iron industry, the information they desired was lodged in British ironworks. The classic relationship for the scientific accumulation of knowledge was reversed, since Americans wanted to extract information from existing centers of accumulated knowledge, experience, and practice. What Latour calls "immutable and combinable mobiles"—the pieces of information that cumulatively lead to new knowledge and even more valuable generalizations and abstractions—were in this case only partially available to American entrepreneurs and the workers laboring at their behest. Unless they spent considerable time in the original places of production they wished to replicate, Americans could accumulate only partial knowledge. As events would demonstrate, the shortcut of hiring experienced

British ironworkers was inadequate, because the techniques used to smelt iron with coal and refine it through puddling and rolling also required physical conditions and social relations that were much more difficult to replicate than the machines, buildings, supply chains, and visible skills that those techniques most obviously required. My case studies bear out Latour's insight that, wherever a new scientific approach (here, industrial technology) fails, some part of its supporting network "has been punctured." What I hope to add to this observation is the crucial significance of place, in all its geographical dimensions, to the efficacy of networks in the service of industrial production.

As this declaration suggests, my approach to the iron industry is predominantly empirical, not theoretical. I strive to reconstruct and then understand the concrete places where iron was made, from the very specific landscapes of production created by individual firms, their workers and managers, to the industrial valleys, transportation systems, and regions where particular modes of production took hold. Latour's explication of network analysis has influenced my interpretation chiefly because his theoretical framework is grounded in a materialist view of history. Similarly, David Harvey's materialist view of industrial capitalism inspired my interest in the shifting balance of power in labor relations and the tension between capital's contradictory requirements for labor mobility and resident labor supply. David Meyer's notion of networked machinists also helped me conceptualize the very uneven results of technology transfer in the iron industry.[19] Implementing British technologies on American soil was as much a social process actualized through networks of variously allied actors as it was a matter of grappling with new physical circumstances. Both the social and physical aspects of modernization involved struggle—hence the subtitle of this book. Company owners, managers, and skilled workers struggled to various extents to implement technologies that had been developed in very different geographical contexts in Britain. They also struggled to create and maintain viable work practices in the face of conflicting cultural expectations, labor scarcity, and differing degrees and kinds of technical understanding.

One last thing that needs explaining is the prominence of Welsh immigrant ironworkers and Welsh iron companies in this book. I knew from previous research that Welsh artisans were an important minority of immigrants at American ironworks.[20] As I researched company records for this book, Welsh puddlers and rollers cropped up at almost every rolling mill. Although the Welsh did not outnumber English iron artisans in the United States, they brought a distinctive ethnic-occupational identity that proved both troublesome and advantageous for their American employers. The presence of Welsh immigrants at so many American ironworks also reflects the prominence of leading Welsh iron companies in the imagination of American entrepreneurs. Time and again in manuscript correspondence and printed reports, I found references to Dowlais and Cyfarthfa, two of the largest iron companies in South Wales. Those firms epitomized modern ironmaking to contemporaries because they were among Britain's first fully integrated ore-to-rail operations. Understanding the geographical circumstances of these prototypical firms is a key part of my analysis of failure and success in the American industry.

Reflecting these various threads and methods, the book is a hybrid of history and geography, thematic exposition and chronological narrative. Chapter 1 explains the iron industry's evolution by retracing the footsteps of J. Peter Lesley and two young men who helped him assess the state of the industry in 1855–58. Lesley published the results of their survey as *The Iron Manufacturer's Guide,* a massive compendium that captured the industry's striking variation and major trends from the end of the eighteenth century to the eve of the Civil War. The company-level information that Lesley and his assistants collected provides the basis for mapping the development of the American industry in unprecedented detail, with particular insights into the spatial and temporal patterns of construction, production, the abandonment of facilities, and technological change. The chapter concludes with a comprehensive description of the country's major ironmaking districts and regions. The analysis is enriched by frank commentary contained in letters the three men exchanged during their survey. This chapter provides crucial geographical and historical context for everything that follows.

Chapter 2 explores the social aspects of the iron industry. Historians have long known that skilled ironworkers were essential to the industry, but we have lacked a comprehensive understanding of workplace environments and the culture of ironmaking. The chapter begins with the perception of ironworks as extreme places, as portrayed by outside observers and by workers themselves. I then pull back to look at ironmaking communities more broadly. Ironmaking in large industrial towns most nearly matched the hellish images of the industry in literature and art, but there were also iron villages and hamlets, frontier iron towns, and iron plantations. Each type of community developed around particular technologies, scales of production, and social relations. Understanding how these interrelated factors formed distinctive industrial places helps explain, among other things, why some kinds of ironmaking communities were more likely than others to experience labor conflict. The chapter closes with a consideration of the wide range of incentives and punishment that managers used to discipline labor at American ironworks.

Chapters 3 and 4 present case studies of four United States companies that variously failed or succeeded in replicating British smelting technologies and large rolling mill operations. Chapter 3 begins with Farrandsville Furnace, one of the first attempts to make iron with coke in direct imitation of the British model. This disastrous foray into modern ironmaking in central Pennsylvania illustrates problems that beset many pioneering iron ventures on the industrial frontier, including managers' ignorance of imported techniques and machinery, the poor quality of local resources, cultural and personal conflicts between managers and workers, the unreliability of waterpower and transport, the distance to markets, and the impact of national economic swings, boredom, drunkenness, and bad luck. The somewhat better outcome at Lonaconing Furnace in western Maryland points up the importance of resource quality, reliable transportation, and more equitable labor relations. Chapter 4 looks at two examples of very successful companies in the Mid-Atlantic region, where the British model was most swiftly and fully replicated. The Lehigh Crane Iron Company was the first American firm to prove the viability of making iron with anthracite, and for years it was one of the largest producers in the United States. The managers of the Trenton Iron Company, one of the country's first integrated operations, showed exceptional entrepreneurial drive. In both cases, collegial labor-management relations, locations along established transportation lines, relative proximity to both excellent raw materials and urban markets, and managers' creative response to economic crisis all contributed to long-term success.

The patterns of regional development in the antebellum iron industry set the stage for chapter 5, which examines the role of iron in the Civil War. The volume of production in the North gave the Union an enormous initial advantage in manufacturing war materiel, and Northern companies were quick to exploit government contracts to accelerate their growth. Confederate chief of ordnance Josiah Gorgas recognized the South's perilous position and moved quickly to expand manufacturing capacity. He was less successful in halting the loss of skilled workers. Wartime experiences at Alabama's Shelby Iron Works illustrate the South's difficulties and show how important it was for Southern companies to retain a skilled workforce. Labor scarcity in both the South and the North inflated wages and gave artisans more leverage with management. The drive to increase output while achieving the precision demanded by the military, however, created strong incentives for firms to wrest control of production from workers' hands. This sparked labor disputes at Northern ironworks that presaged more frequent and violent conflicts after the war.

The conclusion reflects on the iron industry's development from 1800 to 1868. While this period saw tremendous geographical expansion and the adoption of technological innovations that greatly increased production, those changes by no means eradicated older forms of production. The hybridity that characterized American ironmaking in the 1830s and 1840s intensified with time, culminating in wartime industries that used virtually every known kind of ironmaking technology. Ironmaster Abram Hewitt, who led the American delegation to the Paris Universal Exhibition in 1867, was frustrated to see that the US iron industry still could not match the impressive achievements of larger, more advanced European works. The Krupp works in Germany and Le Creusôt in France joined the great British ironworks as the envy of American entrepreneurs. Fundamental changes were afoot, however, that would shift the balance of global production to the United States, give managers the upper hand over labor, and create new processes and landscapes of production that riveted world attention in the age of mass-produced steel.

'Tis astonishing how little the owners & managers of these "one-horse" affairs know about their business.

JOSEPH LESLEY, May 27, 1857

Mapping the Iron Industry 1

In the middle of the nineteenth century, the British iron industry experienced a series of booms and crises. Industrialists greatly expanded ironworks' capacity in the 1830s and early 1840s in response to domestic demand for rails and manufactured goods, then companies suffered economic arrest when overproduction glutted the British market. When depression hit after the British railway boom of 1846–47, iron companies cut skilled workers' wages by up to one-third. In search of other markets, British companies shipped boatloads of Welsh rails and Scottish pig iron to the United States. Shocked to see the American market inundated with cheap British iron, ironmasters held a summit meeting in Philadelphia in December 1849. They launched a lobbying effort to raise tariffs and gathered information from ironworks across the Northeast to document the damage that cheap imports did to domestic producers. Their report helped persuade Congress to raise tariffs, as it had done in 1842, to shield the country's struggling iron companies from foreign competition.

The tariffs stimulated domestic iron production, but by late 1854 supply was exceeding demand. American railroad construction flagged as one project after another encountered cost overruns or simply ran out of money. Continuing overcapacity in Britain once again resulted in large shipments of cheap British rails and pig iron, which further depressed the American market. In late 1855 and early 1856, blast furnaces in many states ceased production, and some of the country's biggest rolling mills ground to a halt. More ironworks closed the following year, as the nation's growing economic malaise turned into a run on banks that became known as the Panic of 1857.[1]

In the early days of this developing crisis, owners of leading East Coast ironworks formed the American Iron Association. They intended to lobby Congress again to increase tariffs, but they also had a broader agenda. They wanted to cure the chronic problems that generally made American iron more expensive than British iron and of poorer quality than Swedish, Norwegian, and Russian iron. The founding members of the association believed that American companies should emulate their British competitors by promoting modern methods of mass production and greater cooperation among ironmasters. In addition to urging association members to embrace these priorities, the preamble to the organization's constitution encouraged ironmasters to recognize the need for more thorough geological surveys and for schools to train young men in the profession:

> WHEREAS, The manufacture of iron, in its various branches, has acquired an importance in this country second only to the great agricultural interest; and whereas, its more rapid and economical development has been retarded by want of unity of action and free intercommunication of opinions and experiences among those interested; and whereas, we believe great advantages are best obtained by united action, we therefore deem it important to form an association in this city, to be called the AMERICAN IRON ASSOCIATION.
>
> The general objects of this Association shall be to procure regularly the statistics of the trade, both at home and abroad. To provide for the mutual interchange of information and experience, both scientific and practical. To collect and preserve all works relating to iron, and to form a complete cabinet of ores, limestones, and coals. To encourage the formation of such schools as are designed to give the young ironmaster a proper and thorough scientific training, preparatory to engaging in practical operations. And, generally, to take all proper measures for advancing the interests of the trade in all its branches.[2]

The "want of unity of action and free intercommunication" referred in part to the gulf between East Coast "iron men," whose large ironworks competed most directly with foreign imports, and the much greater number of ironmasters who ran small operations. Owners such as Henry Burden of the Troy Iron Works and Cyrus Alger of the South Boston Iron Works had vigorously lobbied Congress for tariff protection against European imports in 1842 and again in 1849–50, and protested the downward adjustment of tariffs in 1846. They resented the lack of support for their position from the industry as a whole. As users of the latest iron-making technology, they had a low opinion of the majority of ironmasters who persisted in using old-fashioned methods. It was high time for American iron to become a modern industry.

These ideas also reflected the convictions of the man who wrote the association's constitution, who happened to be one of the few members who was not an ironmaster. J. Peter Lesley (1819–1903), the first secretary of the American Iron Association, was a "topographical geologist" who made his living chiefly by locating and mapping mineral deposits, knowledge that he summarized in 1856 in his *Manual of Coal and Its Topography.* Lesley was editor and virtually the sole author of the *Bulletin of the American Iron Association* from its inception in 1856 through 1859. He had learned geology and surveying in 1837–41 while working as a young field assistant with the First Geological Survey of Pennsylvania under the direction of Henry Darwin Rogers. Lesley left the Survey to attend Princeton Theological Seminary, thinking he would become a Presbyterian minister, but within a few years he returned to work with Rogers as a cartographer. Lesley believed that modern industry required scientific knowledge that only a thorough understanding of mineral geology could provide. In articulating the goals and purpose of the American Iron Association, it was Lesley who called for scientific exchange, who saw the value of amassing a library of works related to iron and storing collections of mineral samples and fossils that would form the basis for scientific classification and the development of geological theory. It was also he who recommended industrial education, several years before he was appointed as the first professor of mining at the University of Pennsylvania.[3]

Peter, as he was known to family and friends, was a tall, thin man with tremendous nervous energy. He was often ill and suffered several collapses over the course of his life, each one brought on by intense exertion, usually on several fronts at once.[4] His enthusiasm and his highly systematic mind worked together to conceive of almost every project on a grand scale. His approach to mapping is typical. While working for Rogers in 1851–52, Peter made two very large maps of areas he surveyed in the anthracite coal region. One "great map" was of the Shamokin, Pennsylvania, coal beds. The second was a topographical map measuring twenty feet by thirty feet showing anthracite deposits and underground mines at Pottsville, Pennsylvania.[5] Giant maps became something of a specialty for Lesley. In 1853–54, he made a huge map of the holdings of the Pennsylvania Railroad as part of his survey of the company's route across the Alleghenies to Pittsburgh. Eight years later he produced a map measuring fifteen feet square for T. E. Blackwell, managing director of Canada's Grand Trunk Railway, which showed all US ironworks on a base map of Appalachian topography.[6] In later years, at the height of his career, Lesley became state geologist and directed the Second Geological

Survey of Pennsylvania from 1875 to 1889. The published reports of the Second Survey ran to 120 volumes of text and atlases. This staggering compilation of geological knowledge eclipsed all other nineteenth-century state geological surveys in scope, content, and cartographic achievement.[7]

When Lesley accepted the position of secretary of the American Iron Association in September 1856, he told his wife, Susan, that the job would oblige him "to travel round a little," but that it would leave him plenty of time to continue his remunerative survey work.[8] He would travel for the association to drum up members and gather statistics on ironworks, which he had begun doing in January 1856.[9] Statistics had been part of government and trade association accounts of industry since the late eighteenth century. A report prepared by secretary of the treasury Louis McLane in 1832 included local and statewide summaries of manufacturing as well as detailed tables of industrial production, capital invested, labor, and wages. McLane's report covered only eleven states in the Northeast and Mid-Atlantic region, however, and like Tench Coxe's 1814 report on manufactures, the amount and kind of information on individual works varied by state and by the willingness of individual ironmasters to answer the survey's questions.[10] The 1850 US manufacturing census captured more information and was more systematic. It recorded the amount of capital invested in ironworks, the amount and cost of materials, labor, and machinery, and the value of products produced. But the census omitted many firms, and published census volumes provided only aggregate statistics at the county or state level. Books and pamphlets published in Europe and the United States offered descriptive accounts that combined histories of the iron trade with travelogue-style commentary and snippets of statistics from the records of individual ironworks, trade publications, and census reports. They typically discussed only the most modern production methods and the largest works, such as Cyfarthfa and Dowlais in Wales and Fourchambault in France. Their main subject was the trade, whose pulse was recorded in the volume of imports and exports, profits relative to input costs, and most important, the fluctuating price of iron.[11]

Lesley's notion of trade statistics differed from his predecessors' approach in character and scope. He saw the iron industry as a collection of types to be cataloged, much like the fossils and ore samples he collected on geological expeditions. As a man of science, Lesley wanted to count everything that could be counted. If he was going to "collect data," as he put it in his journal in 1857, the collection should be as complete as possible.[12] It should also be put to instructive use, offering models of best practice to the many ironmasters who had not yet adopted modern techniques. To comprehend its present difficulties, the industry needed to know which ironworks had failed or were struggling and why. Entrepreneurs needed to know whom to contact for advice on implementing new methods. Potential investors could benefit from learning what kind and quality of products firms produced, their current capital investment, and whether a company was a steady or a fitful producer.

With all these goals in mind, Lesley set out to document the entire industry, from Maine to Alabama, including the most progressive ironworks and the most backward, those currently in operation and those that had gone out of business. To further the association's aim to promote better communication between ironmasters, Lesley would include the names of owners, managers, and superintendents and provide each company's mailing address. The latter is the most unusual piece of information included in the survey. Lesley specified each ironworks' location as precisely as he could—county and state for obscure, long-abandoned works; nearest post office or distance along a river, stream, or turnpike for rural companies; street address for works in major cities. In addition to wanting to encourage communication within the industry, Lesley planned from the outset to map his data. Geographical location was essential to his vision for the project.

In sum, Lesley saw the survey as an instrument for hastening the adoption of modern production methods throughout the industry. He planned to publish installments on each iron-producing region in the *Bulletin of the American Iron Association*. The information thus circulated to association members would advance scientific knowledge, facilitate communication, and promote the trade.

The first difficulty Lesley encountered in carrying out his grand plan was the geographical extent of the area he intended to canvass. The country's approximately two thousand blast furnaces, rolling mills, forges, and bloomeries, not to mention the innumerable foundries, machine shops, tool manufacturers, and smithies, lay scattered over an immense territory. Lesley knew some parts of the East very well. He could have mapped much of Pennsylvania from memory, having surveyed large swaths of the state and visited its "back valleys" on horseback as a colporteur delivering Presbyterian religious tracts in the 1840s. He had surveyed parts of Appalachian Kentucky and Tennessee and had probably traveled in northern Virginia and western Maryland. He spent much of his life in the Greater Philadelphia region, and he knew Massachusetts from frequent visits to relatives in Boston and Northampton and his years as a minister in Milton.[13] In each of those areas, he had at least passing familiarity with ironmasters he could depend on to provide information. But Lesley was acutely aware of the gaps in his mental map of the iron industry and was determined to fill them.[14] He originally hoped to gather most of the information he required by sending out "circulars" (paper survey forms), but he soon discovered the problems inherent in that method of data collection. Some companies never answered his written requests. One respondent sourly replied, "Go to Hell." As the Panic of 1857 took hold, Lesley noted, "the tone of the replies" to his circulars "was almost without exception angry and desponding."[15]

He decided to visit as many ironworks as possible in person. With so much country to cover, Peter called on two assistants to help him in the field. Benjamin Smith Lyman (1835–1920) was his nephew by marriage. Peter was very fond of Ben and rejoiced at every meeting with the young man. The two shared many interests, from geology to the transcendentalist philosophy of Ralph Waldo Emerson and Henry David Thoreau, whom both had met. Well before Ben joined his uncle on the iron survey, Peter had become a committed Unitarian, a transition that disturbed his Scottish Presbyterian family but cemented his intellectual kinship with Ben and other members of his wife's family.[16] When Ben set off to survey southern ironworks for his uncle, he was just twenty-one, a bright fellow with a classical education from Phillips Exeter Academy and Harvard University. He had shown promise as a "common hand" working with Peter during the summer of 1856 on a survey of coal lands around Broad Top City in Huntingdon County, Pennsylvania. "Ben Lyman delights me," Peter wrote in July of that year; ". . . he is remarkably quick in mind."[17] The iron survey would be Ben's apprenticeship in economic geology. He later went to Europe for advanced study in geology as Peter had done, after which his uncle helped him get work on a series of US surveys. Ben's greatest achievement as a geologist was leading the first geological surveys in Japan in 1872–79. In 1887 he rejoined Peter in Philadelphia as assistant geologist of Pennsylvania, serving once again as his uncle's right-hand man for the Second Geological Survey.[18]

Lesley's other assistant was his youngest brother, Joseph Lesley Jr. (1831–89), named after their uncle Joseph, a mining engineer and ardent abolitionist. Eleven years Peter's junior, Joseph was a studious, shy young man. He was very attached to his older brother, for Peter had taken him under his wing after their mother died when Joe was a baby.[19] The two brothers shared many talents and affinities, including their graphic abilities. All the Lesley children had been triply schooled—at fine Philadelphia institutions, by private tutors, and at the family table, where Peter Lesley Sr., a carpenter by trade,

> taught them at an early age to draw from objects. If nothing better offered, he would cut the loaf of bread into shapes, and make them copy that. Then, if they went anywhere, they were expected to tell in detail about what they saw, and he corrected and explained. He taught them the rudiments of architecture, making them observe and draw any architectural detail which he or they thought worthy. He was himself an excellent draughtsman, and usually made use of a pencil in explaining things to his children. The children in turn soon caught the same habit.[20]

Mary Lesley Ames, Peter's daughter, recorded this anecdote from her father's childhood. She went on to say, "In later life I seldom saw my father or my uncles talking together without sooner or later seeing the pencil and paper appear, and any mooted point was made clear by a few strokes." Peter wrote of this remarkable upbringing that what his father taught him "was worth more, in a strictly scientific sense, than what we received from all other sources put together; for it laid a deep and broad foundation for original investigation . . . and started us on our careers equipped for both seeing, thinking, and describing, what we felt to be useful and beautiful, as what we believed to be true."[21] His own training found useful outlets in cartography, diagramming geological formations, and topographical drawing. His younger daughter, Margaret, became an accomplished artist, as shown in a portrait of her father during his work on the Second Geological Survey (fig. 1).[22] Joseph proved to be an excellent cartographer, surpassing even his older brother in that art. The iron survey would be his apprenticeship as well. He conducted geological surveys after several seasons in the field tracking down ironworks. He later worked for the Pennsylvania Railroad and served for many years as that company's corporate secretary.[23]

Figure 1. J. Peter Lesley, painted by his daughter Margaret Lesley Bush-Brown in 1884. Reproduced by permission of the American Philosophical Society.

Peter, Ben, and Joseph also shared a strong family tradition of letter writing. A great deal of what we know about the past is based on the propensity of literate people to write letters to one another. Even among educated nineteenth-century Americans, however, the Lesley and Lyman families were exceptional correspondents. Throughout their married life, Peter and Susan Lesley wrote to one another at least once a day whenever Peter's business took him away from home. Peter sometimes wrote home two or three times on particularly slow or exciting days, in addition to maintaining a voluminous correspondence with other relatives and business associates. With characteristic zeal, Peter had the letters he and Susan exchanged beautifully bound to keep for posterity.[24] Binding them was unusual, but his impulse to treasure the letters was not, for various branches of the family had been keeping letters since at least 1750. Many thousands have survived. They constitute one of the largest and best-preserved collections of family correspondence in the United States.[25]

As they worked on the iron survey, Peter, Joseph, and Ben wrote to one another constantly. The two young men sought Peter's guidance and sent him their reports from the field. They amused him by narrating the adventures and frustrations they encountered during their months of travel. Peter replied with advice, money, lists of ironworks to visit, maps to guide them, and family news. Together with Peter's personal journals and his letters home, the men's correspondence makes it possible to piece together their travels in search of America's iron industry. Their descriptions of the places they visited and people they met provide rare glimpses of the underlying cultural, economic, and geographic conditions that characterized American iron regions on the eve of the Civil War.

The Iron Survey Tours

Peter Lesley logically began his statistical survey with the anthracite iron district of southeastern Pennsylvania (fig. 2). This industrial region was conveniently near Lesley's home in Philadelphia, but more important, it was considered the most technologically advanced and therefore the most interesting iron region by members of the American Iron Association.[26] The association was initially dominated by Pennsylvania and New York ironmasters who represented the country's largest iron companies. The first published membership list included no firms west of Pittsburgh and only one southern firm, Morris and Tanner, which owned the Tredegar Iron Works in Richmond, the largest in the South.[27] The anthracite iron district had attracted attention ever since the Lehigh Crane Iron Company put its first anthracite furnace in blast in 1840 in Catasauqua, on the banks of the Lehigh River. After that signal success, sixty-one anthracite furnaces were built in southeastern Pennsylvania by 1856 and another thirty-five in New York, Maryland, New Jersey, Connecticut, Massachusetts, and Ohio. Southeastern Pennsyl-

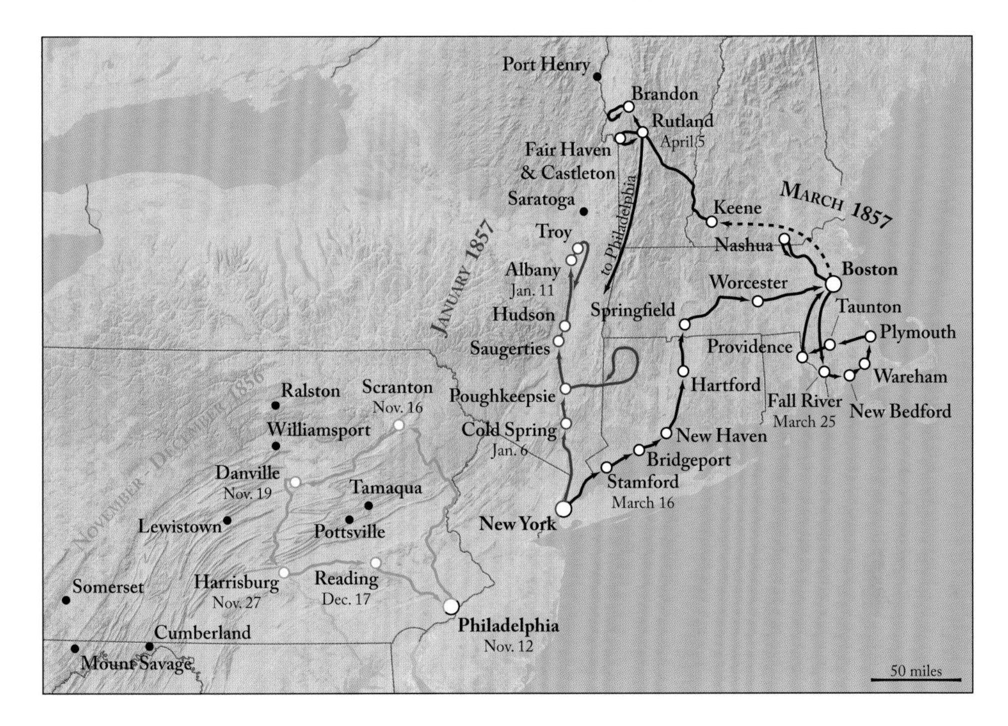

Figure 2. Peter Lesley's tours of eastern Pennsylvania, the Hudson Valley, and New England, 1856–57.

vania claimed both the origin of anthracite iron in the United States and the greatest concentration of anthracite furnaces anywhere in the world. The region also held more than half of Pennsylvania's rolling mills.[28]

Lesley knew the topography and mineral wealth of the anthracite district well from his work on the First Geological Survey. Familiarity sharpened his appreciation of the area's industrial development. "How improved everything is!" he wrote to Susan after riding the train north along the Delaware River. "The country is alive with new enterprise, the villages are ten times as large as when I saw them last. The RailRoad has changed all."[29] A number of Pennsylvania's earliest railways were built to carry anthracite from the mines around Scranton and Pottsville over the mountains to the industrial valleys along the Delaware, Lehigh, and Schuylkill Rivers. It was a sign of the advanced state of rail transportation in this part of Pennsylvania that Lesley was able to "take the cars" for more than two-thirds of the journey.[30]

The landscapes of heavy industry roused Lesley's romantic admiration. After visiting anthracite blast furnaces at Catasauqua, Pennsylvania, he wrote,

> We may cease to regret the absence of European castles from American routes of travel, since these gigantic and most picturesque structures have risen with all the charm of use and ruin combined. Towering 50 and 60 feet into the air, groups of immense furnaces connected by lofty arch ways, stand in line upon the river flats,

> or against the hills, supporting numerous windowed stone or brick buildings from which depend vast iron steam and hot blast tubes, and also high square chimneys which rise like the towers and turrets of Caernarvon Castle as far again above the general summit.[31]

Anthracite furnaces at Phillipsburg inspired "sentiments of wonder & admiration; mountains of ore & coal surround them and perpetual freshets of fire devastate their casting floors." When Lesley found the works manager absent, he explored the site like an eager boy, climbing "over, through, into & out of all the dirty, hot, steamy, gazy, tumbledown breakneck holes & corners I could find, tracing up the innumerable pipes and guessing out the general system according to which this triple cathedral of science & art was constructed." He had a whale of a good time in spite of his "cracking headache and intense weariness."[32] The massive ironworks and coal mines of the anthracite district embodied protean American industry. For Lesley they signified the country's rising to its full potential.

Even in this dynamic region, however, progress was not universal. "The moment one lets go of a rail road," Lesley wrote from Wilkes-Barre, "one is in a wilderness, & helpless."[33] From Wilkes-Barre to Bloomsburg, his pace slowed to three miles an hour in a hired buggy pulled by an old horse. The canal boat would have been even slower. Riding out on horseback to visit charcoal furnaces in the countryside took more time than Lesley had anticipated; he had to get information on a couple of furnaces from locals rather than visiting the works himself.[34] Although the trip yielded a great deal of information, it also began to bring home the magnitude of the task Lesley had set himself, and he lowered his sights somewhat. After the anthracite tour, he rarely noted the number of employees at ironworks, and he published labor statistics in only one number of the *Bulletin*.[35]

Lesley's next tour was a quick sweep through New England in March and April 1857, this time almost entirely by rail. Although the travel was easy, he encountered another problem that came to plague the survey: the struggle to obtain information when managers and owners were absent from the works or unwilling to answer his questions. At a rolling mill near Hartford, Connecticut, Lesley was kept waiting two hours, "cold & hungry and was then told they couldn't be bothered with such things. I was very angry," he continued, "and said so. They grew a little frightened and apologized & gave me a good deal of information all of it false, but I fortunately fell in with an honest old man hoving out the drains in the rain who set me right."[36]

During this trip Lesley further lowered his expectations for the survey by deciding to exclude "machine shops and manufactories properly so called." The number of such businesses in New England exceeded even his statistical appetite.[37] This was the biggest concession Lesley made to the realities of his quest for information. By eliminating machine shops and small manufactories, he could concentrate on documenting primary producers. "Were we to enter upon a summary" of all iron manufacturing, he wrote in the *Bulletin*, "there would be no limit to our tables until we reached the making of needles and watchsprings, and the engrossing of fine wire and sheet iron with wood and other materials in the workshops. It is not the use but the production of iron which we express at present by these statistics." Lesley did decide to include forges, foundries, and bloomeries, whose products included finished manufactured goods as well as, in many cases, bar iron, blooms, and anchonies.[38]

Lesley continued to gather data on Pennsylvania ironworks while traveling on coal survey business. In June 1857, exploratory surveys in central Pennsylvania took him to Lycoming and Tioga Counties, where several ventures in coal-fired iron smelting had failed. The area's coal mines were also doing poorly. Lesley had surveyed the area in 1840. Since then, he wrote, "Lands have risen in value but only 500 tons a day or 150,000 tons a year of coal are sent down the RR from the mines of the entire basin. Two rival companies do all the work. The coalbeds are thin & the coal is inferior & has to be carefully selected." His client, the son of Philadelphia judge and politician John Kintzing Kane, was taken in by a local furnaceman's boasts about the area's economic potential, but Lesley was not impressed. He knew poor resources when he saw them.[39]

Lesley's last major data-gathering trip for the iron survey was the most arduous. He set off for Pittsburgh and the West in late December 1857 (fig. 3). After collecting what information he could from an acquaintance in Pittsburgh, he hired a horse and rode north into the wilds of Clarion County and the Beaver Valley, a sparsely settled area where many new blast furnaces had recently been built. Peter's letter home on Christmas Eve painted a dreadful but comical picture of what he endured for the sake of collecting data:

Figure 3. Peter Lesley's tour of western Pennsylvania and Ohio, 1857–58.

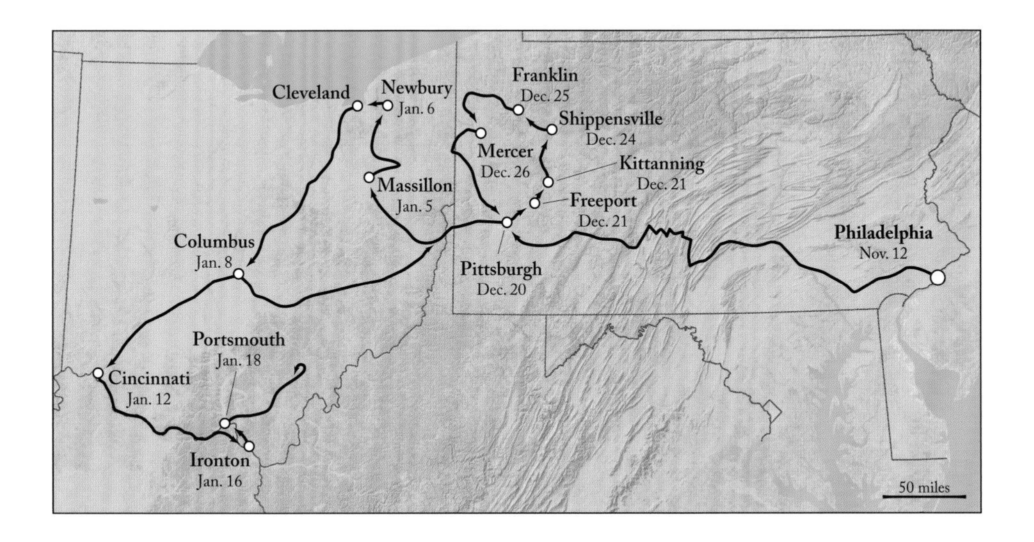

> Ah, such roads! . . . It rained two weeks to reduce the surface to a proper paste; then the charcoal wagons & horses, ploughed and stamped it into the likeness of a smallpox valetudinarian's face & hands; then it froze stiff; and since then it has taken good care to do nothing at all—lest its beautiful work should get spoiled. Such is the travelling that cost me two days to go 35 miles. Such the topography over which I have saddled it & footed it today. Such alas will be the turnpike over which tomorrow the stage to Franklin will have a chance to pound me to a jelly, 24 miles and 12 mortal hours.[40]

In western Pennsylvania he also encountered the suffering of laboring families whose men and boys were out of work. Four hundred people were unemployed at Brady's Bend, one of the West's first and largest rail manufacturers. Lesley particularly pitied a "poor german woman with a three weeks old babe & two little boys" whose "husband had lost his work in the Pittsburg [*sic*] mills."[41] The economic depression was hitting the region hard.

Pittsburgh was a gateway to the West in the late antebellum period. To a Philadelphia scientist who thought of himself as a "noble New Englander," Pittsburgh was also a threshold to a different culture. Down the Ohio River, in the Hanging Rock iron district (fig. 4), Lesley found himself in a region that was "Virginian, a little mixed with Pennsylvania elements—not of a superior kind. Houses are un-

Figure 4. Sketch of the Hanging Rock Rolling Mill, 1857, by J. Peter Lesley. Ames Family Historical Collection. This mill was on the Ohio River, not far from Ironton, Ohio.

comfortable, roads vile, people rough but good natured, beds soft & short, bread hot & doughy, churches few and intermittent." He thought it peculiar that Hanging Rock blast furnaces shut down on Sunday, an inefficiency no eastern furnace company would have contemplated, however religious its owner.[42] What drew Lesley's strongest disapproval, however, were the region's greedy, self-serving, stick-in-the-mud ironmasters. It was bad enough that the lion of Pittsburgh's iron industry, Peter Shoenburger, was "an illiberal entracted mean millionaire." The ironmasters in the rural Hanging Rock district were even worse.

> Ten of the principal iron men of the region have died within a few years worth from 80,[000] to 300,000 dollars and not one has done anything for the town in his will. They all began poor & made their money without foreign help. Obliged to pay but little [for] land at first, they kept the business secret as far as possible, made as little stir as possible, & they were able gradually to get all the timber land they wanted, and also to take possession of the best sites one by one. This policy became habitual & stuck to them after its necessity had passed away. They are ignorant & prejudiced in favor of old regions, old forms, old ores & old methods.[43]

It seemed harder and harder to find informants the farther west Lesley went, and when he found them, the results were often disappointing. A gathering organized for his benefit by the leading ironmaster of the Hanging Rock district came to naught when the man forgot to bring the key to the meeting room and a temperance meeting called away most of the other ironmasters he had invited to come.[44] The effort of the survey, on top of his other work and family cares, had worn out Lesley's enthusiasm. "I am sick of iron," he told Susan.[45]

Bad roads, bad food, unreliable riverboats, and untrustworthy informants should have made Lesley welcome an urgent request to return to Philadelphia, to take up the position of secretary of the American Philosophical Society. Instead, he was "excessively annoyed at leaving my work."[46] He had almost completed his map of Hanging Rock furnaces on the Ohio side of the river but had spent only one day on the Kentucky side. Although the map of the area in Lesley's journal is a rough sketch (fig. 5), it shows the care he took in every aspect of the survey.[47] Lesley hated to leave any job unfinished. He knew by this time that he could not count on ironmasters to mail in the information he required. Even friends in the East had failed to send promised letters. "It is always better to get *double* than to risk having nothing," he had urged Joe in August 1857.[48] Peter had planned to continue as far west as St. Louis, but politics within the newly formed Philosophical Society demanded his presence in Philadelphia. He boarded a riverboat to Cincinnati, took the train across Ohio to Pittsburgh, and was soon back in his study on Morris Street.[49] His assistants would have to visit the Far West and the South in his stead.

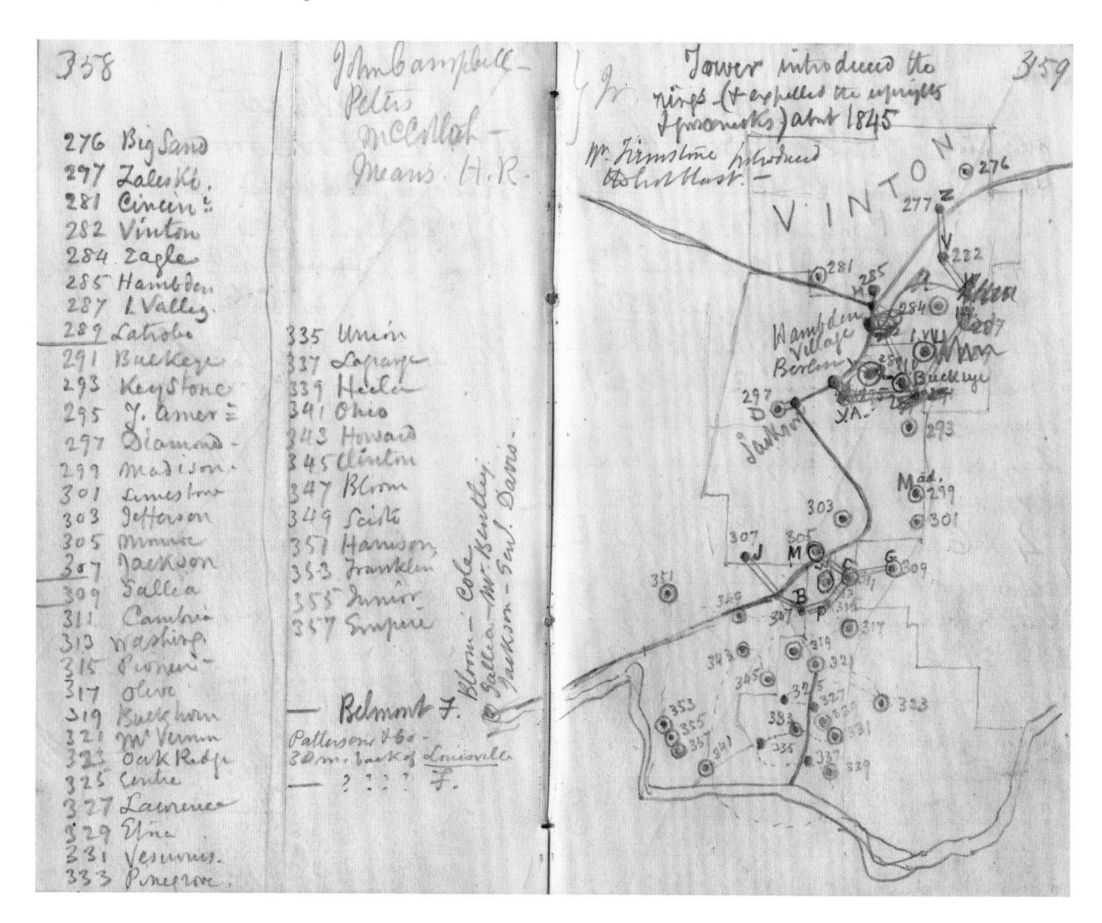

Figure 5. A page from J. Peter Lesley's journal of 1857 showing his documentation of the Hanging Rock iron district in southern Ohio. The sketch map locates blast furnaces in the district, keyed to the list at left. Ames Family Historical Collection.

Joseph Lesley's first contribution to the iron survey was to gather information on forges in the New Jersey Highlands in May 1857. The small scale of operations in this old iron region surprised Joseph, as did the ignorance of its ironmasters. "'Tis astonishing how little the owners & managers of these 'one-horse' affairs know about their business," he wrote Peter. "The complete absence of account books seems quite a matter of course."[50] This was a foretaste of what Joseph was to find during his first long trip, from June 21 to August 3, 1857, when he traveled from Washington, DC, down the Piedmont of eastern Virginia to the uplands of North and South Carolina (fig. 6). Although a few southern ironmasters were deeply knowledgeable about ironmaking in their region, such as William Weaver of Virginia,[51] Joseph generally found it more difficult than he had expected to gather statistics in the South. He believed most of the information he received was exaggerated beyond the overstatement typical of the industry. By the same token, the advance information that Peter gave Joseph was sometimes egregiously wrong. Joseph came to South Carolina expecting to find a rolling mill in Columbia, when in fact the capital city had no ironworks

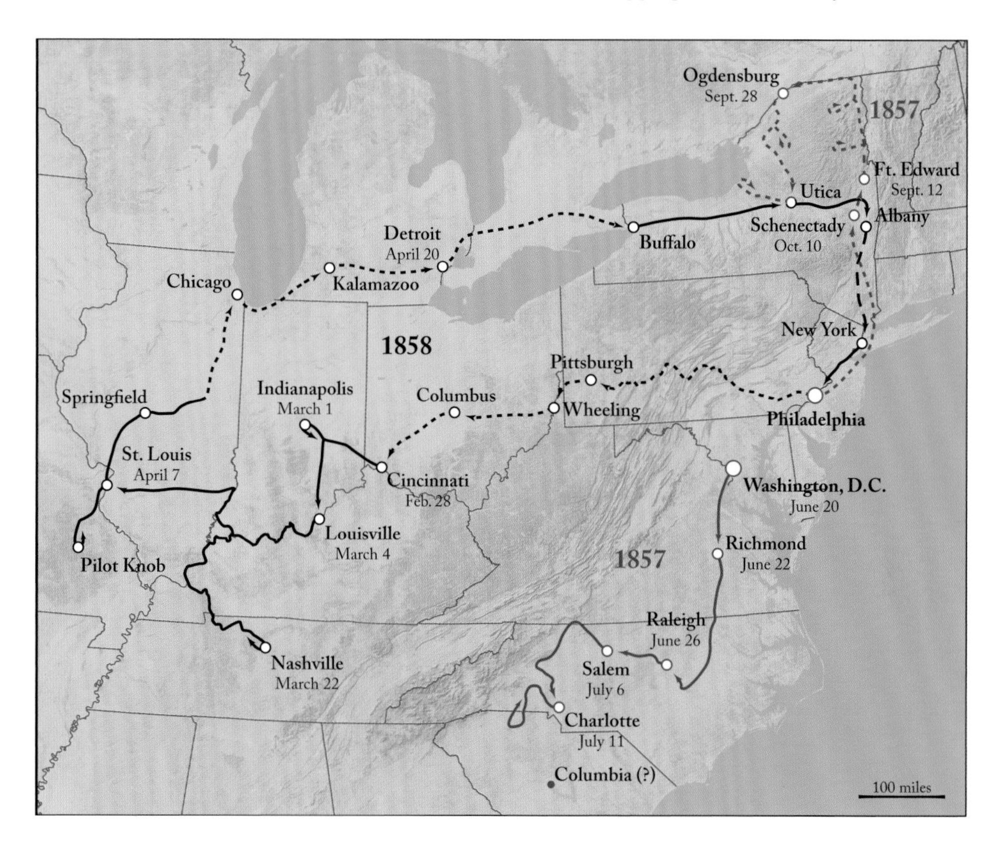

Figure 6. Joseph Lesley's tours of the southern Piedmont, New York state, border states, and the trans-Appalachian West, 1857–58. Dashed lines indicate uncertainty regarding the route Lesley took between places.

at all. The state's three rolling mills, along with a number of blast furnaces, were one hundred miles to the northwest, near Spartanburg.[52] Joseph had crossed into terra incognita, the space on Peter's mental map of the iron industry that was populated by rumors if by anything at all.

Information was hard to come by because of the poor quality of communication, in both senses of the word. Distances were great between ironworks in the South, and transportation was often very slow. Joseph reported that the 180-mile train ride from Richmond to Raleigh, North Carolina, took thirteen and a half hours, longer than it took Peter to travel by rail from Columbus, Ohio, to Philadelphia, a distance three times as far. Means of conveyance were scarce and expensive in the Piedmont and southern Appalachian foothills. In Greensboro, North Carolina, Joseph found no stagecoaches, buggies, or horses for hire. The area's boasted plank road turned out to be "pitch holes, plank turned up & worn out and 'wabbly,' with 2 cts a mile toll." "Oh!" he wrote his brother William, a medical doctor in Philadelphia, "what places—what a country we have passed through—what enormous ticks & bugs, lizards & mocking birds, snakes, toads & dirty table clothes—what porches and architecture, darkies, bad roads & broken-down bridges—Whew! If what we've seen is a sample of the 'chivalrous South' then *I* dont want to live here."[53] Joseph's low regard for the region and its culture could not have inspired confidence in his informants. They might have been more forthcoming had he not been such a snob.

The other side of the coin of denigration was the opinion Joseph shared with many northern businessmen who visited the antebellum South; namely, that the region was ripe for development by intelligent, industrious northerners. He wrote to William, "When I see rich people & rich land & mineral resources such as they have here, I cannot but be struck with the great difference between the north & here. They want Yankees or Pennsylvanians to come & show them how."[54] Pennsylvanians had in fact already been coming to southern iron regions for some time. William Weaver grew up the son of strict ethnic German Dunkers in Flourtown, Pennsylvania, a section of Germantown about fifteen miles from the center of Philadelphia. He moved to the Valley of Virginia in his late twenties to take advantage of the commercial opportunities of iron manufacturing during the War of 1812.[55] Horace Ware, born in Boston and raised in New York State, followed his father in the iron business and established one of Alabama's first blast furnaces in Shelby County.[56]

Little evidence survives about Joseph's second iron tour, from about August 28 to November 4, 1857. This journey took him up the Hudson Valley to Lake Champlain, where he and Peter spent several days examining big open-pit iron mines near Port Henry. He continued north to Plattsburgh, New York, then headed into the Adirondacks to find remote furnaces at Salmon Lake and a few other locations. Once he emerged from that forested wilderness, he followed Lake Champlain up to the moraine fields flanking the St. Lawrence River, turned west to Ogdensburg, then south to Utica. He ended the tour somewhere near Oswego.[57] As in the iron districts of the Carolinas, Joseph found that many upstate New York ironworks were "doing but a miserable business," despite the region's excellent magnetic ore. Some firms had failed repeatedly.[58]

Joseph's final trip for the iron survey covered the territory that Peter had abandoned in his hasty return to Philadelphia. Joseph left Philadelphia on February 17, 1858, armed with letters of introduction and lists of ironworks to visit in Wheeling, Cincinnati, Louisville, Tennessee, Missouri, and Michigan. He had agreed to pay his own way and seek reimbursement later, an expedient necessitated by the association's inability to attract new paying members after the Panic hit full force in late 1857. He reached Wheeling on February 24 and Cincinnati four days later.[59] The information he received there suggested that he was already near the western edge of significant iron production. "There are probably really only 2 fur[nace]s in Indiana & 2 in Illinois," he wrote to Peter. Rumor had it that a rolling mill had been built in Chicago and that a furnace might be operating somewhere north of Milwaukee, but otherwise he heard of little activity in the Great Northwest. He took the train to Indianapolis for a quick inspection of the rolling mill and blast furnace there, then rode it due south to Louisville, where he began a slow journey down the Ohio River on the stern-wheel steamboat *John Gault* to see the iron and coal region of western Kentucky and Tennessee.[60]

Joseph's letters from this part of the journey were more somber. He was entering slave country again, but this time it was a land of slave catchers and sectional conflict. During one long, uncomfortable night on the steamboat, he lay shivering in his cabin, "a glass door alone keeping out the snow and rain." He was unable to block out "the continued clanking of that poor negro's chains." A runaway slave had been "caught *in Indiana* by a couple prowling Louisville policeman." All the Lesleys were abolitionists; Joseph did not have to spell out his revulsion.[61] He apparently kept his feelings in check once he arrived in the heart of the iron district that straddled the Kentucky-Tennessee border, for he was warmly received by George T. Lewis, a former Philadelphian turned slave master whose firm, Woods, Lewis and Company, owned the Cumberland Rolling Mill and a number of charcoal furnaces in the vicinity. During his stay, Joseph persuaded Lewis to join the association. He gratefully accepted Lewis's offer of "a horse, saddlebags & a negro to make my trip south from here among the iron works."[62] The two men must have discussed what Joseph and the whole country called "the great slave insurrection" at the Cumberland mill and furnaces on Christmas 1856, when the attempted escape of fifty-six slave ironworkers met violent white reprisals, ending in the execution of at least nineteen of the slaves.[63]

The greatest difficulties came as Joseph picked his way along the deeply incised valley of the Cumberland River. He rode through rainstorms, struggled to keep his horse's footing on slippery mountain tracks, and forded swollen streams, only to find that "there have been so many failures among iron men that the works have been and are constantly changing hands—the old proprietors moving away & the new knowing little or nothing of the former's business. Out of 42 fur[nace]s on the 2 rivers [the Cumberland and the Harpeth] 20 only will be in blast this year & not over 10 of those in blast during the whole working year. . . . The iron-making business here, for the present, is, as the natives say, *done gone*."[64]

All together, Joseph spent about ten days in the Cumberland Valley. He was proud of his work there but was eager to finish, not least because he had hopes of assuming the lead position on a survey of eastern Kentucky being organized by David Dale Owen, state geologist of Kentucky and son of Robert Owen, who founded the utopian community at New Harmony, Indiana. Joe stopped at New Harmony to finalize the contract with Owen en route to St. Louis, where several rolling mills needed his attention.[65] His final frontier destination was Iron Mountain, a clutch of ore-rich hills on the edge of the Ozarks about sixty miles southwest of St. Louis. Joseph rode the St. Louis and Iron Mountain Railroad down to Pilot Knob, whose charcoal furnace was later targeted for Confederate attack near the end of the Civil War. He may also have visited the furnace and forge at Maremec Spring, thirty miles northwest of Pilot Knob, where entrepreneurs from Ohio had been making iron since 1829. It was as far west as anyone was making iron in 1858, so far as the Lesleys knew.[66] Joseph then headed home, stopping briefly in Chicago, Kalamazoo, and Detroit.[67]

The third iron scout was the youngest and least experienced, but in some ways he was best suited to the demands of work in the field. During more than seven months inspecting ironworks up and down the Appalachian chain, Ben Lyman never complained of illness in his letters to Peter. He was not prone to headaches like his neurasthenic uncle and seemed to accept the rigors of the outdoor life more readily than Joseph did. His acerbic comments on culture, politics, and industry, however, suggest deep-seated regional prejudice. So long as he was in the company of Pennsylvania men or New England ladies, Lyman reveled in the journey, but he found little to like in southern society or southern places.

The first half of Ben's tour passed mostly through familiar territory (fig. 7). Peter wisely started him with well-established ironworks close to home, in Baltimore, northern Virginia, central Pennsylvania, and New Jersey. His conscientious reports to Peter noted the ironworks he did not see as well as those he visited. In his journal Ben transcribed excerpts from geological reports on the areas he was visiting. He was a good student. He was also a Massachusetts Yankee to whom the town of Winchester, Virginia, gateway to the Shenandoah Valley, seemed "the outskirts of Pandemonium."[68] He rode trains when he could, suffered saddle sores on his first long horseback ride from Harpers Ferry to Antietam Furnace, and grew accustomed to the routine frustration of chasing information from ironmasters who did not know he was coming or did not leave word of when they would return. Data collection was, as always, more businesslike in Pennsylvania.[69] Peter and his protégé happened to cross paths in Cumberland, Maryland, in early May. They spent a happy evening poring over Ben's tables of data from Virginia and Maryland furnaces, then strolled around the town, taking note of the "pretty gothic church" with "the spire at the wrong end."[70]

In mid-July, Ben crossed over the invisible line bounding the outer limit of the economy he knew and understood—the economy of northern industrial efficiency. He detected the change as he traveled west from Lynchburg, Virginia, into the Blue Ridge Mountains. "My 'three days more or less' in this neighborhood have proved to be nearly seven days," he wrote Peter from the little town of Christiansburg. "The horseback journey was longer than I expected it to be, and the roads were extremely rough and hilly so that I could average hardly thirty miles a day." He correctly predicted that every stage of the trip they had discussed, through the long wedge of southwestern Virginia and down mountain valleys in eastern Tennessee, would take many more days than they had estimated. The work might also require more hunting than anticipated, if Ben pur-

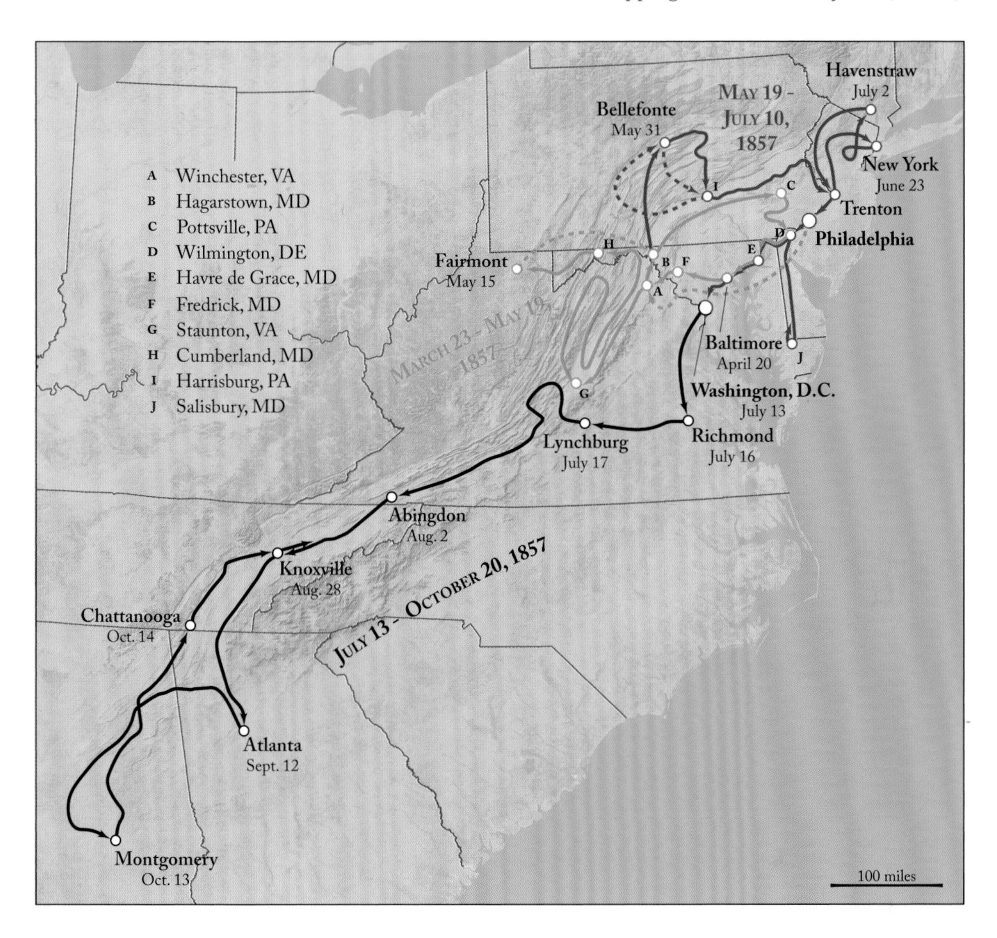

Figure 7. Benjamin Smith Lyman's tours of the central Appalachians, Mid-Atlantic region, and Appalachian South, 1857.

sued all the leads he came across. "There are said to be iron works of some kind at Charleston in Kanawha Co., Va. [in present-day West Virginia]," Ben wrote, but "no one knows where there are any or not in any other part of that great western region of the state."[71] He needed to buy a horse, but the cost of the trip had already drained the association's coffers. "The railroad ends at the Tennessee state line," he warned Peter. The specie changed over the border as well. In Johnson and Carter Counties, he heard,

> forges are so plenty that the current coin is said to be "long dollars," that is bars of iron. This is rather an extension of the Spartan custom, and I expect to meet with a Spartan diet and freedom from luxury—ham broiled, or fried in abundance of fat, and corn bread as hard as a brickbat, and indescribable horrors in the bedroom. What men those Spartans must have been! I mean to stand up for the Persians in future. I suppose that they had mattrasses [*sic*] and clean sheets and washstands furnished with crockery throughout, and wheat bread and perhaps even waffles

> occasionally with wholesome meat. They had their slaves, too, decently dressed, I presume, and not waiting on you in nothing but a filthy shirt, or at most shirt and trousers.[72]

Having found an inexpensive pony for sale in Abingdon, Virginia, Ben headed bravely into Tennessee. He had a pencil sketch map from Peter for the eastern Tennessee furnaces but no map of Georgia or Alabama. He would have to feel his way, following tips from strangers in hopes that they would steer him straight.[73] In Atlanta he "came across a man who said he was as well acquainted with Augusta [Georgia] as with his own dooryard." Contrary to what Ben had been told, this man said that "there were no iron works . . . of any sort" in Augusta. Spared a long trip across Georgia, Ben rode north to Cass County, where he visited a cluster of six furnaces, a forge, and a rolling mill at Etowah. The books "were not kept so well . . . as they are in the North," he noted, perhaps because the owner was a native Georgian.[74] Although the northern-born ironmasters he met provided reliable information, Ben was deeply suspicious of most of what he was told from Knoxville, Tennessee, to Montgomery, Alabama. "I have found that the exaggerations of the furnace-owners of the south are of no ordinary kind," he wrote Peter, "for they state their product to be two, three, and even nearly four times what it actually is, and it is really so small that the exaggeration would hardly be suspected by a northerner."[75] After his "long tramp" through Alabama, Ben headed northeast along the Tennessee River, then retraced his steps from Abingdon back to Philadelphia. He finally reached home on Thanksgiving, November 26. Peter totted up Ben's journey as 225 days at a cost of $444.12. All together, he calculated that the three of them had spent 443 days traveling for the survey. The total cost to the association, including per diem salaries for Ben and Joe, came to about $6,000.[76]

Results of the Survey

Viewed from any rational perspective, Lesley's survey of the iron industry was excessive. It consumed a large portion of the association's revenues in a period of worsening economic conditions. As the survey's expenses mounted, paid memberships in the association plateaued and then began to decline. To recoup some of the cost of the survey, the association's board decided to publish Lesley's final report as a book. Publisher John Wiley of New York City agreed to print a volume of 500 pages "of large print." In Lesley's hands, it grew to 772 pages of small print with dozens of geological diagrams, tables, and four gatefold maps he had specially printed by a new method of lithography, which added further expense to the project. Two-thirds of the book was taken up by long chapters describing the geology of iron ore across the country. The entries describing individual ironworks ran to 262 pages, covering 770 blast furnaces, 497 forges and bloomeries, and 224 rolling mills. In Lesley's imagination, *The Iron Manufacturer's Guide* had become "the great book," like the great maps before it, an all-consuming effort to capture the industry he had surveyed.[77] The first print run was sold and given away to association members. The second printing languished at Wiley's. When Lesley ran into the publisher during a visit to New York, he was told that "the Guide don't sell on either side of the Atlantic."[78]

Lesley meant his book to be a reference work, not "recreation." He explained in the preface that he had imposed "a fixed order" on the ironworks entries "for the sake of easy reference and comparison." This unfortunately robbed the book of the surveyors' personal insights and lively observations. Entries monotonously reel off blast furnace height, width at the boshes, blast pressure, number of tuyeres, kind of iron ore and fuel used at the furnace, mine locations, mode of transport, and so forth—in all, sixty-seven variables related to the four kinds of ironworks. The mass of repetitive detail severely limited the book's commercial appeal. Today, the embedding of all those details in dense text makes the work's historical content difficult to analyze. Although generations of industrial historians have drawn on the *Guide*, most have used just a handful of entries or the statistical summaries Lesley provided at the end of the book. One rare exception, the summary map of ironworks' locations, based on Lesley's maps, was published in Charles O. Paullin and John K. Wright's *Atlas of the Historical Geography of the United States;* it has been reproduced numerous times. Historical geographer Kenneth Warren mapped a few key variables for furnaces and rolling mills.[79] Thus Lesley's survey has been a touchstone but not an influential source in scholarship on the American iron industry.

To bring the survey information to light, it was necessary to extract the data from each textual entry and parse the details into a relational database, then connect the information about each ironworks to its geographic location in a spatial database. The linked databases (point locations and their attributes) became the Lesley historical GIS (see the appendix for more information). I supplemented the entries with information on production, raw materials, and markets that was contained in the *Bulletin of the American Iron Association*, where Lesley published much of the original survey data. I also supplemented Lesley's information with dates of construction and abandonment recorded in local histories and other sources, with the help of fellow historical geographer Richard G. Healey. In the end we were able to document the year of construction for 91 percent of blast furnaces, 96 percent of rolling mills, and 73 percent of forges and bloomeries listed in the *Guide*. We also ascertained the year of abandonment for 83 percent of blast furnaces known to have gone out of business by 1858.[80] Lesley's descriptions of the geographic locations of blast furnaces and rolling mills made it possible to place most of them within a mile of their probable location. He did not describe the location of forges and bloomeries with the same precision, particularly those south and west of the Potomac and Ohio Rivers, so only those in Pennsylvania were placed as individual points in the GIS. For the rest of the country, I aggregated forges and bloomeries to the total number per county.[81]

The correspondence between Lesley and his assistants adds invaluable context and commentary to the data so drily summarized in the *Guide*. The letters suggest that modern ironmaking methods, and what the surveyors regarded as modern attitudes, were confined to a fairly small territory that included southern New England and the Mid-Atlantic states. They portray iron companies in western Pennsylvania, southern Ohio, and Virginia as technologically conservative and behind the times. Those in upstate New York and western Tennessee were struggling to stay in business. The Carolinas, Georgia, and Alabama were downright backward. Joseph's and Ben's letters give the impression that the benighted iron industry of the Deep South was propped up by stalwart Pennsylvania men who were doing their best despite the region's profound disadvantages. The letters provide valuable glimpses of the industry in parts of the country that other sources cover poorly if at all. At the same time, the picture they paint is sometimes strongly colored by class and regional prejudices and by the authors' technological bias. Like the iron men who paid for the survey, the Lesley brothers and Ben Lyman respected and admired the latest technologies, large-scale production, and the accounting mentality that characterized the business culture of East Coast industry. They regarded small charcoal furnaces, rural bloomeries, and water-powered helve-hammer forges as antique, inefficient, " 'one-horse' affairs." To borrow a term from Bruno Latour, the surveyors' attitudes reflect the "great divide" between traveling collectors of information, who consider themselves rational and scientific, and those they observe, whose practices seem local, closed, and culturally determined. Such people's apparent backwardness typically increases with distance from the travelers' place of departure.[82] While larger ironworks in the North clearly produced more iron than the generally smaller works in the South, the judgment that southern works were therefore inefficient or that southern ironmasters were ignorant or unwilling to try new methods reflected more prejudice than knowledge.

The surveyors' attitudes, as well as the difficulties they encountered on the road, influenced the data they collected. Believing that southern ironmasters and workers were liars, they discounted production figures for most southern works. This probably resulted in an underestimate of southern production. More important, their disbelief of what they were told may have disinclined them to pursue leads as energetically as they did in the North. Northern informants, on the other hand, were taken at their word, which might have resulted in at least a slight overrepresentation of iron production, particularly for the Mid-Atlantic states and the eastern seaboard. Checking the *Guide* against other sources suggests that the survey most accurately documented ironworks in the Northeast, including old works that had gone out of business but persisted in local memory and remained visible in the landscape. Transportation improvements in the Northeast made data collection easier, whereas the lack of good roads and railroads in the South and West made it difficult for the investigators to visit all the ironworks they heard about, let alone those beyond the knowledge of their informants. For example, Ben Lyman heard vague reports of blast furnaces in what is now central West Virginia that neither he nor his uncles pursued. As a result, the *Guide* omits at least nine antebellum blast furnaces from that region.[83] Similar omissions almost certainly occurred elsewhere.

Despite its gaps and weaknesses, the iron survey created a remarkably comprehensive and systematic record of a major industry (fig. 8). Unlike most nineteenth-century social or industrial surveys, such as the US Manufacturing Census, it is not a snapshot of a single moment in time. For although the *Guide*'s information is most complete and reliable for ironworks that were active during the survey, it includes dates of construction, modification, and abandonment, as well as a great deal of other information, about ironworks that operated from the

Figure 8. Ironworks visited by Peter Lesley, Joseph Lesley, and Benjamin Smith Lyman. Red dots mark works that one of the survey team definitely or probably visited. White dots stand for works included in *The Iron Manufacturer's Guide* that the team did not reach or did not mention in their correspondence.

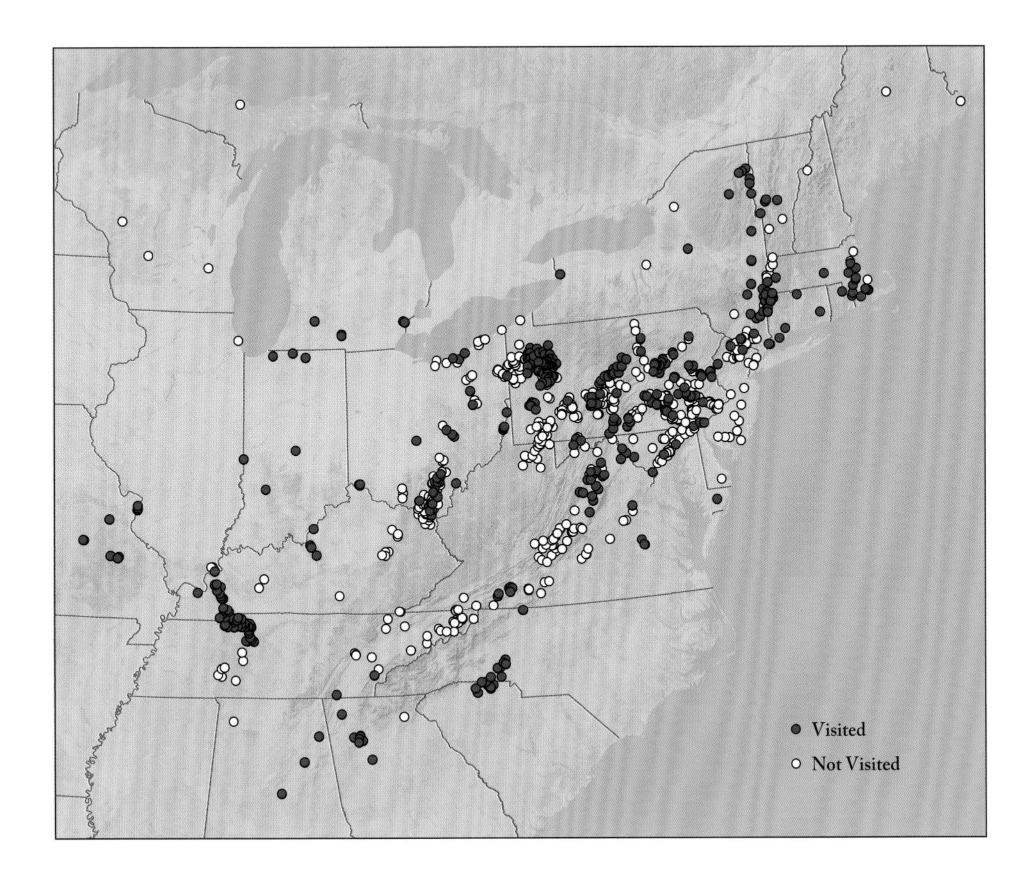

late colonial period through early 1858. It is rare indeed for a single source of geographically located historical information to cover more than seventy years of history.

The rest of this chapter explores the contents of Lesley's survey to answer a number of basic questions about the historical geography of the American iron industry. How did the industry spread and change during the early nineteenth century? Did all iron regions grow or suffer periods of decline at the same time and with the same intensity? How did a region's geological endowment affect industrial development? Where did British methods of ironmaking take root most rapidly or most slowly? Were the various methods of the British model adopted in whole or in part? What was the geography of iron markets, and to what extent were they vulnerable to, or insulated from, outside competition and national economic trends? In sum, how can the iron survey help us understand the industry's development and regional differences?

Spatial and Temporal Dynamics of the Iron Industry

Several studies have tabulated or mapped the basic story of the antebellum industry's growth. Peter Temin summarized nineteenth-century production and average prices for pig iron, steel, and rolled iron, including the way those figures broke down according to the kind of fuel used in blast furnaces and the process used to make steel. The *Atlas of the Historical Geography of the United States* and the *Atlas of Early American History* provided maps of the distribution of ironworks in the colonial period, in 1810, and in 1858.[84] Lesley's data fill the gaps between these benchmarks with annual information and link the details of production and technology to individual ironworks. The iron survey data thus provide a continuous narrative of national industrial change—or, to use a more suggestive metaphor, the survey data are notes in a complex musical composition waiting to be scored for orchestra. With proper instrumentation, it should be possible to pick out the voices of particular regions and kinds of ironworks as they became prominent or lapsed into silence. The overall tempo changed several times, and not every section kept the same rhythm, but underneath these variations the driving theme of concentrated growth in the Mid-Atlantic became more and more insistent.

The idea of a musical score inspired the combined map and graph in figure 9.[85] The map shows the location of every blast furnace listed in the *Guide* as nearly as I could place it according to the geographical description Lesley provides. The graph above the map plots every known year of construction (along the *y* axis) at the furnace's approximate longitude, or east-west location. Thus Cincinnati on the graph is above Cincinnati on the map, Pittsburgh is above Pittsburgh, and so forth. Time on the graph reads from the top down.

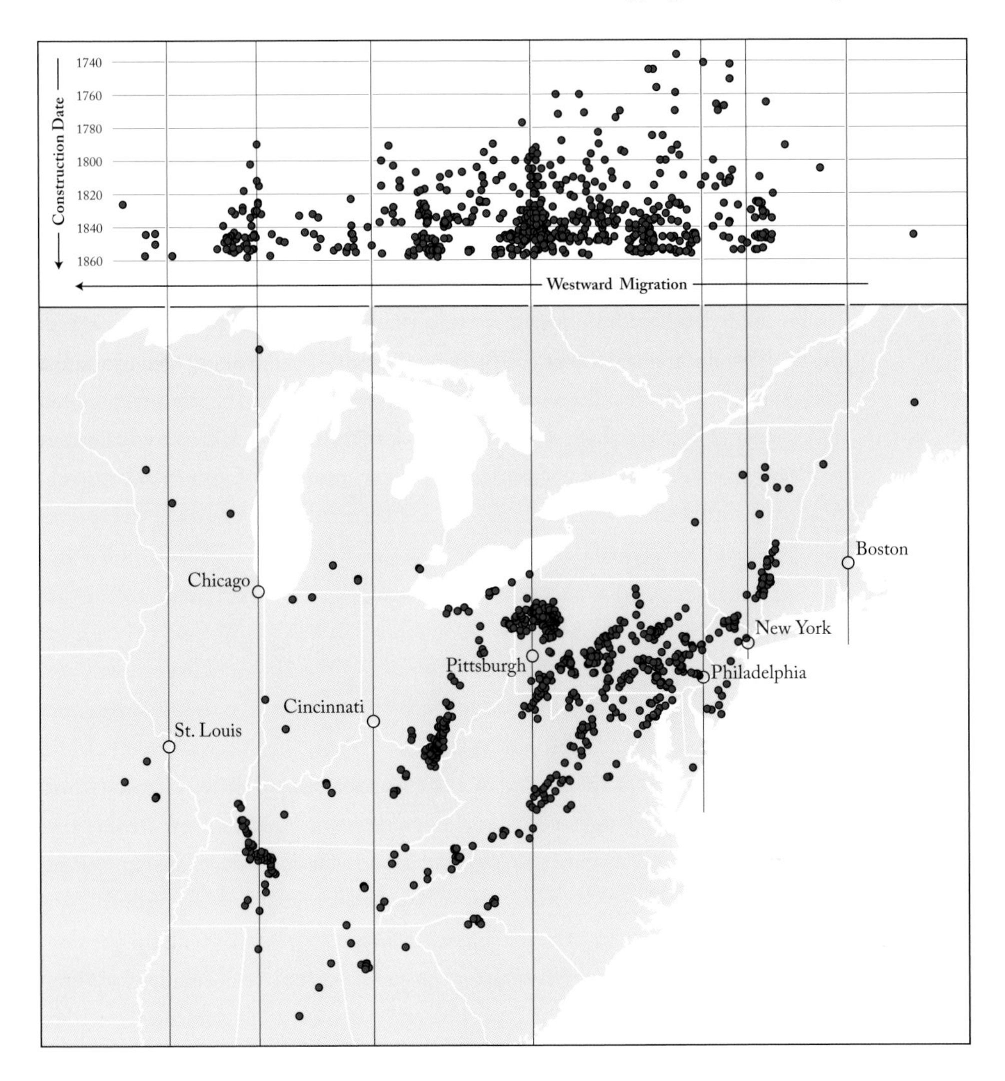

Figure 9. Blast furnace construction correlated with location in space and time, 1736–1858. This map shows only furnaces with a known year of construction. Lesley's *Guide* does not include all colonial furnaces; it is much more comprehensive for the antebellum period.

The graph reveals a number of significant patterns. One is the ubiquity of blast furnace construction, and thus of iron production, across the eastern United States throughout the early nineteenth century. The oldest furnace Lesley recorded was built in 1776 near Pittsburgh, then a frontier outpost. By 1830 most states had at least one blast furnace. Reading from east to west (right to left), one can follow several irregular lines of dots trending to the southwest, down the Valley of Virginia, for example, and from that 1776 furnace near Pittsburgh down the Ohio Valley to the first furnace built in Trigg County, Kentucky, in 1845. These linear clusters indicate that the construction of ironworks followed the general movement of Anglo-American settlement. A rule of thumb appears to have been that a region required a population density of at least six to eighteen persons per square mile to support an ironworks. This level was achieved throughout Tennessee, Kentucky, southern Illinois, and the eastern side of Missouri by 1830.[86]

At the same time, the graph is punctuated by strong clusters of fairly intense, localized furnace construction. The first marked clustering of construction occurred around the turn of the nineteenth century in Fayette County, Pennsylvania, just north of the Virginia border. This was the first of several concentrated bursts of furnace construction in western Pennsylvania. Farther west, the graph captures smaller episodes of localized construction in eastern Ohio and the Hanging Rock iron district, where construction peaked on the Kentucky side in 1845–47 and in Ohio in 1854–56. The Cumberland district in western Kentucky and Tennessee grew gradually for decades, then experienced a flurry of construction in the late 1840s and again in 1853–55. Furnace construction in the Northeast produced a different pattern. New furnaces were built in Philadelphia's hinterland and the district along the Connecticut–New York–Massachusetts border fairly steadily through the 1830s, with no intense activity until 1844–47. Almost no furnaces were built east of the Berkshires and the Green Mountains of Vermont. The map shows one furnace in New Hampshire and one in Maine, the latter a remote outlier near Mount Katahdin.

Figure 10. Cycles of blast furnace construction by primary fuel type, 1820–58.

Figure 10 portrays the timing of blast furnace construction in more familiar form. It shows the aggregate number of furnaces built each year while distinguishing between furnaces by their predominant fuel (charcoal, anthracite,

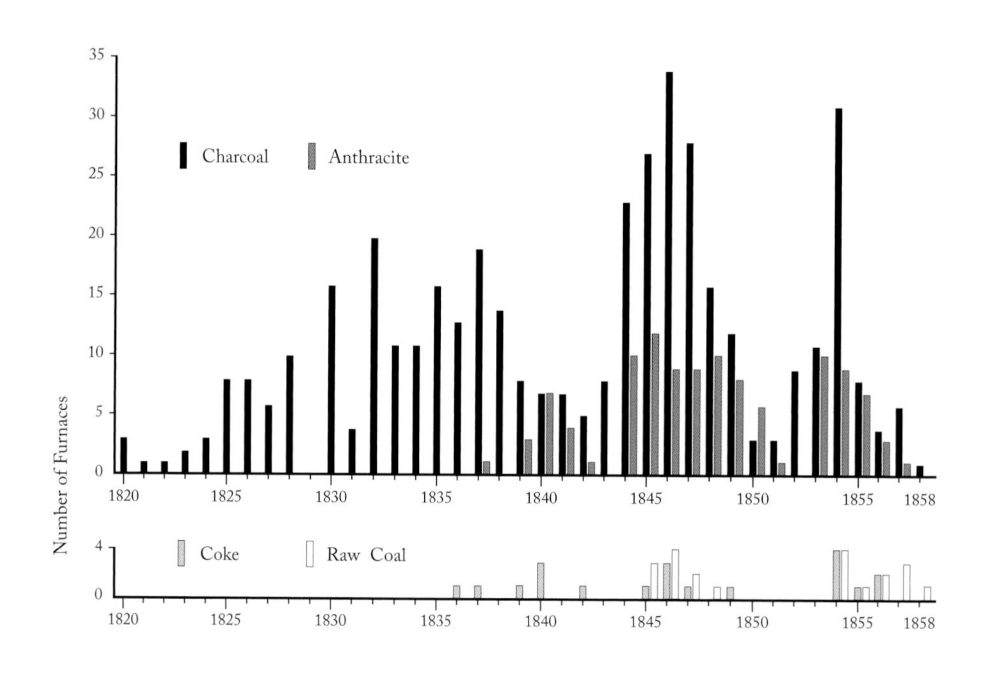

coke, or raw coal), each type implying the use of a particular bundle of smelting technologies. In the United States as a whole, construction of every kind of blast furnace increased during periods of national economic growth and fell off sharply in the wake of economic downturns. The graph shows a sharp decline in new construction after the Panic of 1837, followed by a boom in the mid-1840s, another depression in the late 1840s, and the more gradual rise of investor confidence after 1851. The iron industry was a bellwether for the depression that hit the country in 1857.[87] Investors appear to have responded strongly to the protection offered by the Tariff Act of 1842, which raised duties on imported iron in hopes of stimulating domestic industry. Before lower tariffs went into effect under the Walker Act of 1846, new furnaces were built in many iron districts. Not all districts followed national economic trends, however. In graphing the data by region, one finds that furnace construction was least cyclical in Kentucky, Tennessee, and Virginia. Charcoal furnaces were built in those states almost every year from 1820 to 1858, with no significant building booms before the Civil War. In the mid-1850s, Ohio experienced a more sustained boom than the rest of the country, owing in part to the stimulating effect of railroad construction.[88] Pennsylvania gained many new charcoal furnaces in the 1840s but almost exclusively mineral-fuel furnaces after 1852.

National economic conditions also registered in the timing of rolling mill construction, particularly the favorable impact of the Tariff of 1842 (fig. 11). Before 1842, all imported railroad iron had been admitted duty-free. The 1842 Tariff imposed duties of $17 a ton on bar iron and $25 a ton on rolled iron.[89] These rates were high enough to cut British imports sharply, particularly imported rails, providing a powerful stimulus for the construction of domestic rail mills. According to Lesley, three-quarters of the antebellum rolling mills that specialized in making rails were constructed after 1842. The first to go into operation was the Lackawanna Rolling Mill in Scranton, which opened in 1844. Thirteen new rolling mills were completed each year in 1845, 1846, and 1847, including large ventures such as the Trenton Iron Works (1845), the Montour Rolling Mill in Danville, Pennsylvania, and the Rensselaer Rolling Mill in Troy, New York (1846), and the Bay State Rolling Mill in South Boston (1847). The lowering of duties to a flat rate of 30 percent of declared value in 1846 coincided with the collapse of British railroad construction, and US markets were flooded with low-priced British iron of all kinds. American ironmasters, particularly those who owned large mills on the East Coast, were outraged by the sudden shift of advantage to their commercial adversaries. British companies shipped hundreds of tons of coke pig iron and rails across the Atlantic. This helps explain the decline in new furnace construction in the United States in 1847–51. The impact of British imports is harder to read in rolling mill construction patterns because it took one to three years to build a mill and because advance contracts for rolled iron, particularly

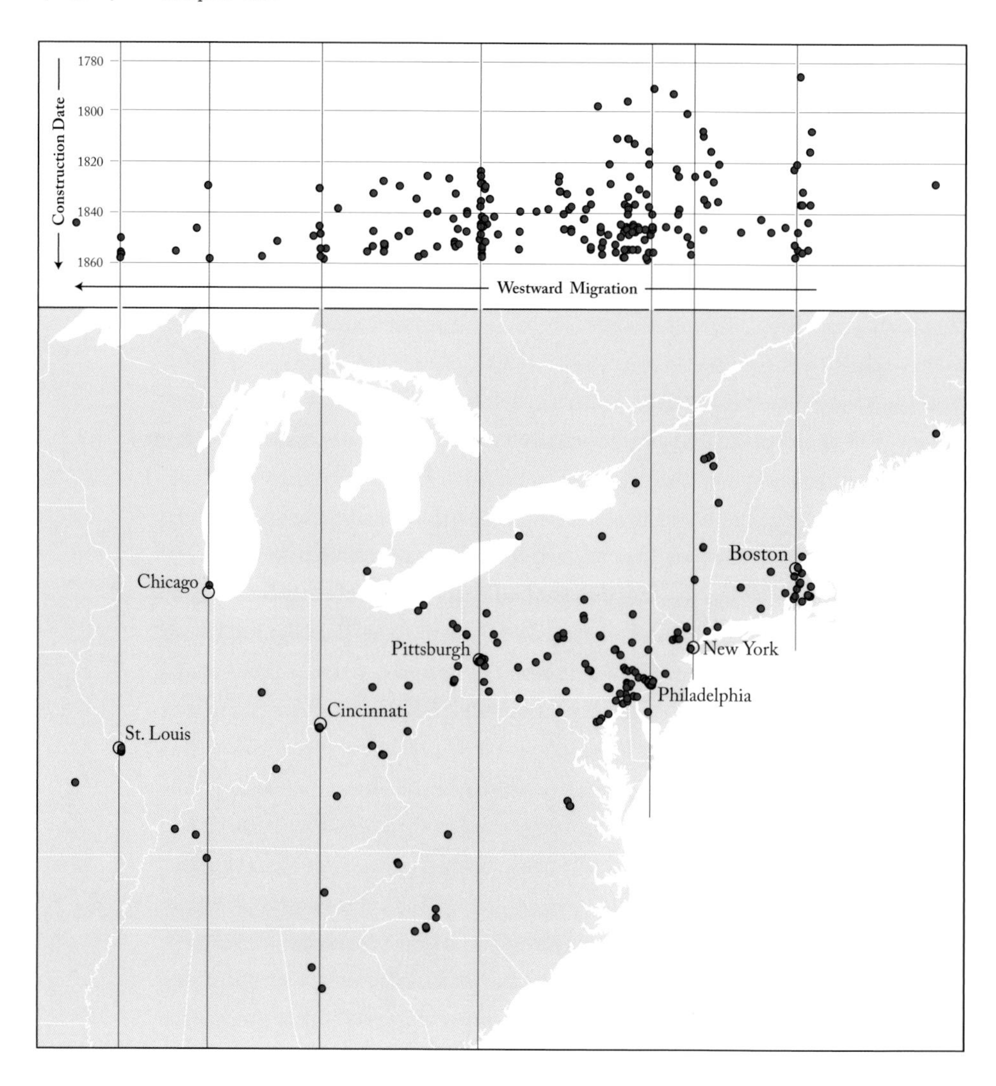

Figure 11. Rolling mill construction correlated with location in space and time, 1785–1858. This map shows only mills with a known year of construction.

the very large contracts for rails, cushioned American mills for a year or more after the tariff reduction went into effect.[90] Farther inland, the construction of rail mills appears to have been more closely tied to particular railroad construction projects than to tariff policy and foreign competition. The Mount Savage mill in western Maryland was built in 1839 to provide rails for the Baltimore and Ohio Railroad. The Cambria rolling mill, the first rail mill west of the Allegheny Mountains, was completed in 1856. When Lesley visited Johnstown in 1858, the Cambria mill was producing rails at a rate of 18,000 tons a year.[91]

The graph in figure 11 also demonstrates that rolling mills became increasingly concentrated in a handful of urban locations as the antebellum period progressed. The trend is visible in the chordlike clusters of dots at Boston, Greater

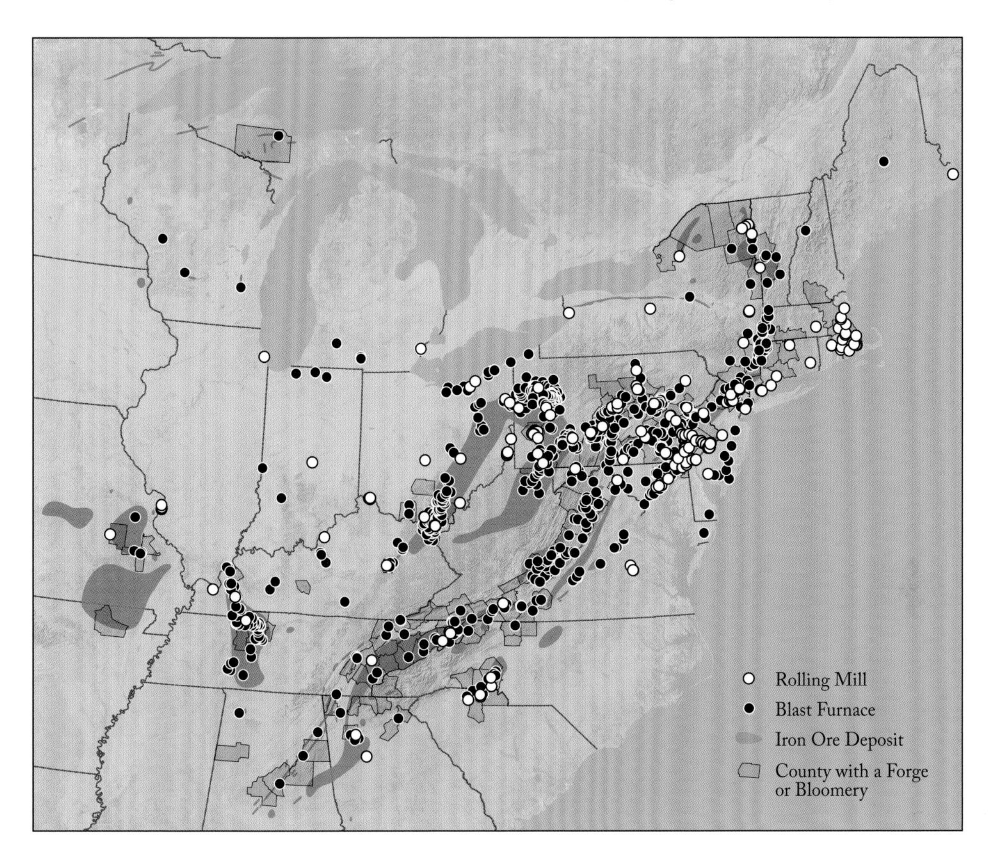

Figure 12. Ironworks in relation to iron ore deposits. White dots are rolling mills, black dots are blast furnaces. Forges and bloomeries, more difficult to map, are represented by the counties where they were located. The map shows all works constructed between 1736 and 1858 as recorded in *The Iron Manufacturer's Guide*.

Philadelphia, Cincinnati, and St. Louis, but most resoundingly at Pittsburgh. By 1857, twenty-one rolling mills had been built within a ten-mile radius of downtown Pittsburgh, compared with eleven mills within the same distance of central Philadelphia.[92]

Figure 12 maps ironworks in relation to major iron ore deposits. As industry observers have long noted, forges and bloomeries (aggregated here by county) were usually close to ore deposits. Proximity to ore was important for early blast furnaces as well. At the beginning of the nineteenth century, the greatest concentration of ironworks lay athwart ore deposits that were within a day or two by wagon from the centers of population at Boston, New York, and Philadelphia.[93] By 1855, most of the large ore deposits east of the Mississippi River were being exploited, including outcrops of high-grade hematite ore in northern Alabama and Georgia, the Iron Mountain area southwest of St. Louis, and the Marquette range in Michigan's Upper Peninsula. Of the two hundred furnaces for which Lesley provided a specific distance from furnace to ore supply, only fifteen were more than ten miles from their farthest source. Many midwestern furnaces that look distant from the nearest ore deposit on the map smelted bog ore from local wetlands, including Kalamazoo Furnace in southwestern Michigan, three furnaces in northern Indiana, and several along the Ohio shore of Lake Erie. Bog ore had supplied the earliest ironworks in coastal settlements in the Massachusetts Bay colony and colonial Virginia. Iron in wetlands was easily accessible, and it was a renewable resource if harvested in small amounts. Where the metallic precipitate from iron-rich groundwater formed consolidated lumps called seed ore (about 45 percent iron) or massive ore (up to 53 percent iron), bog ore was fully adequate for smelting.[94]

The proximity of smelting operations to iron ore followed a compelling economic logic. Ironmaking was the heaviest of heavy industries, and iron ore was the heaviest of the raw materials required to make iron. If a given deposit contained ore that was 50 percent metallic iron (a fairly high percentage), miners had to dig two tons of ore for every ton of usable iron, with some extra to allow for wastage in smelting. About one-tenth that amount of limestone was required for "flux" to remove chemical impurities from the ore. Charcoal, the only fuel used in America until the 1830s, could not be transported long distances because the friable cargo would break down to dust if it was carried in wagons for more than a few miles.[95] Where charcoal was unavailable, coal could be shipped more cheaply than ore because it was lighter per volume. Thus the logic of transport costs made it sensible to locate ironworks as near to the ore as possible.

That logic began to change in the 1850s. Shipments of iron ore from northern Michigan's Marquette range began in 1854. By the end of Lesley's survey in 1858, at least ten furnaces in northwestern Pennsylvania, Ohio, and the Detroit area were making pig iron from Marquette ore. Iron mines in Essex and Clinton Counties, New York, near the western shore of Lake Champlain, did not supply many blast furnaces outside their immediate area in the antebellum period, but Lesley did record one customer in Dutchess County, New York, about two hundred miles to the south, and he noted that Lake Champlain ore was being used in small amounts at rolling mills as far away as Pittsburgh.[96] Iron ores from Lake Champlain and Lake Superior were among the richest in the country. The purest deposits in the Adirondacks were up to 70 percent metallic iron, nearly matching the best Lake Superior ores, which first drew attention at Marquette and later became world famous with the opening of the Mesabi range in northern Minnesota in 1892.[97] Transportation improvements, notably the completion of the Champlain Canal in 1823 and of locks at Sault Sainte Marie in 1855, lessened the cost of transporting the industry's heaviest resource. Quite logically, the ores worth shipping long distances were those with the highest metallic content. They were first used in significant amounts in regions that had relatively small, low-grade iron deposits and good access to shipping via the Great Lakes and canals.

While low-cost water transport greatly extended the markets for the eastern United States' best iron ore mines, iron producers still faced another obdurate geographical problem. Almost nowhere in the country did large deposits of high-grade iron ore coincide with metallurgical coal—that is, coal that could be used in blast furnaces in its original state (fig. 13). This constituted a fundamental difference between the geographical conditions in Great Britain and those in the United States. In Shropshire, Staffordshire, South Wales, and the Glasgow area of lowland Scotland, iron, metallurgical coal, and limestone were found in proximate strata within easy reach of open-pit mines or levels into the hillside. Such convenience was rare in the United States, as it was in most European countries.[98] The anthracite coalfields of eastern Pennsylvania, smaller anthracite deposits in southern Virginia, and semibituminous coalfields in central Pennsylvania, western Maryland, and northern Alabama lay closest to good beds of iron ore, though in most cases they were separated by ridges that posed significant obstacles to transportation by water and rail. The great exception to the separation of iron ore from coal was the vast saddle of carbonate, or fossil, ores running through eastern Ohio, across the Pittsburgh region, and down into western Virginia. Some ore deposits within this formation (called "black band" or "coal measure" ores) were interbedded with bituminous or semibituminous coal that could be used

Figure 13. Major coal and iron ore deposits. In the eastern United States, few large deposits of ore and coal were close together. The one great exception was the coincidence of low-quality carbonate ores and bituminous coal girdling Pittsburgh and the upper Ohio Valley.

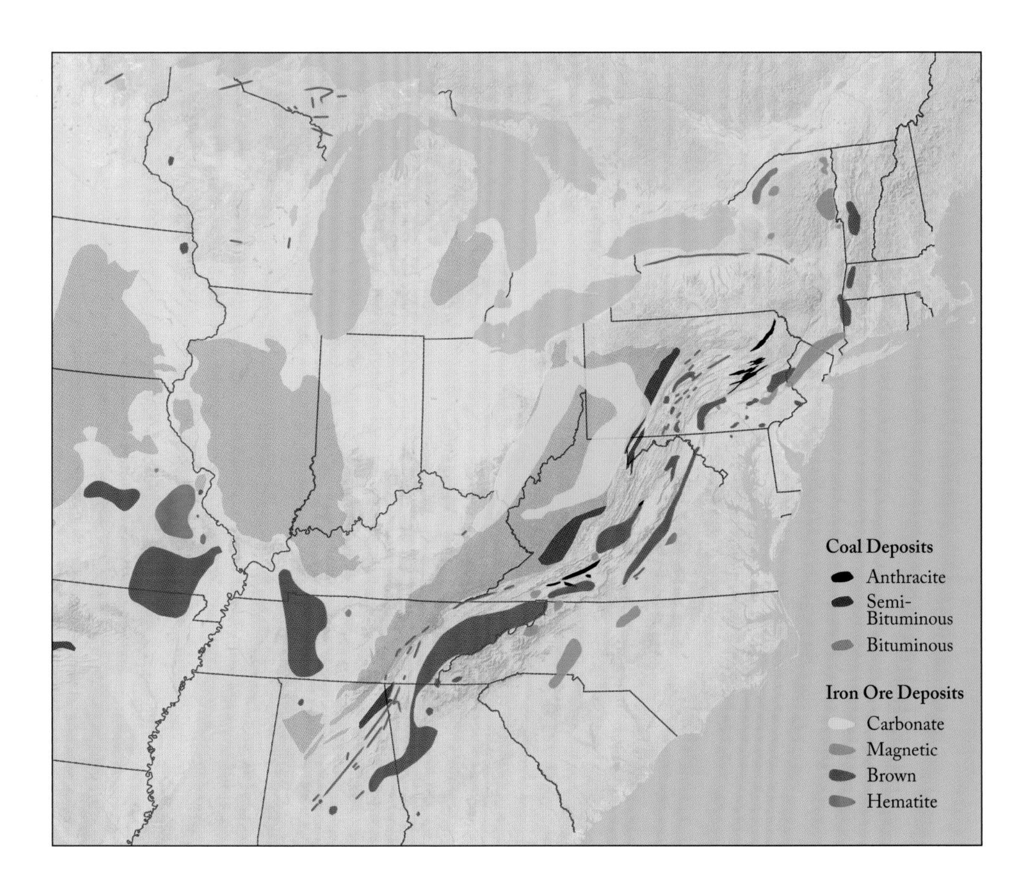

raw (that is, without being coked) in the blast furnace. Some carbonate deposits in this region contained ore that was up to 50 percent metallic iron, but most ores were just 25 to 40 percent iron. Much of the bituminous coal in this region was problematic for smelting as well, since it commonly either contained too much sulfur to make good iron or ran in thin veins between beds of useless shale that made it costly to mine and clean.[99] As we will see in chapter 3, these conditions did not prevent American entrepreneurs from attempting to capitalize on the region's resources, but the quality of these ores and coal sometimes created serious problems.

The location of mineral resources also strongly influenced the development of distinctive subregions where blast furnaces smelted iron with particular kinds of coal (fig. 14). The country's first anthracite iron furnaces were built near substantial deposits of good magnetite and brown iron ores that lay within thirty to fifty miles of anthracite mines to the north, from which the coal was shipped on existing canals and railroads.[100] In the 1840s and 1850s, anthracite furnaces spread down river valleys toward Philadelphia, across northern New Jersey, and

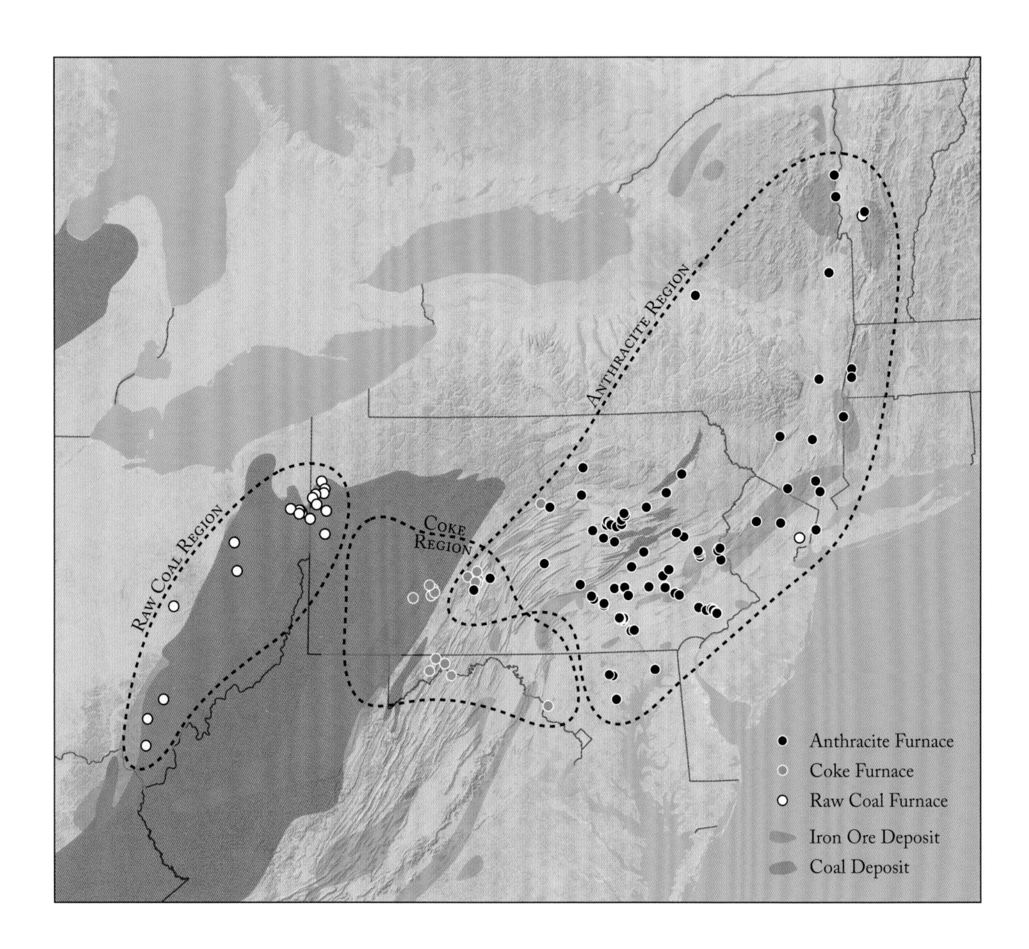

Figure 14. Mineral fuel regions in the Northeast. Black dots stand for anthracite furnaces, gray for coke furnaces, and white for furnaces that burned raw coal. (The map omits the many furnaces that burned only charcoal.) Note how little the mineral-fuel regions overlap. Each fuel required adjustments in furnace design and operation to smelt iron successfully.

up the Hudson River. The expanding network of railroads and canals in the Mid-Atlantic region was key to the spatial diffusion of anthracite furnaces, for relatively inexpensive bulk transport was necessary to get the anthracite to ore locations. Blast furnaces that burned coke or raw coal relied mainly on local deposits of bituminous or semibituminous coal, which was plentiful, though of widely varying quality, throughout western Pennsylvania, eastern Ohio, and western Maryland and Virginia. Throughout the antebellum period, the location of good iron ore deposits exercised the strongest influence on the location of the ironworks that produced "crude" iron, that is, pig iron and blooms. Transportation improvements were just beginning to loosen this constraint at the end of the period.

While the distribution of minerals hindered the quick adoption of British smelting technologies in many parts of the United States, American industrialists did benefit from some natural advantages. One was the abundance of waterpower, which was used to drive the blast engines, which forced air into the furnace to aid combustion, and to run machinery at rolling mills. It was relatively easy, on either side of the Appalachian chain, to find waterpower sites sufficient to run a small ironworks. Although waterpower also influenced the location of rolling mills before steam power became widely available, other geographic factors were equally or more significant in deciding where mills were built. Going back to the seventeenth century, rolling mills were located along established transportation routes near their chief markets in port cities. Colonial "slitting mills" (so called because they ran iron plates through hard-edged grooved rollers that slit the iron into narrow rods, which were then made into nails and other products) were built on the Atlantic coast near Boston, New York City, tobacco ports along Chesapeake Bay, and on tributaries of the Schuylkill River near Philadelphia.[101] From 1785 to about 1830, when most American rolling mills were still fairly small operations, one could find them in manufacturing villages and towns along coastal waterways or inland rivers, where water provided inexpensive transportation to regional markets—places such as Taunton and Fall River, Massachusetts; Stamford, Connecticut; and Covington, Kentucky. In some remote locations with very good iron ore deposits, ironmasters built a blast furnace, finery, and rolling mill or forge, or both, in one industrial complex. After 1800, most of these small integrated works were in the South, where the relative scarcity of ironworks, the distance between settlements, and the availability of river transport argued for concentrating production in single locations from which value-added products such as bar iron could be shipped to customers across a wide territory. Maramec Iron Works in Phelps County, Missouri, was one such operation, the Etowah Iron Works in Cass (later Barstow) County, Georgia, another.[102]

Up to this point, I have considered the dynamics of the industry's development in relation to the location of mineral resources, transportation, population, and national and international economic conditions. It is also important to consider the dynamics of decline. The cessation of iron production recorded in Lesley's survey is one of the best indicators of the volatility of the industry and the varying geography of risk for antebellum industrial investment. The geography of failure provides a fresh perspective on how competitive the industry was when railroads were just beginning, in Marx's phrase, to "destroy space with time" and to expose even frontier ironworks to competition from large-scale, low-cost producers. For all the emphasis on market competition in the literature, we actually know very little about how markets for iron were structured geographically, which products from particular iron districts were sold where, in competition with what other producers. Did small, old-fashioned producers go out of business because of competition from larger firms using more modern methods? Although the data in Lesley's survey do not answer this question definitively, they do provide some clues.

The longevity of blast furnaces suggests that the advent of larger-volume producers in the East did not generally drive small firms out of business. As far as I have been able to determine, furnaces built between 1785 and 1858 had an average life span of about 23 years. Pennsylvania furnaces survived somewhat longer, 29 years on average. The median national figure of 15 years may better represent typical longevity, since the averages are skewed by a handful of venerable firms that produced iron for 50 years or more, including Pennsylvania's Cornwall and Hopewell furnaces, which by 1858 had produced steadily for 113 and 125 years, respectively. These statistics argue against seeing small ironworks as short-lived, ephemeral operations. That about 81 percent of all US blast furnaces burned charcoal as late as 1858 indicates the enduring demand for charcoal iron and the continuing availability of adequate woodlands for making charcoal in many parts of the United States.[103]

The second issue the survey data raise is the meaning of failure in the antebellum iron industry. Lesley and his contemporaries rarely said that an iron company had outright failed. Instead, they spoke of ironworks' being "abandoned." This term literally meant that the works manager had put out the fire, turned off the power, let the workers go, and locked the office door. In some cases, abandonment lasted less than a year. In others, it was permanent: works were dismantled or let fall to ruin. Throughout the colonial and antebellum periods, it was common for ironmasters to abandon their works when the market was slow or when the company suffered any of a long list of difficulties. Furnaces and rolling mills most often shut down when the price of iron dropped too low to sustain production. Brady's Bend Iron Works, for example, "blew out" its three furnaces for a year, from fall 1848 to fall 1849, then resumed production when prices rebounded and orders for western railroads picked up. Carter's Caney Furnace in Bath County, Kentucky, closed in the depth of depression in 1849 but was reportedly back in blast in 1857–58.[104] Blooming Grove Furnace in Montgomery County, Tennessee, suspended operations when the owner died in 1846. It resumed production under a new owner in 1849 and ran until the Civil War. The "want of timber" forced charcoal furnaces to shut down in coastal iron districts in the late colonial period, and charcoal scarcity gradually became more widespread east and west of the Alleghenies in the decades leading up to the Civil War. Some furnaces went back into blast years later on second-growth timber or were refitted to burn coal. Occasionally a furnace exhausted local ore supplies or stopped running when a key piece of equipment failed to operate properly. A number of antebellum furnaces and rolling mills were "washed out by a flood" during spring or fall freshets or halted for long periods to repair damage from fire or explosions. In addition to these hazards, blast furnaces had to be shut down periodically for workmen to clean, repair, and reline the furnace stack. This work was usually done during the winter, when part of the workforce at charcoal furnaces was redeployed to chop wood and make charcoal for the coming year.[105]

These interruptions gave the antebellum iron industry an irregularity or, viewed on a longer timescale, a periodicity, somewhere between the strongly seasonal rhythms of the agricultural calendar and the almost seasonless, perpetual operation of textile factories. The stopping and starting of blast furnaces makes it virtually impossible to gauge precisely the relative strength of any one sector in the industry at any given time. Lesley's data on furnace abandonment are particularly good for Pennsylvania in the 1850s, yet county histories from later dates report that some furnaces that had been abandoned in the depths of the 1857 Panic resumed operations in the floodtide of the Civil War.

Abandonment might best be read as a mixed message that speaks of iron companies' adaptability as well as their struggle to achieve and sustain profitable production. That struggle gradually intensified during the quarter century before the Civil War. The first significant peak in furnace abandonments came in 1837, when ten furnaces were permanently shut down. A number of furnaces were abandoned in 1846, then twenty-five in 1850 and forty in 1855. All but two of the 1850 closures were charcoal furnaces, as were twenty-nine of those abandoned in 1855. Were these furnaces the "mayflies" that Allan Nevins said were put out of business by larger firms—the "tiny ironworks" that "passed rapidly from birth to death," "with poor equipment, and an uneconomic force of men"?[106] The scarcity of labor statistics prevents a decisive answer on this point, but the survey data do permit analysis of the furnaces' technology, scale of production, and age.

Table 1 compares the technological characteristics of furnaces that were abandoned by 1858 and those that remained in operation through that year.[107] The first finding is that the overwhelming majority of abandoned furnaces burned only charcoal. This stands to reason, since charcoal furnaces accounted for 81 percent of all furnaces built up to 1858. At 73 percent of surviving works, they were still by far the most common kind of blast furnace before the Civil War. As

Table 1. Differences of technology and age between abandoned and surviving blast furnaces

	Abandoned by 1858 (%)	Active through 1858 (%)
Primary furnace fuel[a]		
Charcoal (n = 610)	284 (93)	326 (73)
Anthracite (n = 89)	11 (4)	78 (17)
Raw coal (n = 25)	7 (2)	18 (4)
Coke (n = 17)	3 (1)	14 (3)
Mixed fuel (n = 11)	1 (< 1)	10 (2)
Total	306 (100)	446 (99)[b]
Blast type[c]		
Hot	54 (18)	254 (57)
Cold	245 (80)	156 (35)
Warm or mixed	6 (2)	35 (8)
Unknown	1 (< 1)	1 (< 1)
Total	306 (100)	446 (100)
Power		
Steam	61 (20)	243 (54)
Water	3 (1)	13 (3)
Steam and water	6 (2)	26 (6)
Unknown	236 (77)	165 (37)
Total	306 (99)[b]	447 (100)
Annual production (tons)		
7–500	— (11)	— (16)
501–1,000	— (11)	— (19)
1,001–2,000	— (8)	— (28)
2,001–5,000	— (< 1)	— (23)
5,001–9,731	— (0)	— (3)
0 or unknown	— (69)	— (11)
Total	— (99)[b]	— (100)
Longevity (up to 1858)[d]		
0–5	53 (17)	102 (23)
6–10	52 (17)	47 (11)
11–20	62 (20)	143 (32)
21–30	34 (11)	66 (15)
31–50	20 (7)	37 (8)
51–122	16 (5)	31 (7)
Unknown	69 (23)	20 (5)
Total	306 (100)	446 (101)[b]

Sources: Lesley, *Guide* (1859), and [Lesley, ed.,] *Bulletin* (1856–58).

[a] Lesley notes fuel for every furnace in its name and/or description in *Guide* entries. I also consulted all entries in the *Bulletin*, 1856–58.

[b] Shortfall or surplus in totals due to rounding.

[c] In the rare cases where Lesley did not specify blast heat, I assumed that anthracite, coke, and coal-fired furnaces were hot blast and that charcoal furnaces were cold blast.

[d] Calculated as year of abandonment, or 1858 minus year of construction.

one might expect, most abandoned furnaces used cold-blast technology, while a fair majority (57 percent) of surviving furnaces used the newer technology of the hot blast. It is more difficult to draw firm conclusions regarding power and the scale of production because there is little information on these points, particularly for the 135 furnaces (44 percent of those abandoned) that shut down before 1848 or whose year of abandonment is unknown. It is clear that most active furnaces relied on steam power. Because steam engines were still relatively new to the industry at the time of Lesley's survey, it is tempting to conclude that silence on this point meant a works ran on waterpower. However, the number of anthracite furnaces that reported using waterpower, and of cold-blast charcoal furnaces that used steam, warns against making that assumption. The figures for annual production are least complete, particularly for old works in rural areas, but these were almost certainly small producers. Thus surviving furnaces were probably larger on average than abandoned ones. This accords with the generally low rate of abandonment for mineral-fuel furnaces, whose capacity ranged from two to five times that of typical charcoal furnaces. Longevity is the only trait where the differences between abandoned and surviving furnaces appear negligible. If the life span of all abandoned furnaces could be retrieved, I expect the ages of abandoned and surviving furnaces would be even more similar than the table shows.

Nevins's poetic description thus appears to fit Lesley's data quite well on all points except the matter of age. The dominant impression provided by the aggregate statistics is that abandoned furnaces were predominantly small, cold-blast, probably waterpowered charcoal furnaces. It seems intuitively likely that such furnaces were losing ground to larger, more technologically advanced mineral-fuel furnaces, since the latter weathered the late 1840s recession and the early stages of the 1857 Panic much better than did charcoal furnaces. This technological explanation becomes less convincing, however, when we look more closely at the data and at the geography of abandonment, which varied markedly over time.

Exploratory mapping[108] of the survey data reveals that 1847–48 marked a significant dividing line in the characteristics and location of abandoned furnaces. Up to 1847, furnace attrition appears to have been gradual and widely dispersed, as shown by the open circles in figure 15. Furnaces were abandoned in almost every part of the country where iron was made, although concentrations were evident in northern New Jersey, southwestern Pennsylvania, on the Ohio shore of Lake Erie, and along the Virginia-Tennessee border. The New Jersey furnaces were among the first abandonments the survey recorded, dating back to 1818. The New Jersey Highlands had very good magnetic ore, but many deposits were difficult to reach by poor roads through rugged terrain. Lesley noted that old New Jersey furnaces were also among the first to lose customers to the new anthracite furnaces built in the 1840s in eastern Pennsylvania.[109] In southwestern Pennsylvania, furnace companies also faced transport difficulties because of terrain, but a greater problem was the poor quality of local ores. Small or low-grade ore deposits may have precipitated the abandonment of furnaces in northern Ohio, Indiana, and southern New Jersey, although one New Jersey furnace reportedly subsisted on its bog ore reserves for over fifty years.[110]

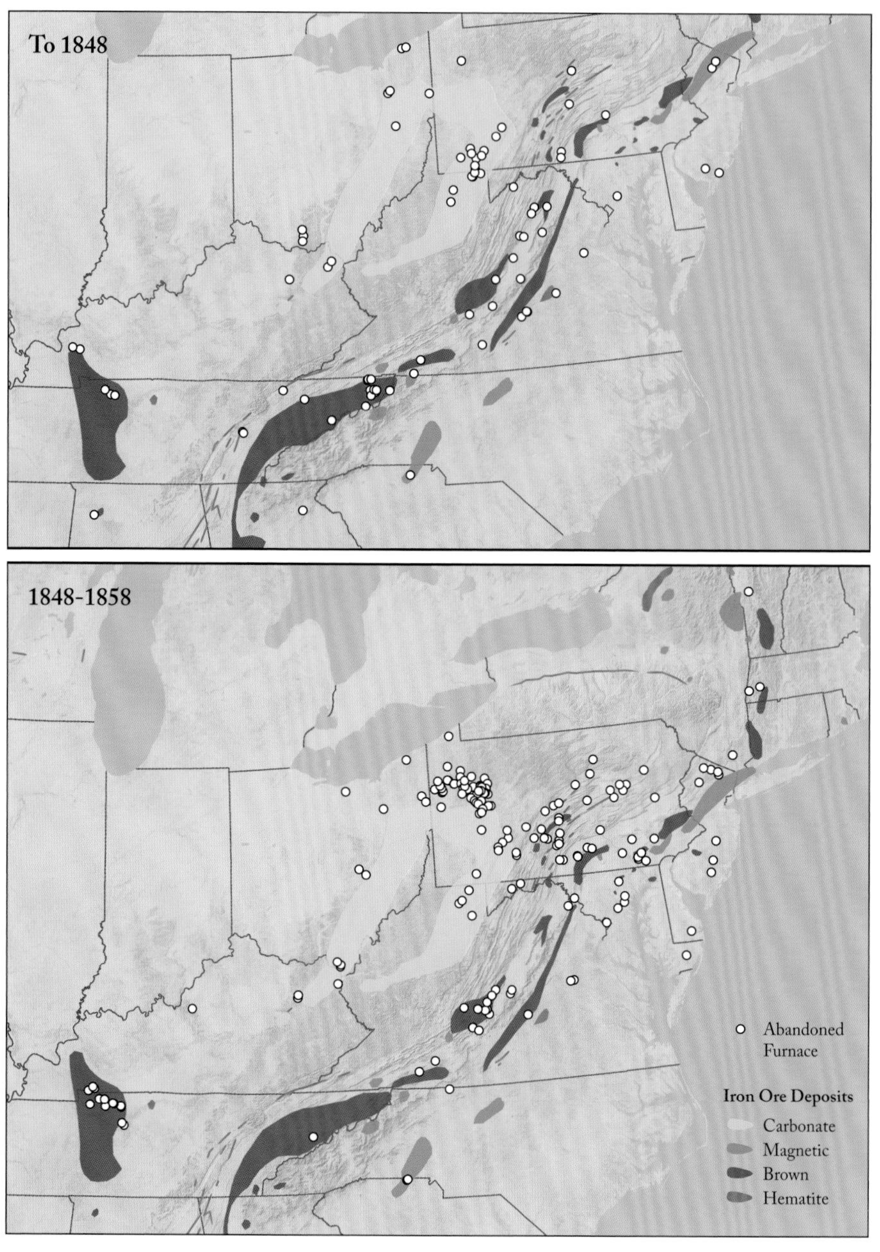

Figure 15. Furnace abandonment. Abandonment up to 1848 (*top map*) was fairly ubiquitous over space and time. Abandonment between 1848 and 1858 (*bottom map*) was much more concentrated, particularly in northwestern Pennsylvania, the Valley of Virginia, and along the Cumberland River in Tennessee.

From 1848 through 1858, the rate of abandonment picked up sharply, and it became much more concentrated geographically. In the Cumberland Valley in western Tennessee and the Shenango and Allegheny Valleys north of Pittsburgh, abandonment in this period was clearly related to localized boom-and-bust cycles. This was not the case in Virginia's iron region, which had gradually acquired blast furnaces and other ironworks from the early 1820s to 1850. There was a marked increase in abandonment in 1849–55, however, when fifteen furnaces closed. The Cumberland Valley of western Tennessee emerged as a significant ironmaking region in the 1830s. It experienced dramatic growth in 1843–55, gaining twenty-six new furnaces. A third of them were abandoned between 1854 and 1856, as were a number of older furnaces. The highest proportion of abandonments occurred in the Shenango and Allegheny Valleys. This area experienced a major building boom in 1843–48, when fifty new charcoal furnaces were constructed to feed Pittsburgh's rolling mills, whose total capacity doubled in the 1840s. From 1849 to 1857, forty charcoal furnaces and five new coal-fired furnaces in the district shut down. All told, 52 percent of the furnaces in the Shenango-Allegheny district went out of operation between 1848 and 1858. Their mean age at abandonment was just 10.9 years, much younger than the state average.[111]

A number of factors contributed to the closure of furnaces in these three districts. Dew credited competition from Pennsylvania anthracite furnaces as the cause of decline among Virginia's pig iron producers. The account books of Virginia's largest rolling mill and foundry, the Tredegar Iron Works in Richmond, show substantial purchases of anthracite pig iron in 1859–60. While it is possible that anthracite pig iron reached the Richmond market before then, the lack of continuous rail lines between Pennsylvania and Virginia's capital meant that heavy freight had to be shifted from canal boats to rail cars or coastal schooners and back to canals, making this route too costly to permit complete substitution of the new kind of pig iron for Virginia's own charcoal iron.[112] The more likely explanation is that most Virginia furnaces were reaching the end of their normal working life of fifteen to thirty years. Not so the western Tennessee and Pennsylvania furnaces that were abandoned in 1849–58. Many of them were built just five to ten years before they were abandoned. Although the great majority burned only charcoal, they did not match the profile of old-fashioned furnaces. Half of them ran at least occasionally on steam power. Nearly 30 percent used the hot blast, a method intended to improve smelting efficiency by heating air before it was forced into the crucible under pressure. The general squeeze on credit that began in 1855 and worsened in 1856 may have hit newer furnaces particularly hard, since they were burdened with debts from the heavy start-up costs required to construct steam-powered hot-blast furnaces.

In the Allegheny Valley, people told Peter Lesley that a few furnaces in the area had shut down owing to scarcity of charcoal, an indication of either poor woodland management or furnace demand beyond the woodland's capacity. He also reported that drought and exceptionally cold weather in 1856–57 had reduced production by curtailing waterpower and impeding shipment of pig iron downriver.[113] GIS analysis of western Pennsylvania furnaces further suggests that distance to improved transportation routes influenced a works' vulnerability to economic downturns. Table 2 shows that mineral-fuel furnaces in Pennsylvania were closer to rivers and channel improvements and much closer to canals and railroads than were the state's charcoal furnaces. Although most charcoal furnaces were built within a short haul of some kind of watercourse, the necessity of spacing them widely within the woodlands that provided charcoal meant some were well away from major rivers. It is not surprising that 60 percent of the charcoal furnaces north of Pittsburgh that were abandoned by 1858 were more than one mile from the Shenango River or Allegheny River, and 88 percent were more than one mile from a canal.[114]

The classic explanation of business failure in the blast furnace sector is price competition. Peter Temin argued that the "downward drift of prices" caused by the shipment of anthracite iron to Pittsburgh, which began in 1852, forced western ironmasters "to change their production techniques and lower their costs or die, as it were, in the midst of plenty."[115] While his argument was chiefly aimed at explaining the late development of coke iron in western Pennsylvania, it also expressed the general notion that competition from mineral-fuel iron put charcoal iron furnaces out of business. One of the key bodies of evidence behind Temin's interpretation was Louis C. Hunter's compilation of newspaper reports of pig iron sales in Pittsburgh from the early 1840s to the beginning of the Civil

Table 2. Distance of furnace from water or rail transport in Pennsylvania

	Distance (in miles)			
Fuel type	Major river	Improved channel	Canal	Railroad
Anthracite (N = 68)				
Mean	3.1	2.8	1.4	0.9
Median	0.3	1.0	0.3	0.2
Coal and/or coke (N = 30)				
Mean	5.3	10.4	3.3	5.3
Median	2.7	11.8	0.6	4.0
Charcoal (N = 210)				
Mean	6.9	10.7	12.7	10.5
Median	4.9	9.6	10.3	8.1

Source: Lesley HGIS and Pennsylvania Transport GIS, analyzed with "Near" function in ArcGIS 9.0, which calculates the shortest linear distance from each point to the nearest feature in the specified feature class.

War. Hunter's tables show that anthracite iron sold at competitive prices in Pittsburgh but rarely below the price of comparable grades of charcoal iron because of the extra cost of transport. Anthracite iron sales jumped in 1854 and again in 1857, when an exceptionally large contract for 14,000 tons of anthracite iron was reported. Most recorded anthracite iron sales in Pittsburgh were of No. 3 iron, which was suitable for foundries but was rarely used at rolling mills because of its hardness.[116]

Hunter's digest of the newspaper reports also shows that Tennessee charcoal iron appeared on the Pittsburgh market by 1841, coincident with the building boom in the Cumberland Valley. That district's industrial growth was led by a number of former Pennsylvania iron men, including Daniel Bell and Lewis Woods. Although newspapers recorded only small lots of Tennessee pig iron and blooms being sold in Pittsburgh in 1846–53, the tonnage suddenly spiked (or was suddenly much better reported) in 1854–55, when Tennessee iron accounted for nearly 11 percent of reported sales, compared with anthracite iron's 19 percent. At prices of $40 to $45 a ton, Tennessee charcoal pig iron was competitive with Allegheny iron. The first big wave of charcoal furnace closures in western Pennsylvania came in 1849–51, the second in 1855–57. In 1855, prices fell as low as $25 a ton and rarely topped $31 for the best charcoal iron. Low-grade anthracite and small amounts of perhaps even lower grade coke iron sold well at these prices, but so did the region's most competitively priced charcoal iron—the pigs shipped upriver by the "old-fashioned" furnaces of the Hanging Rock iron district in Ohio and Kentucky.[117] If price competition undermined western Pennsylvania charcoal iron companies, the pressure came from Tennessee and Ohio charcoal iron as well as eastern Pennsylvania anthracite iron.

Localized problems such as shortages of charcoal may also have shut down furnaces in northwestern Pennsylvania, compelling Pittsburgh iron manufacturers to look farther afield for crude iron.[118] Price competition between charcoal iron producers, whose pig iron was used chiefly in wrought-iron manufacturing at rolling mills and forges, may have been more significant than anthracite or coke iron in undermining marginally profitable enterprises.

Lesley's survey provides scanty but suggestive information on markets. Figure 16 shows the destinations beyond the "home market" to which furnace owners, managers, or workers said they shipped pig iron. Lesley reported such destinations for 93 furnaces, 12 percent of the total. For about half of those, he named more than one destination. The resulting map significantly underrepresents pig iron shipments in the late antebellum period. One particularly glaring omission is shipment of anthracite iron to Pittsburgh. The map also shows no shipment of iron out of the Hanging Rock iron district, which Hunter documented as a major source for iron sold in Pittsburgh and which I found a significant supplier of pig iron to Cincinnati, Wheeling, and smaller cities in central Ohio.[119] We know

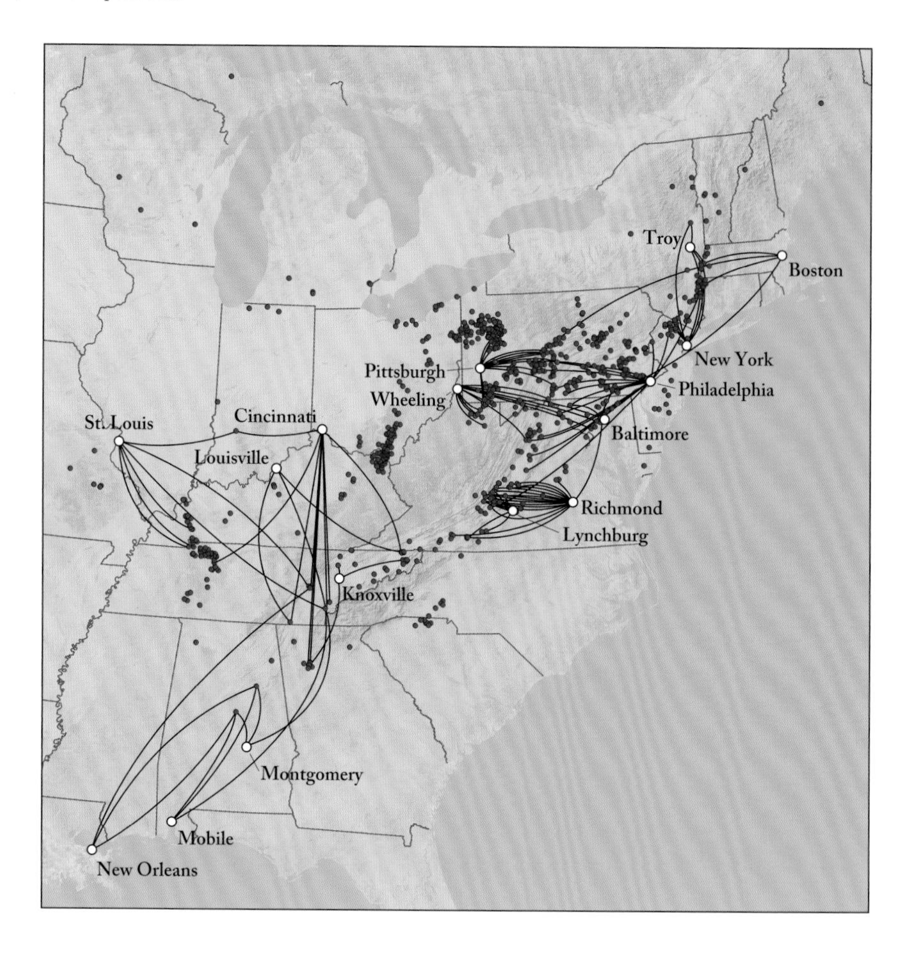

Figure 16. Pig iron shipment beyond local markets, ca. 1854–58. The red dots represent blast furnaces. Differences in topography and transportation systems, as well as the volume of demand in urban markets, produced very different geographical patterns in the Northeast and the South.

that Lesley's trip to southeastern Ohio was rushed and that he missed his best opportunity to meet with local ironmasters when his host forgot the key to the assembly hall. Lesley tended to err on the side of caution: when information was uncertain or incomplete, he omitted it from the *Guide*. He may have assumed that readers understood that furnaces mainly supplied local markets. Further research in individual company records would be required to provide a more thorough mapping of crude iron markets.

Despite these limitations, the map of pig iron markets shows intriguing patterns of regional segmentation of markets by major topographic barriers. Lesley recorded only two western furnaces shipping iron east over the Appalachian divide. Wheeling, connected to the East Coast by the National Road, the Chesapeake and Ohio Canal, and the Baltimore and Ohio Railroad, drew some pig iron from furnaces in eastern Maryland, including Antietam Furnace and Baltimore's Elk Ridge Furnace. As a whole, however, pig iron markets in the northeastern part of the United States appear to have been regionally delimited. Another strongly

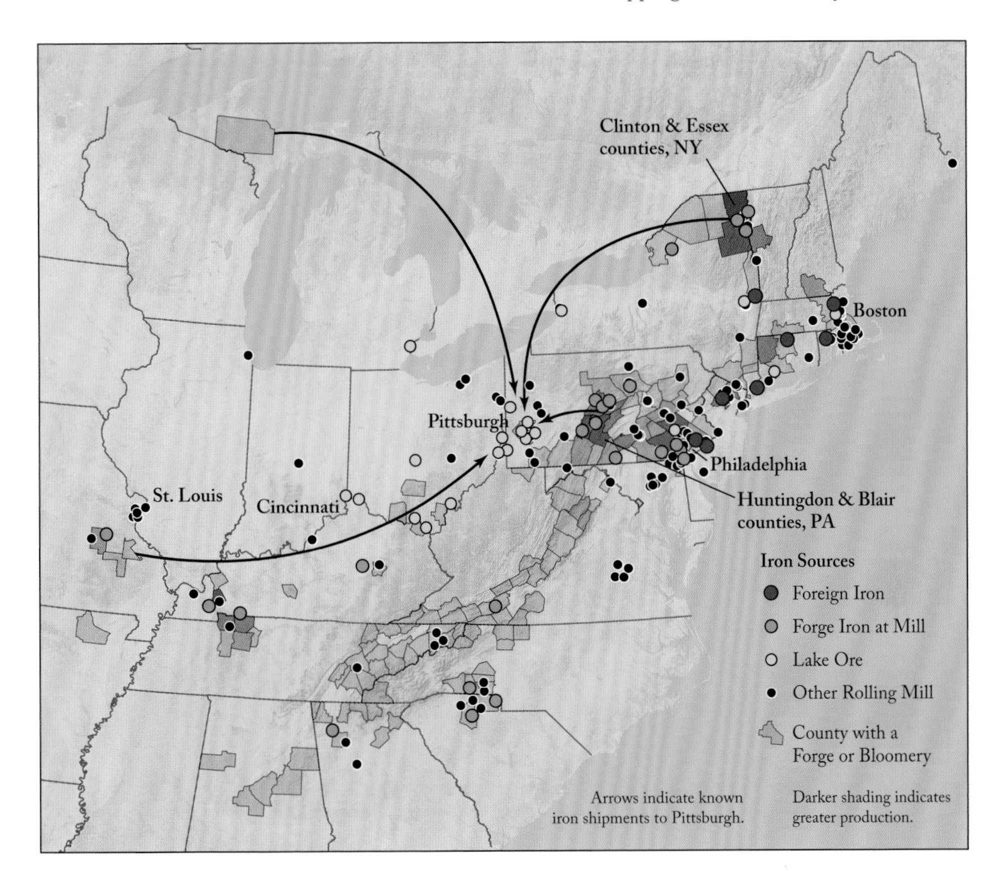

Figure 17. Sources of semifinished iron and iron ore for rolling mills, ca. 1854–58. Forges in upstate New York and Pennsylvania provided significant amounts of blooms and bar iron to rolling mills in greater Pittsburgh and Philadelphia. In the South, mills were more likely to use iron refined at their own forges. Shipping on the Great Lakes and canals carried rich ores from Lake Superior and Lake Champlain to Pittsburgh and Ohio Valley mills.

focused region was the Valley of Virginia, whose charcoal furnaces sent almost all their pig iron to Richmond and Lynchburg. To the southwest, a very different geography prevailed. Before the great dams and reservoirs of the twentieth century, southern rivers were true highways that made long-distance shipment of freight safer and more affordable than shipment on the more steeply graded rivers of New England and the Mid-Atlantic. If the straight lines on the map followed the actual routes of flatboats, barges, and paddle wheelers from Tennessee and Georgia furnaces to their customers in Cincinnati, Louisville, St. Louis, and New Orleans, some would snake for hundreds of miles.

Most antebellum forges and bloomeries met local demand for domestic iron wares, farm implements, horseshoes, wheel rims, and nails. By 1840–50, however, some forges in the Mid-Atlantic and upstate New York attained a new scale of production and specialized in high-quality merchant bar for rolling mills and foundries. Forges in Huntingdon and Blair Counties, Pennsylvania, Essex and Clinton Counties, New York, and in northern New Jersey were foremost in this development (fig. 17). The counties where these forges were located (shown in the darkest tone on the map) possessed deposits of exceptionally good iron ore.[120] Figure 17 also maps several other important kinds of iron used at antebellum rolling mills. The red circles mark rolling mills that used foreign iron, mostly bar iron imported from Sweden. Note that most such mills were on or near the East Coast. The orange circles are mills that included a forge for refining pig iron. Like the few companies that built one or more blast furnaces adjacent to their mills in this period, the integration of forge and rolling mill operations marked a step toward the spatial agglomeration of production. The yellow circles indicate rolling mills that used ore from Lake Champlain (Essex or Clinton County) or Lake Superior (Marquette County, Michigan).[121] These "lake ores" were shipped great distances. Lesley recorded both kinds as being used at Pittsburgh, presumably coming by way of the Erie Extension Canal (completed in 1844). The map does not show the ubiquitous use of scrap metal at rolling mills.

New questions arise from seeing the lineaments of pig iron markets and the contrasting geographies of supply from foreign and domestic forges and new iron mining districts. Company records and other sources tell us how much pig and bar iron American ironworks produced, but far less about where crude iron was used. We still have much to learn about the connection between supply and demand, from iron producers to the foundries and machine shops that manufactured the engines and implements of the Industrial Revolution.

The Geography of Antebellum Iron

Lesley's survey yields several insights into the character and development of the antebellum iron industry. First, it shows that the industry was technologically diverse (fig. 18). Previous characterizations of the industry as divided between modern and antiquated ironworks have not appreciated the great variety of hybrid combinations of technology and experimentation in pursuit of mineral-based technologies at American ironworks before the Civil War. The early and sustained success of anthracite iron companies in southeastern Pennsylvania stands out as a lucky exception, for only in that region were deposits of metallurgical coal and good iron ore economically proximate and suitable for making iron of marketable quality using the methods and machines available in the 1830s. Everywhere else, the distance between resources, the often less than ideal quality of coal and ore, and other problems delayed the sustained, profitable production of iron made with coal or coke. Technologies that could be adopted regardless of the quality of mineral resources were tried in every region, though not by every firm. By the 1840s, steam engines and the hot blast were being widely adopted wherever coal was readily available or could be shipped cheaply, although waterpower was used at many ironworks throughout the period. Rolling mills varied least from region to region. They all embodied the same basic bundle of technolo-

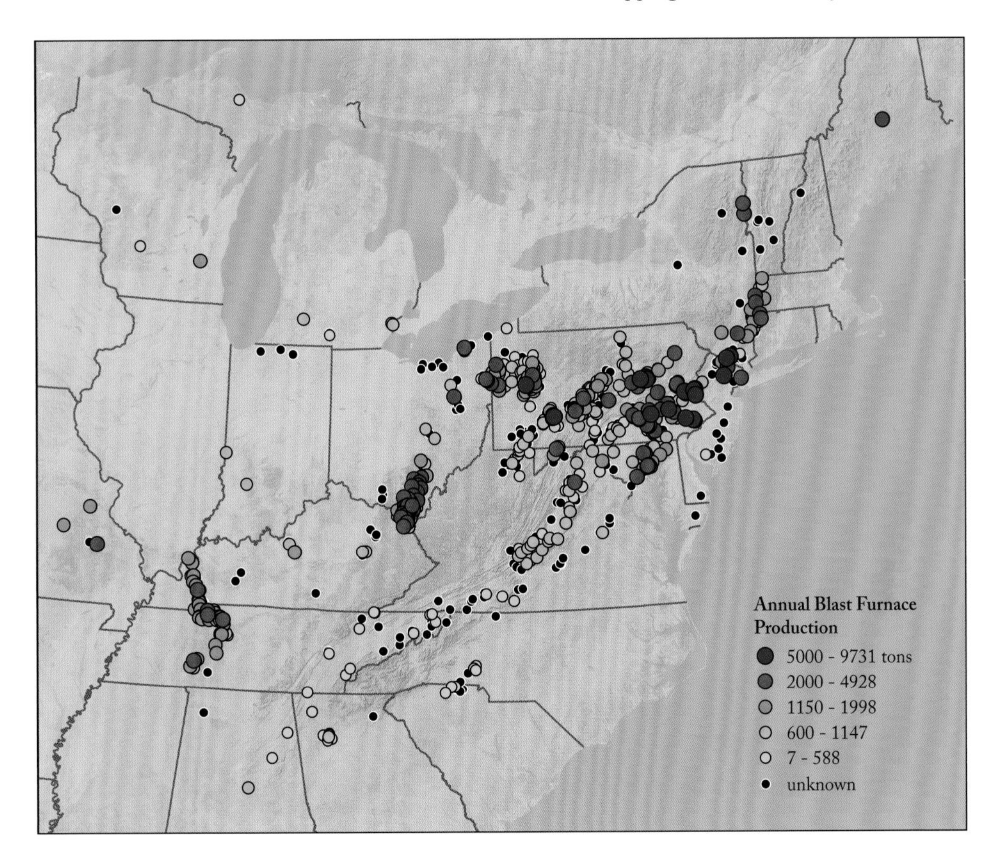

Figure 18. Annual blast furnace production, ca. 1856–58. One indication of the diversity of the antebellum iron industry is the wide range in furnace capacity, as shown here for all furnaces with production data in *The Iron Manufacturer's Guide*.

gies: roll trains, hammers, and puddling and heating furnaces. They varied chiefly in the scale of operations, which was determined by the number and size of each kind of equipment. Lesley recorded the use of the very large, steam-powered Nasmyth hammer, for example, at only a few large rolling mills with heavy forging works.

Second, the survey reveals significant regional variation in the cycles of growth and decline, as well as varying regional sensitivity to changes in international tariffs and national economic cycles. Iron companies in the Mid-Atlantic states were least apt to close during national economic downswings, whereas they responded most strongly to the protection against foreign competition offered by the tariff acts of 1842 and 1849. Investment in ironworks increased everywhere during buoyant economic times and decreased sharply in all regions in 1855–58, but the recession of the late 1840s hit western Pennsylvania and western Tennessee harder than other regions. What had been slow but steady growth in Virginia's iron district also came to a halt about 1849. A few years later the Marquette range, the first new iron region to be developed in forty years, began shipping ore and pig iron via the Great Lakes to works in Pittsburgh, Ohio, Michigan, and Illinois.

To borrow a metaphor from William Cronon, the discovery of excellent iron ore in northern Michigan began the shift of the east-west hinge of the iron industry from the eastern side of the Allegheny Mountains to Pittsburgh.[122]

The third insight that emerges from Lesley's survey is that the antebellum iron industry functioned primarily within a set of distinct regions. Their geographical boundaries were permeable and imprecise, but for most of the antebellum period, six major regions were evident in the spatial clustering of certain technologies; the scale of production and the extent to which smelting, forging, and rolling operations were integrated; the geographical extent and urban foci of pig iron markets; the dominance of certain modes and routes of transportation; and distinctive patterns of historical development. In delineating these regions I also took into account the quality of resources, which largely determined the potential for implementing certain technologies, and major topographic features that influenced transportation routes and markets. Chapter 2 will explore two other important characteristics that differed between regions: their dominant labor regimes and the kinds of iron communities they contained. For the sake of drawing broad regional distinctions at this point, I considered only the basic question whether a region's ironworks employed exclusively free labor or a combination of slave and free labor.

Figure 19 and table 3 lay out the summary characteristics of six antebellum regions and a seventh, the Great Lakes region, that was just beginning to take shape at the end of the period. The four regions covering industrial New England, eastern New York State, and Pennsylvania incorporate twelve smaller ironmaking regions that Robert B. Gordon identified in his study of the American iron industry. Gordon was the first to suggest that natural and economic resources and localized cultural preferences collectively formed distinctive iron regions in the antebellum period.[123] I have included additional factors in an effort to capture the way ironmaking regions functioned as contexts for production (see table 3 for the details). Here I want to emphasize the factors that most strongly influenced each region's development from 1830 to 1860. I do not mean to reify iron regions as if they possessed any explanatory power in themselves or existed concretely the way buildings or mountains do. All regions are human inventions born of our desire to make sense of the world. Other regional encapsulations are certainly possible. I propose the following regional descriptions as summaries of the quite different historical geographies of ironmaking that were manifest in the American landscape from 1800 to 1860.

New England was the first region in North America to develop an iron industry. From the release of entrepreneurial energy after the War of 1812 through the Civil War, ironworks in New England and the Champlain and Hudson Valleys focused strongly on providing high-quality pig iron, blooms, and forge iron to the region's proliferating manufacturers, from Cyrus Alger's ordnance foundry in South Boston to the countless machine shops in New York City and New England manufacturing towns. Charcoal furnaces in Connecticut's Salisbury district provided much of the region's pig iron, while forge iron and merchant bar came increasingly from rolling mills and anthracite furnaces using ore brought by canal schooners down the Champlain Canal to the Hudson River. As the earlier map of pig iron markets suggested, however, the region's furnaces did not keep pace with demand. Crude iron had to be shipped in to New England by boat, rail, and canal, just as coal was. This region's rolling mills were the main consumers of European pig and bar iron.[124]

The Mid-Atlantic region was the powerhouse of the iron industry. It too had colonial origins, but here the industry grew most rapidly during the antebellum period. By 1858, the area had 31 percent of the country's rolling mills, including many of the largest operations. This was Philadelphia's hinterland, a region of synergistic industrial development between the city's dynamic metalworking sector and the anthracite coal trade. The particular needs of heavy equipment manufacturers in Philadelphia, such as the Baldwin Locomotive Works, fostered

Figure 19. Antebellum iron regions and districts. Each dot represents one ironworks. Regions are based on differences in technology, scale, transportation, markets, products, and historical development. For a full explanation, see table 3.

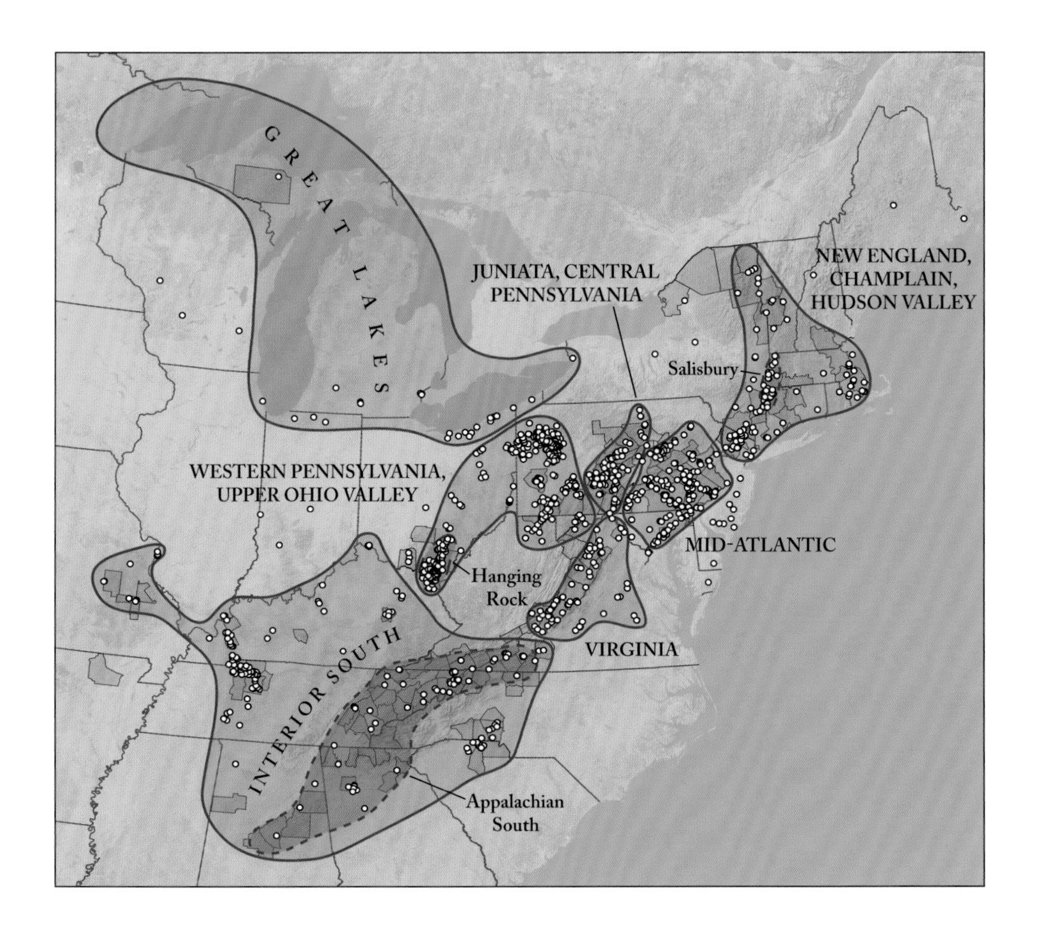

Table 3. Characteristics of antebellum iron regions

Region	Main transportation routes	Technology	Scale and integration	Development	Product specialities	Market foci
New England/ Champlain/ Hudson Valley	Hudson River, Lake Champlain, Champlain Canal, Delaware and Hudson Canal, railroads, coastal shipping	Cold-blast charcoal and anthracite furnaces; puddling and nail mills; steam and waterpower	Small to very large	Colonial roots; steady growth	High-quality charcoal pig, nails, rods and shapes, steel, engines	New York, Boston
Mid-Atlantic	Delaware, Lehigh, Schuylkill, Susquehanna Rivers; railroads, canals	Anthracite furnaces, some charcoal (hot and cold blast); bloomeries and forges; puddling and nail mills; Nasmyth hammers; heavy foundries; steam and waterpower	Small to very large; high concentration of large ironworks; some large firms becoming integrated	Colonial roots; constant growth, with anthracite iron and mill booms 1844–50, 1853–55	Blooms and forge iron, anthracite pig, sheet and boilerplate, locomotive iron, rails, nails	Philadelphia, Baltimore
Juniata/ Central Pennsylvania	Juniata River, Main Line Canal, railroads	Charcoal furnaces (hot and cold blast), a few anthracite, coke; forges; steam and waterpower	Small to moderate	Strong, steady growth 1818–40; mills built 1827–50	Forge and pig iron	Pittsburgh, Philadelphia
Western Pennsylvania/Upper Ohio Valley	Allegheny, Ohio, Monongahela Rivers; Pennsylvania Railroad, Erie Extension Canal	Charcoal (hot and cold blast), raw-coal furnaces from 1845, a few coke; puddling and nail mills; heavy foundries; steam and waterpower	Small to moderate furnaces, large rolling mills, some with integrated furnaces	Episodic growth and decline in furnace sector; steady growth of rolling mills	Charcoal pig iron, nails, bar, sheet iron, rails, ordnance	Pittsburgh, Wheeling
Virginia	Shenandoah, James Rivers, roads, railroads, James River and Kanawha Canal	Charcoal furnaces (cold and hot blast); forges; limited puddling; steam and waterpower	Generally small	Colonial roots; steady growth arrested ca. 1849	Charcoal pig iron, bar, nails, ordnance	Local markets, Richmond, Lynchburg
Interior South	Tennessee, Ohio, Cumberland, Mississippi Rivers	Charcoal furnaces (cold blast outside Cumberland district), bloomeries, some puddling and nail mills; steam and waterpower	Small to moderate, some rural works integrated	Furnaces and bloomeries from ca. 1800, mills post-1826; strong growth 1830–54	Charcoal pig iron, bar, sheet and plate, nails, one large rail mill	Local markets, St. Louis, Cincinnati, Pittsburgh, New Orleans, Mobile
Great Lakes	Great Lakes via Sault Sainte Marie locks, canals	Hot-blast charcoal, raw-coal furnaces	Small to large	Post-1852, aside from a few small furnaces	Iron ore	Chicago, Detroit, Cleveland, Pittsburgh

specialization at rolling mills that produced sheet iron and boilerplate. And the rapid development of rail lines connecting New York, Philadelphia, Baltimore, and Washington to points north, west, and south supported nine rail mills. Compared with other parts of the Northeast, this region was exceptionally well supplied with navigable rivers as well as waterpower, though the obstacle posed by the long ridges of the Allegheny foothills prompted early development of railroads and canals to get anthracite to market, which later served the development of the region's signature anthracite blast furnaces. A number of the region's largest rolling mill companies, including the Trenton Iron Works, began to build their own blast furnaces to regulate supply and to capitalize on the region's excellent ores. Big operators did not obliterate older, smaller producers, however. A number of furnaces dating from the colonial or early federal period continued to smelt local deposits of high-grade iron ore through the Civil War, and their pig iron commanded top prices on the Philadelphia market. Baltimore was a secondary market hub for this region. The rolling mills and nail factories along its harbor and northeast into Delaware used iron from a plethora of furnaces in Pennsylvania, New Jersey, Maryland, and northern Virginia.[125]

The Juniata district spanned the great S-curve of ridgelines through central Pennsylvania, a topographical divide that was not fully breached by the Pennsylvania Railroad until 1852. The region's iron industry began along outcrops of good hematite ore, but the lack of metallurgical coal meant that the Juniata developed only a few mineral-fuel blast furnaces on its eastern and western edges, where anthracite and coke, respectively, could be shipped economically to ore locations. The region thrived in the middle of the antebellum period, when its forges and charcoal blast furnaces shipped high-quality crude iron to Philadelphia and Pittsburgh. After about 1845, increasing amounts of its production were consumed by the region's own rolling mills. By the 1850s, however, the Juniata was beginning to be eclipsed by iron regions to the east and west.

The iron industry in western Pennsylvania and the upper Ohio Valley focused on Pittsburgh from its earliest days, since the city's location at the confluence of three major rivers and its connection to western territories via the Ohio River linked it to a large and rapidly populating hinterland. Pittsburgh was a major industrial crossroads where the traffic of development was delayed by a long yellow light, partly because the region's rugged topography slowed development of the Connellsville coke district southeast of the city.[126] Pittsburgh had no blast furnaces until 1859. Rolling mills and foundries, however, increased steadily from the early 1820s to the end of the antebellum period. A handful of rolling mill owners in Pittsburgh and the upper Ohio Valley sought to ensure their own supply of pig and forge iron by acquiring furnaces and forges in the Allegheny and Juniata districts. As the volume of iron processing increased, most mills sought crude iron from an expanding network of suppliers. By 1857, furnaces and bloomeries from Lake Champlain to Iron County, Missouri, were sending iron to Pittsburgh, most of it charcoal iron. The Pittsburgh region's blast furnaces went through marked cycles of boom and bust. First Fayette County to the south and then the Shenango and Allegheny Valleys north of Pittsburgh experienced concentrated bouts of furnace construction and abandonment. The localized boom in the 1840s was matched only by the rapid addition of charcoal furnaces in Ohio's Hanging Rock district in the mid-1850s. The growth cycles suggest plenty of entrepreneurial ambition in this region, but the number of abandonments calls for further research to understand localized volatility. We particularly need a better grasp of just how good or bad the coals and ores of western Pennsylvania were for making iron with the available technologies and the varying costs of transport within Pittsburgh's extended hinterland.

The South contained at least two distinctive iron regions. In Virginia, the Valley of Virginia and the Piedmont stood out as a historically productive region that made good iron. Its earliest ironworks dated back nearly as far as the first in New England. By the early nineteenth century, the region had a tradition of planter-industrialists, exemplified by David Ross of the Oxford Iron Works and William Weaver of Buffalo Forge. Theirs was a paternalistic model of slave-based agriculture and industry in which small, well-run ironworks were a linchpin in the self-sufficiency of the Virginia economy. As Richmond developed into a manufacturing center in the late 1830s, demand for iron increased and Virginia's mineral resources were harnessed for somewhat larger-scale production, though most furnaces remained small. The region began to flag in the late 1840s and added no new furnaces or rolling mills in 1850–60.[127]

The interior South diverged markedly from Virginia and the rest of the country in the geographical extent of its markets. Owing to the length of its navigable rivers, this region's furnaces and bloomeries were able to ship pig iron and blooms as far north as Pittsburgh and as far south as Mobile and New Orleans. The survey provides relatively little information about production in this region. We know that Ben Lyman and Joseph Lesley Jr. were not always warmly welcomed here. Their heavy discounting of southern ironmasters' reports also makes it difficult to draw firm conclusions from the production data they gathered. Their reports do suggest that a larger proportion of crude iron may have been made at bloomeries in the Appalachian South than was true for any other region. The few rolling mills in this mountainous area were small, excepting only the Gate City rail mill built in Atlanta in 1858. A number of mills were part of integrated rural ironworks that included one or more blast furnaces and a forge, like the Etowah Iron Works.

The Cumberland district in western Kentucky and Tennessee was quite different, with a big, steam-powered rolling mill supplied by a large number of mostly hot-blast charcoal furnaces that were built in bursts of development in

1845–47 and 1853–54. The technology and developmental dynamics of the Cumberland district resembled those of the Hanging Rock district, but Tennessee lacked Ohio's relatively easy access to a major industrial market. St. Louis had five rolling mills by 1858, the Cincinnati–Covington area had four, and small mills in a few other Ohio River towns helped spur growth in the Cumberland district, but the interior South as a whole had no industrial engine comparable to Pittsburgh. This situation would change after the Civil War, with the development of the rich iron ore and bituminous coal beds in what became known as the Birmingham district in central Alabama.[128]

The last region, the Great Lakes, was a mineral frontier in the 1850s. When ironmaster John Fritz visited Marquette in 1852 in the company of prospective investors from Cleveland, Ohio, just a few surficial iron mines had begun extracting the wealth of iron ore embedded along the shores of Lake Superior. Almost as soon as locks were completed through Sault Sainte Marie in 1855, permitting freighters to pass from Lake Superior to Lake Huron and beyond, mine companies began shipping Marquette ore to western Pennsylvania and northern Ohio. Great Lakes ore became the staple source of iron for postbellum ironworks in Detroit, Chicago, and along the north shore of Indiana. In the antebellum period, however, the Upper Peninsula was a remote, hardscrabble frontier. Fritz wrote years later that he wished he had been able to buy the half share of the Jackson Mine on offer in 1852; it would have made him a millionaire. Joseph Lesley had the same pipe dream. "I shall be obliged to stop in N.Y. City," he wrote Peter from Detroit in 1858, ". . . to learn what I can of 2 blast furnaces just erected at Jackson Mnt. Lake Superior. In my opinion that is to be a great iron making region. What think you of setting up there?"[129]

Identifying the differences between antebellum iron regions clarifies the geographical conditions that influenced the development of the industry between 1800 and 1860. Considering how much variation there was in terrain, resource quality, distance to markets, market size, demand, and competition, it is no wonder the American iron industry was so diverse. Peter Lesley, his assistants, and the members of the American Iron Association interpreted that diversity as a problem. In the spirit of national improvement and modernization, they wanted companies across the country to adopt large-scale methods of iron smelting and manufacturing. Like later historians of the industry, they discounted the obstacles and economic disincentives that limited adoption of the British model's component parts almost everywhere except in the Mid-Atlantic states.

Lesley's survey says least about labor in the iron industry. As a geologist, Lesley naturally focused on the mineral resources needed to make iron. His fascination with technology, and his fealty to the interests of the American Iron Association, explain the *Guide*'s emphasis on machinery, power, physical plant, and production. The *work* of making iron interested Lesley far less. Iron company owners and managers, however, were acutely aware of the importance of labor, for whatever technology a works employed, ironmaking in the antebellum period was labor intensive. Just as the physical characteristics of an iron region influenced the kinds of technology a works employed, the choice of technology influenced the structure and composition of the workforce. To see these relationships, we must change scale to situate ironmaking in the local context of industrial communities and look inside the working environments where labor took place.

We cam' na here to view your warks,
In hope to be mair wise,
But only, lest we gang to Hell,
It may be nae surprise.

ROBERT BURNS, August 1787

2 The Worlds of Ironworkers

The Scots poet Robert Burns scratched the poem above on the window glass of an inn opposite the gate where he had been denied access to Scotland's largest ironworks, the Carron Company, one August morning in 1787.[1] Like many other visitors, Burns had been curious to see the huge foundry's fiery industrial landscape. Had he come another day, he might have given us a rich poetic description of the Carron works. Or perhaps not. Ironworks managers and ironworkers were jealous of their business. In the late eighteenth and early nineteenth centuries, Carron and other leading British iron manufacturers screened visitors carefully to fend off industrial spies. British law prohibited the emigration of skilled ironworkers as part of the general effort to prevent loss of the technical expertise that gave Great Britain much of its advantage over other industrializing nations. Those restrictions were lifted after the Napoleonic Wars, but many firms continued to guard the craft of ironmaking into the middle of the nineteenth century. Skilled workers also protected their privilege by maintaining apprenticeship as a common if not always formally observed system to restrict labor supply and ensure that kinsmen had access to jobs and advancement.

Figure 20. Coalbrookdale by Night, by Philip James de Loutherbourg, 1801. Oil on canvas, 680 × 1,067 cm. Science Museum, London. In this painting of the birthplace of coke iron, industry is mysterious and grim. The extremes of blazing red coke, sulfurous yellow smoke, and deep shadow suggest a dirty, dangerous place.

To outsiders like Burns, large ironworks were fabulous, extreme places whose operations were difficult to comprehend. Painters commonly portrayed British iron towns like Coalbrookdale and Merthyr Tydfil as "smoking ruins . . . prey to the devouring element" of fire, as in Philip James de Loutherbourg's painting of Coalbrookdale, Shropshire (fig. 20). From miles away one could see the lurid glow of burning banks of coke and the flames shooting from the tops of blast furnaces, which one contemporary writer described as volcanoes "breathing out their undulating pillars of flame and smoke."[2] Depictions of American iron landscapes are more varied. Some have a much more bucolic cast. Fannie F. Palmer's 1862 lithograph of West Point Foundry, for example (fig. 21), shows industry nestled comfortably in the countryside, the ironworks a quiet pocket of productivity. These views hint at the diversity of ironmaking environments in the nineteenth century and the ways they were perceived. Literature and art give us outsiders' perspectives on the industry and the places it created. Men who worked on the shop floor portray their lives and workplaces in a very different light.

The Devil's Place

Rebecca Blaine Harding (1835–1901) grew up in Wheeling, Virginia, during its transition from a quiet market town to the country's leading manufacturer of cut nails. Between 1832 and 1853, seven rolling mills were built in Wheeling, plus an

Figure 21. Foundry at West Point, by Fannie F. Palmer, 1862, lithograph published by Currier and Ives, 15.75 × 19 in. West Point Museum, United States Military Academy, West Point, NY. Two gentlemen in the foreground take their ease overlooking the tamed wilderness of the Hudson Valley. Through the cleft in the trees, one glimpses the roofs of West Point Foundry, a railroad train heading south over the causeway, and beyond, a steamboat and sailing craft on the Hudson River.

eighth just down the Ohio River in Benwood. All the mills made cut nails, and three of them specialized exclusively in that product. By the early 1850s, Wheeling was third among American iron towns in number of rolling mills, with one fewer than Philadelphia and half as many as Pittsburgh.[3]

Harding wrote a novella about ironworkers in Wheeling titled "Life in the Iron Mills," which was published anonymously in the *Atlantic Monthly* in April 1861. It is the tale of a Welsh puddler named Hugh Wolfe and Deborah, the saintly Welsh woman who loves him. Unlike the coarse men he works with at the rolling mill, Hugh is an artist who carves wild figures from waste material at the mill. The dramatic turning point in the story comes when the works manager brings visitors to observe the night shift. One of the visitors stumbles over Hugh's sculpture of a woman. Distracted by the discovery, he does not notice when Deborah steals his wallet. She hopes the money will enable Hugh to buy wood and proper sculptor's tools so that he can become a real artist. Instead, Hugh is accused of the theft and put in jail. He commits suicide while awaiting transport to prison.

In this melodramatic story, Hugh is imprisoned by industry long before he lands in the Wheeling jail. His work at the rolling mill brutalizes his finer sensibilities as it sickens his body with consumption. The environment he works in is a living hell. Harding describes it through Deborah's eyes as she walks along the cinder path to the mill in the dead of night to bring Hugh his meal of bread, salt pork, and ale:

> The mills for rolling iron are simply immense tent-like roofs, covering acres of ground, open on every side. Beneath these roofs, Deborah looked in on a city of fires, that burned hot and fiercely in the night. Fire in every horrible form: pits of flame waving in the wind; liquid metal-flames writhing in tortuous streams through the sand; wide caldrons filled with boiling fire, over which bent ghastly wretches stirring the strange brewing; and through all, crowds of half-clad men, looking like revengeful ghosts in the red light, hurried, throwing masses of glittering fire. It was like a street in Hell. Even Deborah muttered, as she crept through, " 'T looks like t' Devil's place!"[4]

Harding's description contains all the elements of bourgeois observers' response to heavy industry in the early nineteenth century: fascination with scenes that seemed Shakespearean or biblical in their metaphorical richness; fear of dangerous work that exposed men to tremendous heat and violent chemical reactions; revulsion mixed with concern for the health and morality of the workers who labored in a brutal and irredeemably masculine environment; and the allure of being in the presence of craft skill beyond one's comprehension. If railroads embodied the aggression and excitement of the Industrial Revolution, ironmaking reached deeper into observers' imaginations, calling up mythical, primordial connotations. John Stilgoe captured the tone of anxiety that modern iron manufacturing roused in some Americans when he wrote that the artifice of ironmaking

> embodies rape, and abortion and transmutation too. Artifice thrusts into the very womb of mother earth, into infernal dark, and wrenches living rock from living rock. Smelting, forging, and casting torment the aborted fetuses with fire. Earth, air, fire, and water combine in an unholy alchemical alliance from which husbandmen stand away, shielding their eyes. . . . All the mystery of making obscures the places of artifice. Only the artificers understand. Onlookers stand back and hope for success, and are afraid.[5]

Similar anxiety imbued the warnings of Welsh preachers who urged their rural congregants to stay away from industrial South Wales, which they saw as an inhuman wasteland that God had already judged and condemned. Where once there had been mountains " 'swept bare, with no residence save human, rural dwellings, and the scattered flocks of mountain sheep,' " one preacher wrote, "one now saw . . . 'valleys crowded with expansive works and populous towns; every valley sounding as if it housed a million blacksmiths' shops, and every valley smoking like the plain of Sodom and Gomorrah.' "[6]

Not all artists responded to ironworks or iron towns as hellish places. In Palmer's lithograph of West Point Foundry, the Union's leading manufacturer of heavy cannons is nearly invisible in the embrace of woodland (see fig. 21). The foundry is little more than an initial focal point to guide the viewer's gaze down the wooded embankment and out over the Hudson River. A painting of a forge in Canton, Massachusetts (fig. 22), similarly shows iron manufacturing as a benign presence that sits comfortably in the midst of a New England village. An advertising poster for the Ausable Nail Company of Keeseville, New York, makes explicit connections between America's rural, agrarian past and modern industry (fig. 23). A muscular blacksmith bends to his task while a horse, burro, and hound calmly wait; the tools of the smithy's trade lie close at hand. This part of the composition, except for the boxes of Ausable nails on the floor and the advertising placard near the hearth, could be a genre painting from the eighteenth century. What makes the scene modern is the rolling mill complex visible through the open door, across the river—the site of one of the country's largest nail manufacturers.

Figure 22. *View of the Iron Works, Canton, Massachusetts*, Museum of Fine Arts, Boston, gift of Maxim Karolik for the M. and M. Karolik Collection of American Paintings, 1815–65. This anonymous view, painted ca. 1850, probably depicts Bridgewater Forge. Ironworks were often built on the edge of a village, but as the settlement grew, houses, other businesses, and cultivated fields developed in a close mosaic around them.

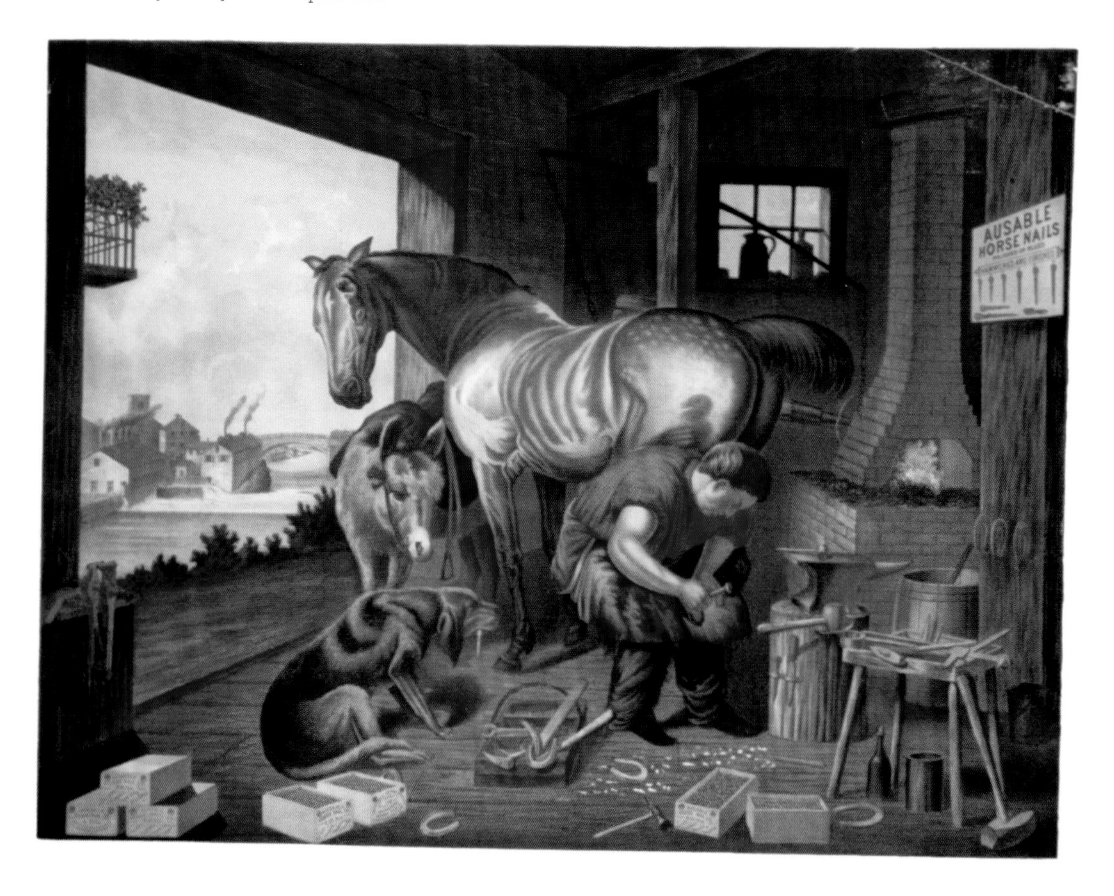

Figure 23. Blacksmith shoeing a horse with Ausable nails. Courtesy of the Library of Congress. Smoke rises from the Ausable Iron Company's rolling mill and nail factory across the river. A sign on the post at right displays some of the company's leading products—horseshoe nails to suit every size of horse and kind of horse-powered labor.

The drama of the inferno, however, created the great art associated with the iron industry. In *Coalbrookdale by Night* (see fig. 20), Alsatian artist Philip James de Loutherbourg portrayed the Bedlam Furnaces in Shropshire as a mystery obscured by odd structures and smoke glowing from half-hidden fires. The intense heat at the center of the image (presumably coal being burned to make coke) pulls one's gaze into the painting. Sulfurous smoke rises behind the silhouetted structures as a draft horse in the foreground labors to haul a loaded wagon. This image from the basement of hell was painted in 1801, just a few years before William Blake gave the phrase "dark satanic mills" to the British imagination.[7] De Loutherbourg's painting is one of the finest examples of what David E. Nye calls the technological sublime. "The English were prone to view industrialization in terms of satanic mills, frankensteinian monsters, and class strife," Nye writes, while "Americans emphasized the moral influence of steam, and often sought to harmonize nature and industrialization."[8] The difference partly reflects the scale of heavy industry in the two countries. At the turn of the nineteenth century, the largest America ironworks were scarcely one-quarter the size of the industrial complexes at Coalbrookdale. The British industry also made a more potent

Figure 24. The Cyfarthfa Ironworks, by Penry Williams, 1824. Cyfarthfa Castle Museum and Art Gallery, Merthyr Tydfil. Cyfarthfa was one of the largest ironworks in the world when this painting was done. Hills in the background held the coal and iron ore used at the works, while the river Taff (flowing in the foreground) or an adjoining canal carried finished iron to the docks at Cardiff. The many human figures at work in this scene suggest the constant activity in such places, which ran shifts around the clock.

impression on observers because it predominantly burned coal. "The Industrial Revolution in the iron and steel industry," writes historian Alan Birch, "was symbolized by the smoking, country-devouring slag-heap; for the blast furnace—no machine—was a sacrificial fire taking its toll from its servants, the iron-workers, and the surrounding countryside. The field by the stream would be changed very rapidly from a green, rural site with its mantled, soft contours to a harsh, stark gash of grey and black, enlivened only by the lurid glow of fire and the billows of steam."[9]

Urban industry was a magnet for British artists such as J. M. W. Turner, who recognized how "the mills, forges, and mines and . . . the new urban sprawl" around them "began to mark the map of Britain."[10] Penry Williams (1802–85), a native of Merthyr, painted some of the most sympathetic and knowledgeable landscapes of ironmaking. Although Williams made his living chiefly as a portrait painter,[11] two of his finest portraits are of the ironworks at Cyfarthfa. Figure 24 shows the works' expanse along the river Taff in 1824.[12] In the left foreground, the slender chimneys of puddling furnaces blaze above the roof of an open-sided rolling mill. In the distance, in a broad arc, are the company's workshops, seven blast furnaces and casting sheds, engine houses, and another rolling mill. Behind

Figure 25. *Cyfarthfa Ironworks Interior at Night*, by Penry Williams, 1825. Watercolor, 157 × 210 cm, Museum of Welsh Life/St. Fagans National History Museum. This rare view of puddlers (*right*), rollers (*center and left*), and straighteners (*left foreground*) at work in the open space of a rolling mill suggests the independence of iron artisans and the whiff of mystery their labor held for outside observers.

the furnaces, thick plumes of white smoke rise from coke ovens, obscuring the bare-backed mountains behind. In 1825 Williams painted an even more remarkable interior landscape of a Cyfarthfa rolling mill at night (fig. 25). Pulses of hot light from the puddling furnaces and clouds of steam and sparks from the roll trains highlight the confident poses of men working in the mill's vast room. The arched roof and pillarlike furnace chimneys suggest a quasi-religious space where the rituals of labor serve some higher purpose. The men work under the watchful eye of Cyfarthfa's owner, William Crawshay, whose castle windows glow in the distance.

Williams's paintings lack the drama and metaphorical suggestiveness of de Loutherbourg's image and Harding's prose, but they provide exceptionally revealing visual evidence of ironmaking environments. Williams knew ironworks well. He sketched and painted the works in Merthyr Tydfil many times, first as a boy learning to compose landscapes and then as a mature artist. William Crawshay became Williams's patron and helped him gain admittance to the Royal Academy in London.[13] As a friend of the ironmaster, Williams had access to the shop floor and informed guides to explain what he saw. Williams's paintings visually express the insider's perspective of engineers and craftsmen. It was much more difficult for a middle-class woman such as Rebecca Harding to gain entry to the world of ironworkers. This male world was very familiar to American poet Walt Whitman, whose poem "A Song of Occupations" catalogs "Iron-works, forge-fires in the mountains or by river-banks, men around feeling the melt with huge crowbars, lumps of ore, the due combining of ore, limestone, coal, / The blast-furnace and the puddling furnace, the loup-lump at the bottom of the melt at last, the rolling-mill, the stumpy bars of pig-iron, the strong, clean-shaped T-rail for the railroads." In "A Song of Joys," Whitman echoes the feeling of Williams's painting when he observes "Foundry casting, the foundry itself, the rude high roof, the ample and shadow'd space."[14]

The last passage by Whitman may refer to West Point Foundry, whose proximity to New York City made it popular among industrial tourists. The foundry's most devoted artist was John Ferguson Weir, who painted several striking scenes of men at work there, including *The Gun Foundry* (fig. 26). Weir was born and raised across the Hudson River from West Point Foundry in the town of West Point, home of the United States Military Academy, where his father taught drawing. Weir was influenced by the Hudson River school of romantic landscape painting that had given national prominence to the beauty of his home region. As biographer Betsy Fahlman notes, Weir was not a landscape painter by temperament. He found his métier in complex interiors, first studio compositions, then the foundry. The process of casting a cannon provided inherently dramatic material. Weir wrote that he was intrigued by "the primitive and massive appliance employed in this heavy work in the dingy shops, with the men sometimes stripped to the waist toiling vigorously."[15] *The Gun Foundry* combines attention to technical detail with a rare grasp of the teamwork, judgment, and strength required of ironworkers. Four men strain to push the tipping bars of the massive ladle while a fellow worker holds back the dross with a long bar. Another group of men wind the massive winch. In the background, a worker reheats iron in preparation for the next cast. Fahlman notes that the painting "records simultaneously the entire array of processes involved" in casting a cannon, including a recently cast cannon curing in its mold and a finished cannon lying on the sand floor.

The painting is also an allegory of ironmaking. Instead of the necessarily shipshape condition of a typical shop floor (compare fig. 26 with fig. 25), Weir gives us an image of Vulcan's cave, a tense, dangerous place of masculine creation. To the far right stand five visitors, tucked safely away from the action. Their still observation amplifies the awe we are meant to feel looking upon the scene. The figures include the aged Gouvernour Kemble, owner of the ironworks; Robert Parker Parrott, inventor of the kind of gun being cast; Parrott's wife and a female friend; and an anonymous Union soldier.[16] From Robbie Burns's day to the middle of the nineteenth century, ironworks provided spectacular diversion for educated

Figure 26. The Gun Foundry by John Ferguson Weir, 1864–66. Oil on canvas, 46.5 × 62 in., Cold Spring, New York, Putnam History Museum. This image emphasizes the teamwork, physical strength, and skill of foundry workers.

tourists. Charles Russell Lowell, one of the investors in Farrandsville Furnace, organized "a party of six, ladies & gentlemen" of Boston to visit the anthracite coal region along the Schuylkill River in 1835. They planned to cap off their excursion with an overnight stay in the rustic accommodations at Farrandsville.[17] Roughing it at industrial sites was an amusing break from urban sophistication.

Neither visiting nor living in an iron town necessarily conferred knowledge of the industry. Rebecca Harding's descriptions of puddling and other ironmaking processes are almost pure fiction.[18] As a resident of Wheeling, however, she was acutely aware of the effects of iron manufacturing on the physical environment and local society. She opens *Life in the Iron Mills* with sensory impressions drawn from years of personal experience.

> A cloudy day: do you know what that is in a town of iron-works? The sky sank down before dawn, muddy, flat, immovable. The air is thick, clammy with the breath of crowded human beings. I open the window, and, looking out, can scarcely see through the rain the grocer's shop opposite, where a crowd of drunken Irishmen are puffing Lynchburg tobacco in their pipes. . . .

> The idiosyncracy [*sic*] of this town is smoke. It rolls sullenly in slow folds from the great chimneys of the iron-foundries, and settles down in black, slimy pools on the muddy streets. Smoke on the wharves, smoke on the dingy boats, on the yellow river,—clinging in a coating of greasy soot to the house-front, the two faded poplars, the faces of the passers-by. The long train of mules, dragging masses of pig-iron through the narrow street, have a foul vapor hanging to their reeking sides. Here, inside, is a little broken figure of an angel pointing upward from the mantel-shelf; but even its wings are covered with smoke, clotted and black. Smoke everywhere! A dirty canary chirps desolately in a cage beside me. Its dream of green fields and sunshine is a very old dream—almost worn out, I think.[19]

Local physician James E. Reeves confirmed this description in his survey of public health, *The Physical and Medical Topography . . . of the City of Wheeling* (1870). Reeves noted that ladies in Wheeling carried a handkerchief on social calls "for the express purpose of preventing their gloves from becoming soiled in opening gates and pulling at the door-bells." "Grass grows with difficulty in Wheeling," he continued,

> and many of the green yards in front of the houses are the result of much care. Neither do tender plants live in summer without constant washing; the leaves become coated with soot, the stomata choked, and respiration ceases. Indeed, Wheeling has acquired almost as much fame for its coal smoke and soot as for its mud, fogs, and manufactures. With every breath, the sooty particles enter the lungs and discolor the bronchial secretions; and housekeepers in the vicinity of the foundries, mills, and similar establishments are compelled to keep their windows continuously closed to keep out the soot. Some of the furnaces are positive nuisances from the quantities of carbon they emit as smoke.[20]

Industrial archaeologists Robert Gordon and Patrick Malone note that "nineteenth-century reformers were almost entirely concerned with smokes and stinks, nuisances that could be detected directly by the senses of sight and smell."[21] The smokiest antebellum ironmaking environments were not around blast furnaces, which were usually in the countryside and separated from one another by some miles, but where coal-burning iron manufacturing plants were concentrated in a city or town nestled in a valley. The most notorious example in the United States was Pittsburgh (fig. 27). Its exceptional concentration of coal-burning rolling mills and foundries, as well as glassworks and other manufacturing plants, put out huge amounts of smoke and grit that were trapped by the surrounding hills. Such air could be deadly. Reeves's report on Wheeling included a statistical summary of the causes of death from 1854 to 1868. It showed that aside from cholera, typhoid fever, scarlet fever, puerperal fever, and eclampsia (a deadly syndrome

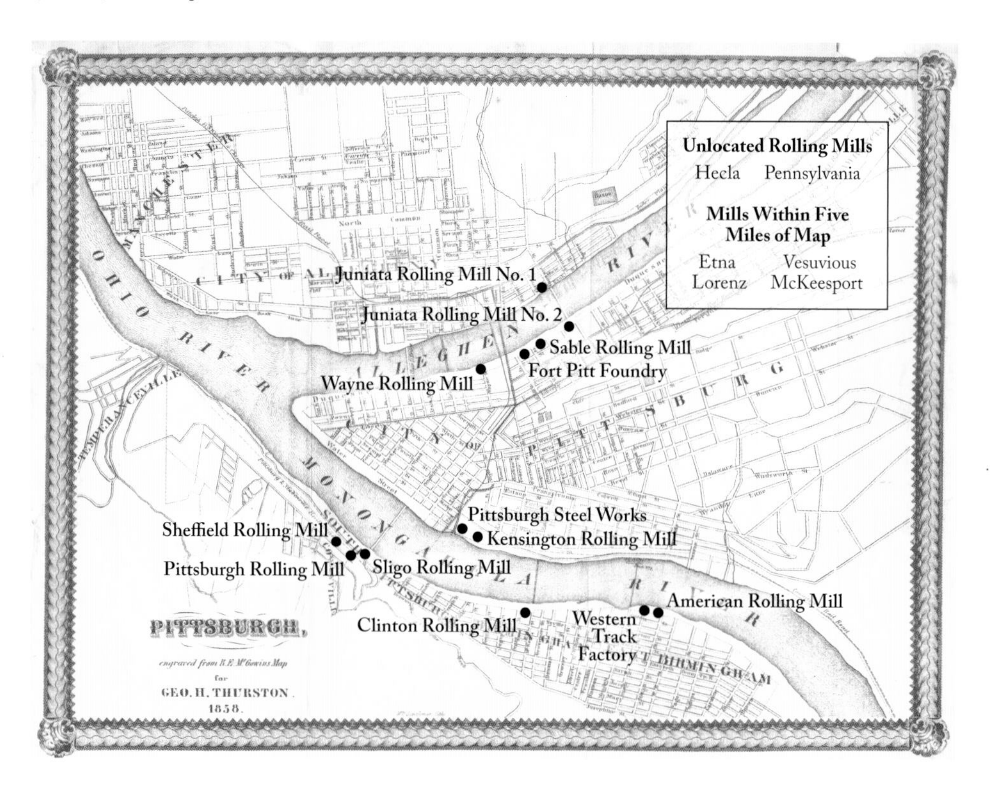

Figure 27. Pittsburgh rolling mills, ca. 1858. Pittsburgh had more rolling mills than any other American city at this date but no blast furnace within the city limits. The Monongahela and Allegheny Rivers enabled low-cost delivery of crude iron and coal, while their conjoined flow down the Ohio River connected Pittsburgh to the rapidly growing markets of the Midwest and upper South. These natural advantages supported Pittsburgh's industrial development while also making it a site of significant industrial pollution in the antebellum period.

affecting pregnant women), the most deaths were credited to respiratory diseases, including pneumonia, pulmonary phthisis, and the "indefinite" killers called "'Consumption,' 'Inflammation of the Lungs,' 'Lung Disease,' etc."[22]

Similar topography penned in the air pollution produced by ironworks in Merthyr Tydfil. Landscape painter George Childs captured the city's heavy atmosphere in a watercolor of the Dowlais Iron Works painted in 1840 (fig. 28). Two men in the foreground, probably furnace fillers, strain to lift chunks of ore or coal that they will load in a tram car and dump down a chute at the top of a furnace stack. The woman standing with them could have been a filler as well, for the 1851 manuscript census recorded a few women in that occupation. Behind them, ranks of coke furnaces pour smoke into the air. Although Childs's composition conflates Dowlais with the phalanx of conjoined blast furnaces at Cyfarthfa, [23] his depiction of the atmosphere is one of the best likenesses we have of concentrated urban iron manufacturing in the early nineteenth century.

Figure 28. Dowlais Ironworks, by George Childs, 1840. © National Museum of Wales. The relaxed observers in this image are incidental in the artist's depiction of labor conditions and the smoke and flame of iron production.

Generally, however, ironworks had a much less devastating impact on the natural environment and on human health than did the postbellum steel industry. Steelworks operated at a much larger scale than ironworks. Unlike the disaggregated American iron industry, most late nineteenth-century steel plants were fully integrated, with smelting, casting, rolling, and finishing operations clustered in vast industrial complexes. Steel production used far more iron ore, it burned mostly "dirty" bituminous coal, and its finishing processes used hazardous chemicals that produced toxic waste.[24] In the annals of industrial environmental history, steel was a major villain, iron a bit player.

Contemporary observers of the antebellum industry complained most about deforestation. Up until 1835, every ironworks in the United States burned charcoal, and charcoal remained the fuel of choice at most blast furnaces throughout the antebellum period. Charcoal blast furnaces consumed from 150 to 1,500 acres of timber per 1,000 tons of pig iron produced, depending on the efficiency of the furnace and the quality of the charcoal.[25] Extrapolating from these figures in conjunction with production data from Lesley's *Guide*, charcoal blast furnaces required clearing from 58,000 to 580,000 acres of forest a year in the late 1850s. The higher estimate sounds destructive, but geographer Michael Williams puts it into perspective by noting that woodland cleared for iron production amounted to "only 1.3 percent of the land cleared for agriculture" in the antebellum period. Nevertheless, charcoaling did produce impressive scenes of destruction. Land cleared for charcoal was usually left idle for twenty years to allow the trees to regenerate. It looked ravaged until new growth restored a "natural" appearance.[26]

The mining of coal and iron ore affected smaller areas. Much of the iron ore used at furnaces in the Hanging Rock iron district in southern Ohio, for example, came from surface deposits, where men scraped the ore from fairly small, shallow beds (fig. 29). At deep-pit coal mines, winding mechanisms and breakers marked the location of pit heads while the actual mine workings were underground, hidden from view. Coal levels, which followed more horizontal seams into hillsides, had few external markers except the tracks laid down for trams and heaps of fairly inert waste rock near the mine opening. Coal tips remained essentially sterile ground for decades, as did the ground where glassy furnace slag and cinders were dumped, but neither kind of waste heap became large until the advent of high-volume mechanized production in the late nineteenth century.

Labor at ironworks did expose workers to some serious hazards (table 4). Smelting iron ore with charcoal produced sulfur dioxide, carbon dioxide, carbon monoxide, and airborne particulates, and smelting iron with coal produced the

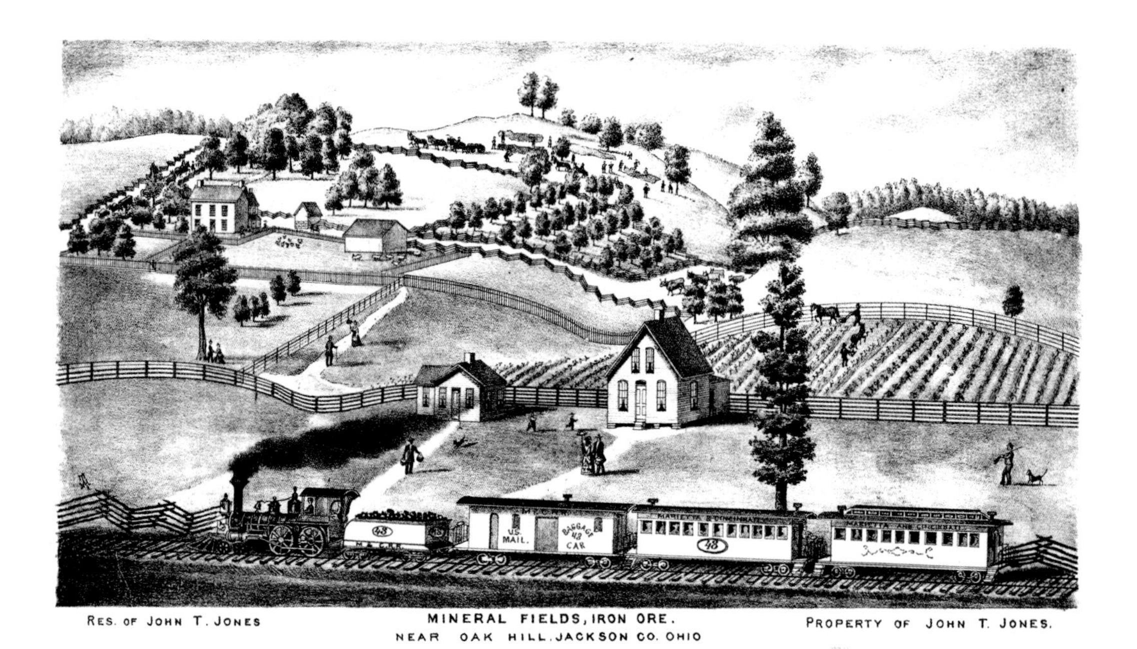

Figure 29. Iron in the rural economy of the Hanging Rock iron district. D. J. Lake, *Atlas of Jackson County, Ohio* (Philadelphia: Titus, Simmons and Titus, 1875), 16. In this image from the farming community in Jackson County, Ohio, agricultural prosperity is bracketed by ore mining at the crest of the hill (*top center*) and the railroad that connected the nearby rural-industrial village of Oak Hill to settlements throughout Ohio and beyond.

Table 4. Workplace hazards at antebellum ironworks

Environmental hazards	Health risks
Blast furnaces	
1. Exposure to poisonous gases, fumes, vapors, esp. carbon monoxide, carbon dioxide, and sulfuric acid	Gradual poisoning (esp. where ore contains lead), headaches, dizziness, nausea
2. Exposure to extremes of temperature	Heart problems, anemia
3. Close proximity to molten iron and processes of combustion	Burns, injury from explosions (rare).
4. Exposure to inorganic dust, fine sand	Respiratory diseases (asthma, bronchitis).
Finishing plants	
Rolling mill operatives	
1. Exposure to excessive heat	Heat stroke, exhaustion, cramps, anemia, diarrhea
2. Hard labor	Exhaustion, heart problems
3. Handling hot metal	Burns
4. Flying sparks and metal chips	Burns, eye injuries
5. Moving large masses	Muscle strain, crushing injuries
6. Exposure to inorganic dust	Respiratory diseases
Puddlers	
1. Exposure to extremes in temperature	Rheumatism, lumbago
2. Close proximity to intense, dry heat	Chronic eye disease: cataracts, conjunctivitis, retinal and choroidal lesions; "sunburn" of arms, hands, and face
3. Intense labor	Chronic muscle strain, exhaustion, heart problems
Workers at foundries and forges	
1. Flying bits of metal	Eye injuries, splinters
2. Exposure to inorganic dust and fine metal bits from grinding and boring	Respiratory diseases (asthma, bronchitis)
3. Repetitious use of heavy tools	"Hammerman's paralysis" of arms; "striker's arthritis" of wrist, elbows

Source: George M. Kober and Emery R. Hayhurst, eds., *Industrial Health* (Philadelphia: P. Blakiston's Son, 1924), 176–95, 614, 621, 636, 856–58, 933–41.

same kinds of pollution in even greater amounts. Where concentrated ironworks burned large amounts of coal, as in Wheeling and Pittsburgh, the air was dangerous for all living things. The risks of respiratory disease may have been especially high for ironworkers who shoveled coal or who lived closest to ironworks. These were, ironically, usually the most skilled workers, who also faced occupational hazards from the extreme heat and weight of the iron they worked. Table 4 lists the hazards and health risks that an early twentieth-century report on industrial health identified as ailments common to ironworkers. Although the report was written decades after the period covered in this book, the hazards specific to work in blast furnaces and rolling mills were much the same.

Blast furnace workers employed in the casting shed, an open-sided structure, were exposed to extremes of temperature, particularly when they tapped molten iron from the crucible in winter. Fatal and disabling accidents were fairly rare at blast furnaces, though fillers were exposed to the poisonous gases that vented out the top of the furnace and sometimes suffered facial disfigurement because of them.[27] At rolling mills, foundries, and forges, accidents were more common and often more serious because men handled hot metal, used heavy tools, and

worked with massive, rapidly moving machinery. Rollers used tongs to guide welded bars of red-hot iron weighing up to two hundred pounds into the grooved rollers of the mill. The catcher at the other side had to snatch the ropy, lengthening tongue of metal and either quickly flip it back up to the higher rolls for the return pass (on three-high mills) or carry it back to the other side (fig. 30). With steam-driven roll trains running at up to sixty revolutions a minute, the iron shot out at such speed that the catcher had to be deft indeed to avoid being injured while preventing the extruded metal from hitting the floor, where it might break or pick up impurities that would send the whole piece back to the heating furnace.[28] Forgemen shifted chunks of hot metal under the blows of massive helve hammers that struck as rapidly as once every two or three seconds. Flying sparks caused burns and eye injuries. Grinding and boring equipment at foundries cast off fine shavings and chips that could easily blind an unlucky or careless operator.

Of all iron occupations, puddling was unquestionably the most arduous. This was the job of manually stirring a mass of white-hot iron at close range to rid it of impurities, as oxygen in the air combined with sulfur, carbon, and other unwanted elements while the puddler stirred. Kenneth Hall, an interpreter at the

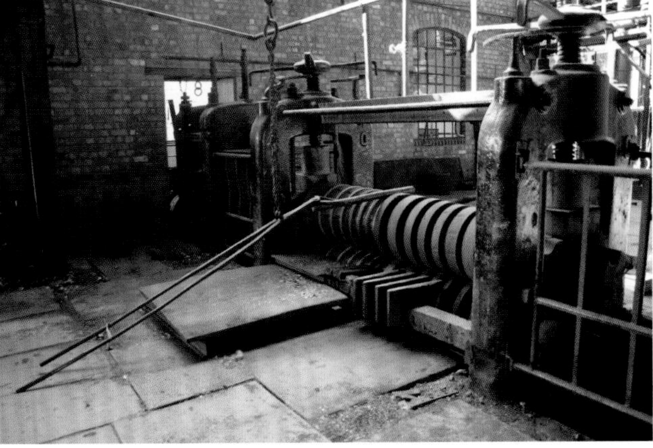

Figure 30. A roller at work. Blists Hill Victorian Village, Ironbridge, Shropshire, 2005, photos by the author. The first three photos show Graham Collis and his catcher working on an 1832 rolling mill run by steam power; the last photo shows the roll train and catcher's tongs. During this demonstration the mill was running at about 30 rmp, half the usual working speed, according to Collis. The bar of iron he was working weighed about 125 pounds. In the nineteenth century, a roller would have worked with bars up to twice as heavy.

rolling mill at Blists Hill Victorian Town in Ironbridge, Shropshire, calls himself the last puddler in England. He learned the craft as a young man and worked several years at the Inglis Ironworks in Dudley before it closed in 1974. When I asked him what puddling was like, Hall said, "Bloody awful! Don't know why I ever did it," he added with a smile. Working the iron in the furnace was "hotter than the bloomin' hobs of hell."[29] Chris Evans, the leading historian of the Welsh iron industry, writes that

> contemporary observers who rated puddling as the most grievously exacting occupation in the industrial world, one which led its practitioners towards premature decrepitude and death, do not seem to have been wide of the mark. . . . To stand at the furnace door and stir about the molten metal and then to haul out incandescent balls of refined iron with nothing more than hand tongs made impossible demands upon human endurance. Puddling, one commentator announced, "taxes the muscle and strength of the operator to a greater extent than [. . .] any other workman engaged in the coal and iron trade." John Percy, the great Victorian metallurgist, was even more emphatic: "The majority [of puddlers] die between the ages of forty-five and fifty years."[30]

Work environments that seem brutal or unacceptably dangerous to present-day sensibilities looked different to nineteenth-century artisans. I have never seen an account by a puddler that complains of the dangers of the job; rather, its physical and mental demands were a source of pride, and the occupation itself was something worth fighting to preserve. As Gordon and Malone note, "Our perception of the risk from accident is strongly conditioned by social factors that are only now being explored by scholars. Through much of the nineteenth century, public perception of industrial accidents [and hazards] was much like the mid-twentieth-century perception of automobile accidents. They were accepted as the price of benefits thought to be worth this cost."[31] Men who endured the demands of working a puddling furnace received some of the highest wages in the iron industry. They enjoyed high esteem among their peers and were leaders in organizing ironworkers, first in Britain and later in the United States. Pittsburgh puddlers organized the first US ironworkers' union, the Sons of Vulcan, in 1858.[32]

The best account of puddling comes from James J. Davis, who represented Pennsylvania in the US Senate and served as secretary of labor from 1921 to 1930. He was born James Davies in Tredegar, Monmouthshire, an iron town a few miles east of Merthyr Tydfil. After emigrating as a boy, Davis worked as a puddler in Pittsburgh for many years. In his autobiography *The Iron Puddler* (1922), Davis berated those who stressed the hardships of labor at iron mills. "The gases, they say, will destroy a man's lungs, but I worked all day in the mills and had wind enough left to toot a clarinet in the band. . . . It was no job for weaklings, but neither was tree-felling, Indian fighting, road making and the subduing of a wild continent." Davis's retrospective heroism may be suspect, but his vivid descriptions of the craft of puddling are convincing. Like many puddlers, he learned the process from his father. "None of us ever went to school and learned the chemistry of it from books." (By the turn of the twentieth century, the chemistry of refining iron was known, but puddling remained an occupation in a small number of American mills and forges even then.) "We learned the trick by doing it, standing with our faces in the scorching heat while our hands puddled the metal in its glaring bath." Davis ran five "heats" a day, as did puddlers in the early nineteenth century. No doubt exposure to the extreme heat scorched their palms and fingers, as it did Davis's hands, which he said "became hardened like goat hoofs, while my skin took on a coat of tan that it will wear forever."[33]

Ironworks generally carried positive connotations in the minds of working people. For agricultural laborers and tenant farmers, work in a dynamic iron town offered a way out of rural poverty and the tedium of rural life. That was the view of young B. L. Coombes when he was growing up on a struggling farm in Herefordshire in the late nineteenth century. To his boyish eyes, the iron towns at the head of the valleys of South Wales were a beacon on the distant horizon. "I was fascinated by that light in the sky," he wrote in his autobiography. "Night after night I watched it reddening the shadows beyond the Brecknock Beacons, sometimes fading until it only showed faintly, then brightening until it seemed that all the country was ablaze."[34]

Coombes was one of tens of thousands of young men who came to work in the Welsh industrial valleys in the nineteenth century. In 1850, Merthyr had a population of more than 46,000, roughly half of whom worked in the iron industry, including some women and many children who helped relatives at the furnaces and rolling mills. Thousands more labored at ironworks in neighboring Monmouthshire.[35] A man who earned one shilling a day as an agricultural laborer could pocket twenty-two to twenty-seven shillings a week hewing coal, or up to thirty-five shillings as a puddler. Although industrial wages were volatile and employment was unpredictable, ironworkers could reasonably expect a higher standard of living than most farmers or rural craftspeople enjoyed.

With skill and higher wages came a new kind of pride and working-class community, and the birth of industrial class consciousness. Ironworkers felt much the same pride as artisans in other metalworking trades. Thomas Wright, a journeyman engineer in England, wrote two books in 1867–68 about "the inner life of workshops." He aimed to part the veil that obscured the secrets of labor in heavy industry or, as he put it, "a life behind the scenes—that is known only to the initiated."[36] Wright described metalworkers' pranks and the culture of drink as reflections of industrial camaraderie. He explained how men and boys at ironworks and machine shops tested one another's wits and physical strength, both of which were necessary traits in skilled artisans. Wright depicted metalworking shops as places built on mastery, trust, inclusion, and strong occupational identities.

Such identities naturally varied according to workers' level of skill, ethnicity, and race. But identity and work culture were also affected by the characteristics of the communities where workers lived and labored. As I moved back and forth between Lesley's *Guide* and the company records that provide the basis for my case studies of individual ironworks, I began to see relationships between the structure, size, population, and social relations of ironmaking communities on the one hand and their geographical setting and industrial technologies on the other. Five distinct kinds of communities became apparent. Recognizing the differences between them clarifies the role geography played in shaping workers' experience.

Iron Communities

Peter Lesley describes three basic settings for ironworks: rural areas, villages, and cities. Assigning each ironworks to one of these categories based on the locale's approximate population provides a general sense of the regional distribution of ironmaking communities (fig. 31).[37] The map shows that many blast furnaces were in rural areas while most rolling mills were in or near cities, such as the mills along the Schuylkill and Lehigh Rivers in southeastern Pennsylvania and the cluster of mills in eastern Massachusetts and Rhode Island. Ironworks in villages and small market towns were concentrated in New England, the Hudson Valley, and the Champlain Valley, though they also occurred across Pennsylvania, New Jersey, and Ohio. The map does not show forges or bloomeries because I am unable to pinpoint most of their locations. Those whose location Lesley describes were almost all in rural places or villages.

This three-part typology, however, does not fully represent the character of places where iron was made. The process of mapping the information in Lesley's *Guide* revealed that some of the most remote ironworks were outposts of heavy industry that employed the most advanced technologies of the era. At the other extreme, some works using the oldest technologies were in tiny industrial hamlets not far from major cities. The variety of ironmaking communities I found was very different from the view promulgated by James M. Swank and Arthur C. Bining, who declared that the rural iron plantation was the archetypal setting for antebellum ironworks. Swank and Bining described nineteenth-century iron plantations as remnants of a decaying, if romantic, social system that had outlived its usefulness. They likened iron plantations to the "small feudal manors of medieval Europe." Plantation ironmasters ruled like feudal lords, exercising paternalistic control over almost every aspect of workers' lives. Their wealth made

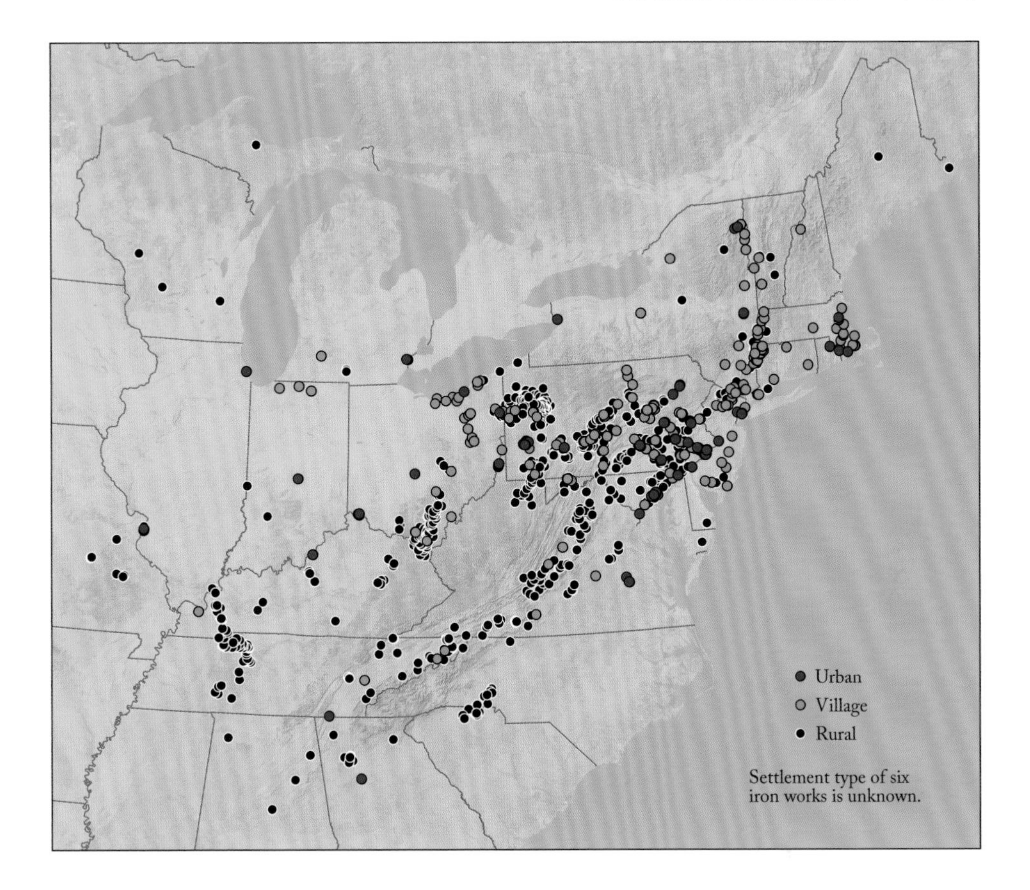

Figure 31. Geographical distribution of ironmaking communities by settlement type. Categorization is based on Lesley's place descriptions and on examining locations on topographic maps. Most rolling mills were in urban locations, most blast furnaces in the countryside.

them members of the rural elite, but they were a very different sort of businessman or woman than the aggressive capitalists who owned modern iron and steel corporations.[38]

Seeing rural ironworks as lingering representatives of outmoded methods and archaic social relations accords with the view that masters of charcoal iron furnaces and forges resisted new technologies and were doomed to fail in competition with the more efficient modern, coal-fired technologies embodied in the British model. I argued in the previous chapter that the dichotomy between old-fashioned and modern ironworks omits much of the experimentation and hybridity that characterized the antebellum industry. Similarly, while the contrast between iron plantations and modern urban, proletarian industrial communities is a fair summary of how much changed between the colonial era and the age of steel, it greatly oversimplifies the diversity of antebellum ironmaking communities. Understanding the variety of these communities matters because the social and geographical contexts of iron production significantly influenced labor relations, which in turn affected how swiftly and successfully new technologies were implemented and how prone ironworks were to labor conflict.

Iron Plantations

In the antebellum period, "plantation" came to mean a commercial farm that focused on producing one crop for export, such as cotton or sugarcane, while also growing food crops and livestock to feed a large resident workforce of slaves. In the context of the iron industry, however, the term continued to hold the colonial connotation of a self-sufficient rural settlement,[39] and it was used to refer to places in both the North and the South. The owner, called the ironmaster, usually lived on the plantation and was often directly involved in day-to-day management. The best-known Pennsylvania examples are Hopewell Furnace in Berks County, now a National Historic Site administered by the Park Service (fig. 32), and the Curtin Iron Works in Centre County. Southern iron plantations, which mainly employed slaves, included David Ross's Oxford Furnace and William Weaver's Buffalo Forge and its associated furnaces in the Valley of Virginia.[40]

Figure 32. A bird's-eye view of Hopewell Furnace. Courtesy of the National Park Service, Harpers Ferry Center Commissioned Art Collection, artist L. Kenneth Townsend. This 1983 rendering of the charcoal furnace shows the ironmaster's home (*center foreground*), the furnace stack smoking above the broad roof of the casting shed, various company buildings, and the lane leading to skilled workers' houses in the distance. The fields and surrounding woodland also belonged to the furnace company.

The main commercial export from iron plantations, of course, was iron, in the form of pig iron or refined iron, such as blooms, anchonies, or merchant bar that was shipped elsewhere to be made into horseshoes, barrel hoops, and nails, as well as military ordnance, machine parts, and railroad rails. Like most charcoal iron furnace companies, plantations mined iron ore, quarried limestone, and cut timber for charcoal on their own land. They differed in also growing their own food and fodder. That may seem a small thing, but the fact that some ironworkers also did agricultural work, and that agricultural laborers were available to assist at the ironworks, gave workers and the business a number of advantages. Plantation workers benefited somewhat from the change of pace that field work provided. They may have had a more varied diet than ironworkers in other kinds of settlements. Skilled white artisans were often allowed a garden plot and milk cow to supplement the basic diet of salt pork, bacon, flour, cornmeal, and coffee.[41] The company benefited from being able to deploy labor where and when it was needed. For example, if prices slumped, plantation managers could stockpile iron and put workers to good use in the fields or forests. If the mill dam broke in a spring freshet, dozens of field hands and woodcutters could quickly rebuild the power system.[42]

Southern ironmasters had the power, and the flexibility, to decide how labor was deployed on every part of their property. Slaves resisted unreasonable and cruel treatment, but ultimately they had little choice in the work they did. Where works employed both slave and free labor, as at the Maramec Iron Works in the Missouri Ozarks, slaves "provided a more dependable source of manpower for unpleasant tasks."[43] On northern plantations, metalworkers and those employed at the furnace or forge were usually contracted to do specific jobs; they rarely did agricultural labor beyond tending their home gardens. Skilled workers lived in company housing near the works. All workers could earn a little extra money, or credit at the company store, by doing extra tasks when time permitted or necessity required.[44] The same kinds of work were done on plantations in the North and the South. The chief differences were the greater volition of northern workers and the division between industrial and agricultural employment.

The difference in control exerted by southern and northern ironmasters was subtly reflected in the layout, or settlement morphology, of iron plantations in the two regions. The plantation master's "big house" always had pride of place at the center of the settlement, usually on a rise that gave a good view of the ironworks from the front porch. On southern iron plantations, such as Buffalo Forge, the view included slave quarters near the ironworks (fig. 33a). The dwellings of free white workers on northern iron plantations were more widely scattered. Down a lane leading away from the ironworks, one would often find neatly built stone or wooden houses for skilled workers along "Puddlers Lane," "Forge Street," or "Furnace Row." These two- to four-room structures provided housing as good as

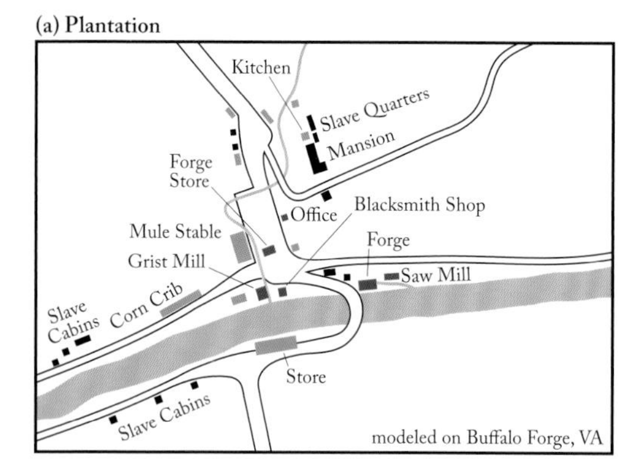

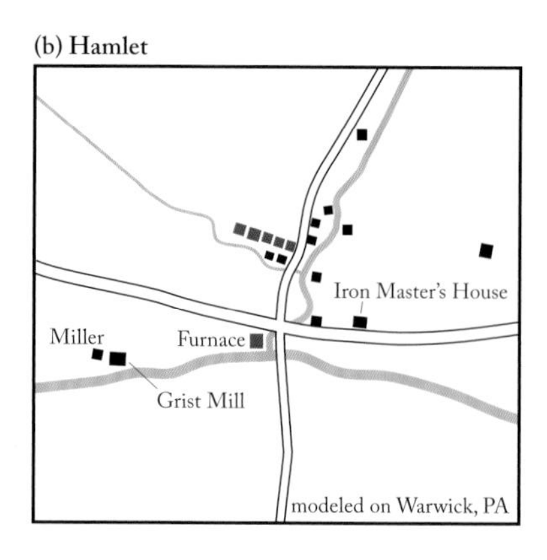

Figure 33. Typical morphology of iron plantations (*a*), hamlets (*b*), and villages (*c*).

or better than most frontier farm families enjoyed.[45] At northern works, proximity to the furnace or forge was meant not so much for surveillance as for workers' convenience. Less skilled workers lived in cabins built by the owner or themselves.[46] Southern ironmasters built simple, sometimes crude houses for slave artisans, though their homes were more spacious and private than the tenements and log cabins provided for slaves of lesser skill.[47] Housing was a key component in the package of benefits ironmasters used to secure labor on their plantations. As in the colonial period, many iron plantations were far from the coastal cities and market towns that most readily attracted industrial workers. Rural ironmasters had to provide housing and food to attract and retain the most mobile members of their workforce, whether they were free white artisans or slave artisans hired on annual contracts.

A typical iron plantation operating a charcoal iron furnace and bloomery forge required from 150 to 200 workers over the course of a year. As few as six to ten were skilled metalworkers: a founder to run the blast furnace, two keepers to assist the founder, and one or two pairs of finers and hammermen and their assistants to convert the pig iron into wrought iron. If the plantation had a small rolling mill, another ten to fifteen men were required to puddle and roll iron. Even with a rolling mill, 75 percent or more of the total workforce at iron plantations did the ancillary labor of chopping down trees for charcoal, tending crops and livestock, hauling materials to the furnace and forge, hauling finished iron to market, building roads and making waterpower improvements, and so forth.[48]

The pool of skilled ironworkers was much smaller in rural areas than it was along the eastern seaboard or in rapidly industrializing inland cities such as Pittsburgh and Cincinnati. Labor markets were further limited by regional preferences. Rural ironmasters hired local men, or workers from their state or region, when they could. This was particularly true of southern ironmasters, who found northern and immigrant white artisans difficult to manage. As Virginia's iron in-

(c) Village

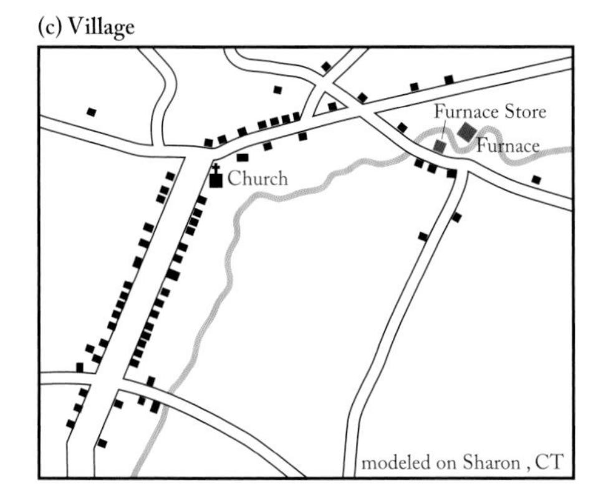

dustry developed in the early nineteenth century, ironmasters invested heavily in acquiring skilled slaves, whom they regarded as more tractable and reliable than free labor.[49] Stability, continuity, and control came at the cost of restricting artisans' skills and inhibiting the adoption of new ironmaking technologies. Virginia ironmaster David Ross noted in 1812 that slaves at his forge "had no chance to improve—they have not [had] an opportunity of travelling to see other works and the annual improvements."[50]

Southern iron plantations were not alone in "doing things in the old, familiar ways."[51] Furnace and forge masters of the Juniata Valley in central Pennsylvania used proven, traditional methods, as did ironmasters in Connecticut's Salisbury district and companies in the Adirondack iron region. In each of these regions, ironmasters persisted in making and refining charcoal pig iron in part because it sold well and competed favorably with all but the very best iron imported from Europe. Every study of an iron plantation shows that the owners were aware of technological changes and competition from domestic and foreign producers who used newer technologies. They were equally aware of the risks posed by new technologies. When Roland Curtin's sons took over the family business in 1848, they built a new charcoal blast furnace on the same design their father had used in 1819. They did not take out loans against their extensive property to attempt a more ambitious coal-fired operation, though they did install hot-blast equipment in the new furnace and experimented with it, unsuccessfully, for ten years.[52] Charcoal was the Juniata Valley's only readily available fuel source for many years (anthracite shipments did not reach the valley until the early 1850s) and the Curtins' forge iron was in demand in Pittsburgh and Philadelphia throughout the antebellum period.[53] The family's conservative approach kept their operation on an even keel and avoided damaging the reputation of the Curtin brand. It also limited the volume of production and hence the potential profitability of their business.

The plantation system was based on a social paradigm of family relations. The ironmaster was patriarch or matriarch of a family business that depended on long-term, personal relationships with workers and their dependents. While slave artisans literally belonged to the master's family, they and their white counterparts on northern plantations were also the heads of locally respected families who raised their sons in the craft. This social system may have been conservative, but that did not mean plantation ironworks were simple businesses. Quite the contrary. Iron plantations were among the most complex enterprises in the rural American economy. Their owners included exceptionally wealthy men whose long careers demonstrated business acumen and managerial ability.[54] Testimony from court cases portrays William Weaver, Roland Curtin, and David Ross as men with driving personalities. At the same time, they understood the iron business and their workers better than many of their younger, more progressive competitors. An ironmaster needed a certain degree of tolerance to handle the minor infractions of discipline that were routine at rural ironworks. Hot weather made furnace and forge work just about impossible. In the steamy Valley of Virginia, for example, slaves claimed they were sick, or they "broke down to loaf," when July temperatures hit ninety degrees or more.[55] Experienced plantation owners and managers knew that the long-term productivity of a good worker was far more valuable than strict obedience. At Buffalo Forge and Oxford Furnace, only slave artisans who set dangerous examples by openly rebelling or trying to escape were physically punished or sold.[56]

William Sullivan observed that up to 1840, ironworkers on Pennsylvania plantations were spared the destitution and strife that periodically afflicted urban industrial workers. Generally stagnant wages lessened the incentive for rural workers to migrate, and geographic isolation shielded iron plantation workers from economic downturns and political agitation. More recent studies depict iron plantations less peacefully. In furnace communities in the New Jersey Highlands, for example, fighting and drunkenness were common problems despite regulations against such behavior. The scarcity of skilled labor, however, and free men's "privilege of quitting," as one artisan put it, compelled plantation ironmasters to negotiate their differences with workers they could not do without.[57]

Iron Hamlets

"Hamlet" is an old word for "a little cluster of houses in the country." In geographic terms, hamlets are the smallest settlements in an urban hierarchy.[58] As industrial sites, iron hamlets sometimes included both a furnace and a forge, foundry, or bloomery, though many consisted of a single smelting or refining facility. Peter Lesley and his assistants found such places in every ironmaking region. The charcoal furnace owned by Vermont's Green Mountain Iron Company was in the hamlet of Forestdale, "3.5 mi northeast of Brandon Village, 17 mi north of Rutland, on a stream . . . under Mr. Royal Blake's house." Retreat Charcoal Furnace was "on Purgatory Creek . . . nine miles north of Buchanan," in Botetourt County, Virginia. Cranberry Forge was somewhere on "Cranberry Creek, 12 miles east of Jefferson," in Ashe County, North Carolina.[59]

On twentieth-century topographic maps, iron hamlets appear as no more than a few scattered buildings, if they have not disappeared entirely. As sites of iron production 150 years ago or more, each one was a tiny industrial settlement surrounded by woodland and farms. Benjamin Henry Latrobe's watercolor of Warwick Furnace in Chester County, Pennsylvania, shows the typical geographical context for this kind of operation (fig. 34). When Labtrobe visited the furnace in 1803, it had been producing iron for nearly seventy years. The furnace was on a tributary of French Creek near a gristmill, store, and schoolhouse (fig. 33b). This community was in the heart of "the best poor man's country," in Philadelphia's fertile hinterland.[60] The 1850 manuscript census of Warwick Township notes the local residences of ironmaster David Potts Jr.; Nathaniel and R. H. Potts, both clerks; Robert Potts, manager of the furnace; and mill owner John Miller. According to the census, the furnace founder, keepers, and molder were native Pennsylvanians, as were all other male workers aside from a few Welsh miners

Figure 34. Warwick Furnace in 1803. Pencil and watercolor on paper by Benjamin Henry Latrobe, Maryland Historical Society, Baltimore, Maryland.

and Irish laborers.[61] Skilled industrial workers were a small minority in this primarily agricultural community.

Warwick Furnace may originally have produced iron for local needs, but when Peter Lesley visited in 1856, it was specializing in cold-blast charcoal pig iron for boilerplate.[62] Warwick illustrates the continuing significance of ironworks that used older technologies in the antebellum period. Located "13 miles west of Phoenixville, . . . ten miles southwest of Pottstown," the furnace was within easy reach of rolling mills along the lower Schuylkill River that produced boilerplate for steam engines. Just thirty feet high, smelting ore with charcoal and a water-driven, cold-blast blowing engine, Warwick Furnace was the first link in a chain of iron production that made southeastern Pennsylvania a major center of heavy manufacturing.[63]

In less fertile rural areas, ironmaking hamlets more nearly resembled plantations out of sheer necessity. One striking example is Adirondac-Tahawus, an iron community fifty miles west of Lake Champlain in the heart of the Adirondack Mountains. In 1826, Archibald McIntyre and David Henderson staked a claim to a rich deposit of iron ore in the southwest corner of Essex County, New York. They opened a mine and charcoal furnace in Adirondac in 1827. The hamlet of Tahawus was to be a train depot for shipping out ore and pigs, but the hoped-for railroad was long delayed. McIntyre and Henderson tried to make their remote industrial community self-sufficient by establishing a three-hundred-acre farm. In 1842, it counted for half of all improved land in the town of Newcomb. They raised the usual northern mix of "sheep, hogs, cattle, poultry, hay, oats, and potatoes." The two men complained bitterly over the cost of trying to farm land that was so hostile to agriculture, but they had no choice if Adirondac-Tahawus was to be a viable community. Workers carted "but little iron" eight miles to the nearest railroad at Sacketts Harbor until the company folded in the economic crash of 1857.[64]

Although some hamlet ironworks were important suppliers of iron, they tended to operate more sporadically than did more heavily capitalized iron plantations. They commonly shut down when farmhands were needed for planting and harvest or to await repairs of broken machinery. Hamlet ironworks may have been what Allan Nevins had in mind when he referred to short-lived, tiny iron works "with poor equipment, and an uneconomic force of men."[65] Lesley noted that some of the furnaces, forges, and bloomeries in hamlets operated intermittently, shutting down when prices dropped and resuming production when conditions improved.[66] Although long spells of idleness might cost a company its founder and keepers, the rest of its workforce could be reconstituted without great difficulty. Unlike plantations, whose owners typically secured both ironworkers and agricultural labor with year-long labor contracts, companies that produced only iron were not obliged to employ workers if metal production ceased for some time. This may help explain why Sullivan found no examples of violence or coercion in early antebellum Pennsylvania, many of whose rural ironworks were in hamlets;[67] discontented or unemployed free labor could move elsewhere.

Iron hamlets were among the most homogeneous of ironmaking communities. Their ironworks tended to hire local men, since the technologies they employed rarely required skilled immigrant labor. The patterns of work and daily life in ironmaking villages may give us the best indication of ironworkers' experience in hamlets, since many villages were hamlets writ large: communities whose economies were rooted in agriculture but whose resident elite invested their surplus capital in iron, giving their localities a distinctly industrial cast.

Iron Villages

In *The Texture of Industry*, Gordon and Malone identify the manufacturing village as the archetypal ironmaking community in New England. New England ironworks proprietors "did not build iron plantations," Gordon and Malone note, "and they sometimes placed their furnaces within existing communities" or on the outskirts. Sharon Furnace in Litchfield County, Connecticut, is an example of the latter situation (see fig. 33c). On most days the prevailing westerly winds would have blown the furnace's smoke away from the village, in the direction of Beardsley Pond, whose water provided power to the furnace.[68]

Like Warwick Furnace, Sharon Furnace was in a long-settled area where small-scale industry had existed alongside agriculture since the early eighteenth century. A bloomery forge produced small amounts of iron from local ores as early as 1736. The first blast furnace in the area was erected in 1762. In the 1770s, Salisbury ironworks produced cannons for the Continental Army. The district's iron industry grew much larger in the 1820s, when five furnaces were built in Litchfield County, including Sharon Furnace in 1825.[69] By that time, three generations of Anglo-American families had established farms and villages in the county. New industrial ventures had to fit in where space and water privileges allowed. The proximity of the furnace to the village center hints at its integration into the local economy and local society—a connection suggested even more strongly in a painting from about 1850 depicting the ironworks in Canton, Massachusetts (see fig. 22). Southern New England farmers had begun to diversify their economic interests by the 1820s, investing in mercantile trade and industry. Many prosperous Litchfield County farmers invested their surplus capital in the iron business. Profits from ironmaking did not produce extremes of wealth, however, because the iron companies remained small and continued to use tried-and-true technologies that produced modest amounts of high-grade iron.[70]

According to Robert Gordon, this conservative approach to industry was congruent with local cultural values. Close community ties and common schooling of a fairly high standard lessened the social distance between resident owner-managers and workers at Salisbury ironworks, many of whom were born and raised in the Connecticut Valley. As the region's iron industry gradually grew, local residents acquired considerable technical expertise.[71] Salisbury iron companies' aversion to risk was reflected in their overwhelming preference for cold-blast charcoal furnace technology and water-driven helve hammer forging techniques. The predominantly family-owned iron companies in this part of New England produced high-quality iron for niche markets. Salisbury forges, for example, provided much of the iron plate used for rifle barrels at Springfield Armory throughout the antebellum period. Salisbury pig iron and blooms also supplied manufacturers of edge tools and machine parts in southern New England.[72]

It is telling that the Salisbury district's largest ironworks and its one technical innovation were introduced by outsiders from Massachusetts. Horatio Ames, John Eddy, and Leonard Kinsley built a large foundry forge at the Housatonic Falls in Salisbury Township in the mid-1830s. The Ames Iron Works was one of the first in the country to use wood-fired puddling furnaces, an innovation that may have been developed by American inventor Peter Cooper to adapt puddling to wood-rich, coal-poor New England. The conservatism of local culture frustrated Horatio Ames, who found his neighbors stubbornly opposed to his own equally stubborn ambitions to expand the Ames works.[73] Their unwillingness to adapt also showed when larger mineral-fueled furnaces in eastern Pennsylvania and the Hudson Valley began to encroach on their markets in the 1850s and 1860s, as a number of Salisbury ironmakers decided to abandon their businesses rather than adopt new techniques to remain competitive. The same qualities that supported consensus in the region may have hardened local attitudes against technological change. There was so little social distance between ironworks owners, managers, and artisans in the Salisbury district that it was not unusual for them to change places as need and opportunity arose. Such harmony and shared knowledge made for smooth working relationships and diminished the likelihood of conflict, but also helped unite the community in opposition to change.[74]

Iron villages were not confined to New England. In parts of the Midwest where Euro-American settlers found both arable land and mineral resources, ironworks and agricultural communities developed side by side. The Hanging Rock iron district of southern Ohio and northern Kentucky was one such area. Oak Hill, a village in Jackson County, Ohio, developed in the 1830s and 1840s as the trading center for a predominantly Welsh immigrant farming community. In 1852–53, several American companies in the vicinity built charcoal iron furnaces along newly constructed rail lines. In 1854, Oak Hill's leading Welsh farmers and merchants decided "to take a stab at what they could do" in the iron industry. They organized three furnace companies, one of which, Jefferson Furnace, survived the Panic of 1857 and returned handsome dividends to its investor families for decades, providing the basis for a diverse local economy.[75] In the first few years of Jefferson's operation, local residents from all walks of life joined the workforce in various roles. Once it was well established, a few skilled workers at the top of the labor hierarchy became settled members of the Oak Hill community. Manual laborers, colliers, and miners constituted a landless proletariat of trans-Appalachian migrants, most of whom worked at Jefferson for a brief while and then moved on.[76]

The small number of artisanal positions at charcoal iron furnaces made it difficult for most furnace laborers to rise into the ranks of skilled labor. As the only salaried workers possessing skills peculiar to the iron industry, keepers and founders stood apart from the workers they supervised. Although semiskilled laborers carried the industry on their shoulders, the burden could be easily shifted to new workers. The organization of labor, the generally low wages at village ironworks, the local availability of alternative employment, and the extent to which skilled and managerial personnel were laced into the local community all dampened conflict and worked against organized protest.

Rural Company Towns

Iron plantations, hamlets, and villages were similar in a number of ways. They shared the loosely clustered focus of small nucleated settlements. They combined agriculture and industry, as integrated businesses on iron plantations and as mutually beneficial pursuits in hamlets and villages. The industrial workforce was similar in number and skills in the three settings, though workers' race and the range of tasks they performed varied. Social tensions certainly existed in these communities, but they were muted by the small size of the industrial workforce, the diversity of the local agro-industrial economy, and the relative security of top skilled workers' employment. Ironmaking company towns, which began to appear in the American countryside in the 1830s, had quite different characteristics, from settlement morphology and workforce to the technologies they employed and their experience of economic swings.

Iron companies typically built workers' housing near the furnace or rolling mill and ran a company store that sold provisions and rented tools to workers. While these traits made almost any ironmaking community a company town, the rural places that require that label were governed by company rules and were devoted to year-round industrial production, with the resident workforce employed full-time in making iron or extracting the resources necessary for making iron. Workers often included coal miners as well as iron artisans and laborers, for many companies that built rural company towns intended to make iron with anthracite, coke, or raw bituminous coal. Because good ore and coal deposits rarely occurred in close proximity, company towns were located when possible along navigable rivers or existing, or planned, railroad lines.

One example is the community of Ashland Furnace in Baltimore County, Maryland (fig. 35a). Ashland's owners built their first hot-blast anthracite furnace in 1844 about fifteen miles north of Baltimore Harbor. The anthracite was shipped in by rail from Pennsylvania. By 1850, the works had two hot-blast furnaces powered by steam and water, with forty-five hands making a total of 4,300 tons of pig iron a year. The 1860 census recorded two hundred employees at the works making 8,000 tons of anthracite pig iron.[77] The layout of the settlement was a classic railroad "T town" in miniature, with a main street coming to a dead end at the railroad.[78] Across the tracks, set off from the workers' housing, was the manager's mansion. A boardinghouse dubbed "Stone Row" faced the tracks on the north side, with two ranks of smaller houses nearby. As the company grew, it added more houses. By the end of the nineteenth century, when Ashland Furnace had become part of Pennsylvania Steel, "solid, livable houses, 65 in all, lined the streets of the company town."[79] Farrandsville Furnace (fig. 35b) was on the West Branch of the Susquehanna River in Lycoming (later Clinton) County, Pennsylvania. The furnace, built in the mid-1830s to burn coke, was intended to supply pig iron for an adjoining rolling mill. The small industrial community included a boardinghouse, cabins for skilled workers, a hotel for visitors, a house for the resident manager, and a small mining settlement laid out in a row on the hill above the furnace. Where Ashland Furnace was built along a railroad line, Farrandsville's owners planned to take advantage of an expected canal along the river.

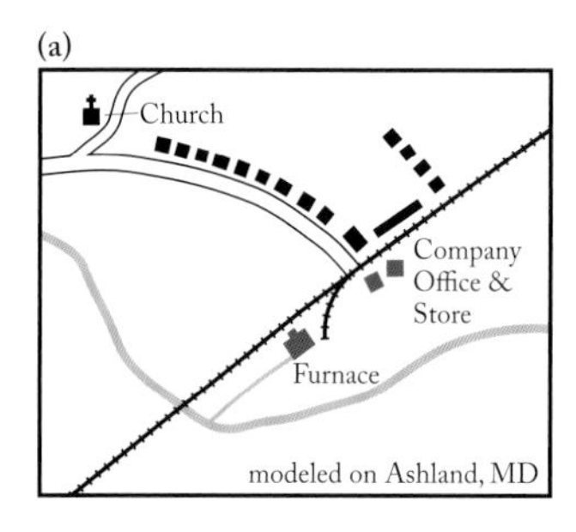

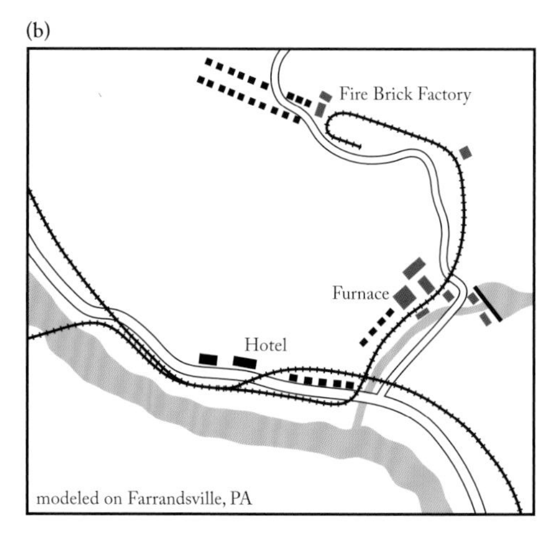

Figure 35. Typical morphology of rural iron company towns.

The Etowah Manufacturing and Mining Company in Cass (later Bartow) County, Georgia, offers a southern example. There were virtually no coal-fired blast furnaces in the antebellum South. The region's coal deposits were not extensively exploited for iron smelting until the development of the Birmingham iron district in the late 1870s and 1880s.[80] The first charcoal furnace at Etowah was built in 1837 about thirty miles northwest of Atlanta, in a hilly region that is now flooded by waters behind Allatoona Dam. The first furnace was replaced by another charcoal furnace in 1844. A bloomery forge built in 1838, supplemented or replaced by a second forge in 1841, was up Stamp Creek a mile or so from the main site. A rolling mill was added along the Etowah River in 1849. By 1856, when the owners commissioned a detailed map of their property (fig. 36), the Etowah complex included a sawmill, gristmill, flour mill, corn mill, granary, barrel factory, two stores, warehouses, storehouses, and office. The mills ran on waterpower stored behind the company dam. Worker housing included boardinghouses, homes for artisans, and log cabins. A company prospectus noted that the store had "the exclusive monopoly of the trade with the operatives." In 1858–59 the company built a branch line to the Western and Atlantic Railroad, which ran from Atlanta to Chattanooga.[81] The Etowah Company owned thousands of acres of timberland and mineral deposits. Resident owner-manager Mark Cooper's personal property, colored red on the map, exceeded 7,500 acres. The company's holdings included 200 acres of farmland that provided food for the working community.[82]

The physical description of Etowah bears a striking resemblance to the largest iron plantations in Virginia and Pennsylvania, but the composition of its workforce was very different. In 1850 the Etowah community housed about seventy white ironworkers and employed dozens of white woodcutters and colliers. Most of Etowah's free white workers had been born in the South, but there were also fourteen men from the British Isles: four Irish laborers and ten machinists, refiners, rollers, and puddlers from England and Wales. Men from Virginia, South Carolina, and Georgia also held skilled positions at the rolling mill. An English founder supervised a dozen molders from South Carolina and Tennessee and a lone Scot. A Pennsylvanian named Daniel Welch, listed as ironmaster (in charge of the blast furnace), was assisted by two young relatives who clerked for the company. The census records only one other northerner at Etowah, a rolling mill worker from Massachusetts.[83]

At least half of the white workers at Etowah were single men; twenty-three of them lived in one boardinghouse. The predominance of fairly young men in industrial occupations was typical of rural iron company towns. It gave places like Etowah, Ashland, and Farrandsville a different tone than iron plantations, hamlets, and villages, where more workers lived with their families. The concentration of male workers in rural iron company towns reflected the interest of the

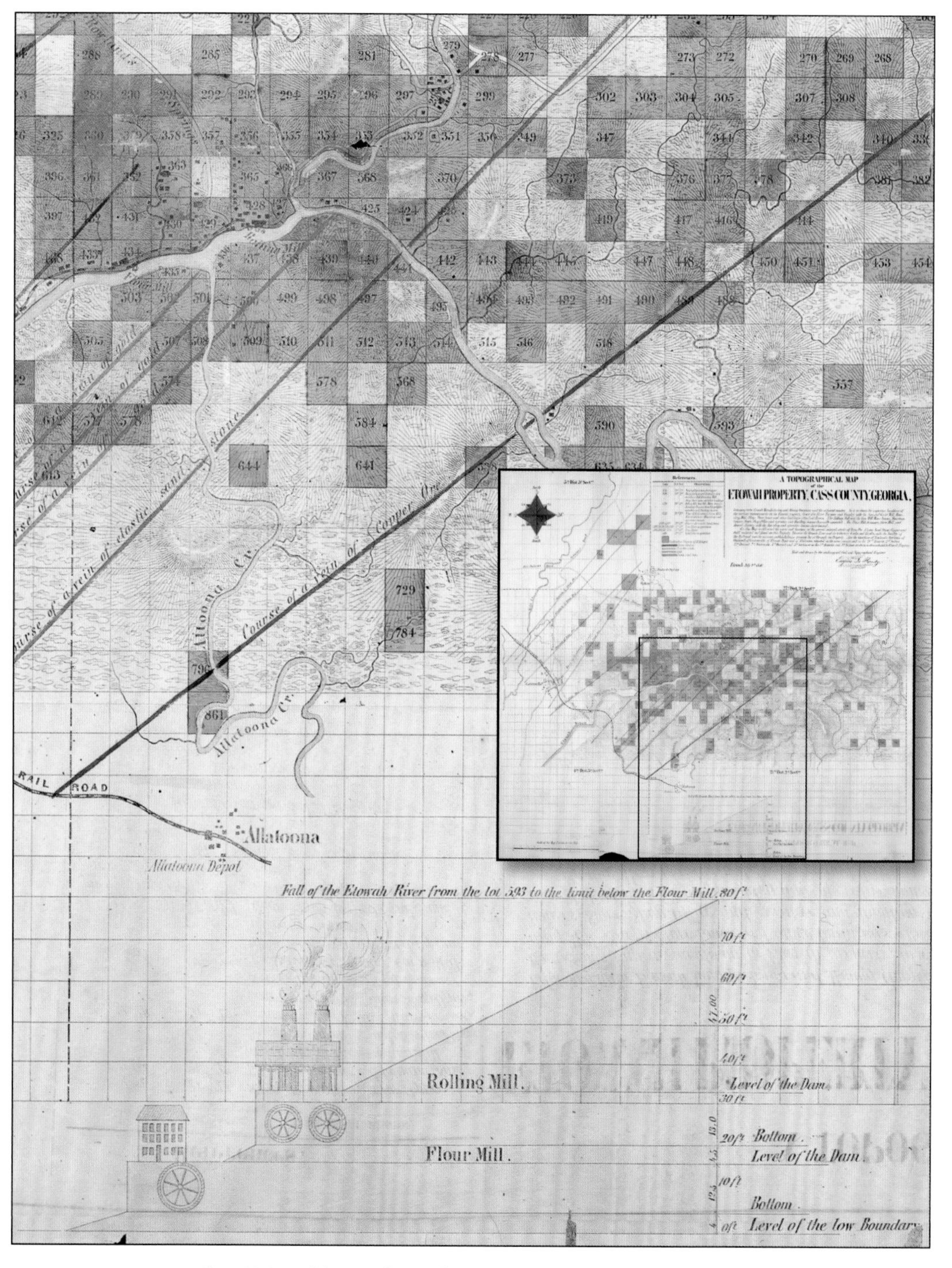

Figure 36. Map of the Etowah Manufacturing and Mining Company and its lands. Eugene T.[?] Hardy, "A Topographical Map of the Etowah Property, Cass County, Georgia," National Archives and Research Administration. This rare printed, color map of a rural antebellum iron company's holdings shows how much land was deemed necessary to fuel a substantial charcoal-iron furnace, forge, and rolling mill operation—thousands of acres. The detailed view below the map shows a cutaway diagram of the fall of water that provided power to the rolling mill and the settlement's flour mill. The long diagonal lines across the map mark veins of copper (*green*) and sandstone (*orange*) that also ran through the property.

owners. Although married men might be more settled, they could also be more demanding about living conditions. If workers came with families, the company had to build more housing. Given the number of workers required to run an operation like Etowah, it was more economical to hire as many single men as possible.[84]

The predominance of men in rural company towns was further heightened in the South by iron companies' preference for slave labor. In 1850, the Etowah proprietors owned forty-two slaves, all but three of them males of working age.[85] Like Etowah, Tennessee's Cumberland Rolling Mill employed both white and slave workers. The rolling mill proprietors, whose industrial holdings included three blast furnaces, owned 341 male and female slaves. The company also hired 63 slaves in 1850, all young men. The proprietors of three other furnaces in Stewart County owned 52 slaves, all male, most of them in their twenties—prime working age.[86] The disproportionately male population of rural iron company towns contributed to their combustibility. Most of the incidents I have found of labor-management conflicts, strikes, and violence at antebellum ironworks occurred in company towns, about half of them in rural settings.

The owners of rural iron company towns brought extensive business and financial experience to their ironmaking ventures. Etowah owner Mark Cooper, a Georgia lawyer, opened a cotton mill and later ran repeatedly for political office. His first partner at Etowah, Moses Stroup, had been raised in the iron trade by his father Jacob Stroup, who built the first Etowah Furnace. Moses went on to become one of the South's leading ironmasters.[87] The incorporators of Ashland Furnace included Philip and Samuel Small, "who owned several flour mills and engaged in the wholesaling of dry goods, at one time accounting for one-sixth of the freight on the Northern Central Railroad." A later partner in the business was John Merryman, who went on to become president of the Baltimore County Board of Commissioners and treasurer of the state of Maryland.[88] Lesley omits the names of owners for many ironworks, but those he does list for rural coal-fired blast furnaces and rolling mills included other merchants, financiers, and industrialists who saw iron as a promising investment connected to their other interests. Perhaps the most striking set of investors in this mold were the members of Boston's "enterprising elite" who financed the venture at Farrandsville. For capitalists such as Thomas Handasyd Perkins, Patrick Tracy Jackson, Edmund Dwight, and George W. Lyman, who had already made fortunes in the textile industry, producing iron was a logical next step in building their investment portfolios.

The involvement of wealthy capitalists had important consequences for the communities they created. For one thing, a great deal of capital was infused into iron company towns—$500,000 or more into Etowah, perhaps as much as $700,000 into the mines, furnace, and rolling mill at Farrandsville.[89] While much of the capital went to purchasing land and mineral rights, it also facilitated the transfer of British technologies, for heavily capitalized firms could afford to import hot-blast and rolling mill machinery and to recruit and house dozens of skilled ironworkers and coal miners. Second, investors in these ventures were often absentee owners. This was the case for most coal-fired furnaces and rolling mills in central Pennsylvania, western Maryland, and northern New York in the antebellum period. Owners hired the best managers and superintendents they could find, trusting them to run ironworks whose technology they themselves rarely understood. They expected a good return on their investment and rewarded managers who met their expectations. Because the geographical circumstances of production were often problematic in rural iron and coal regions of the United States, hard-driving firms that were trying to implement new technologies often encountered frustrating and costly delays, generating tensions that became tinder for labor-management conflict.

Urban Iron Communities

Urban iron communities ranged from New York City, where iron was one of many industries, to Pittsburgh, where the country's greatest concentration of rolling mills dominated the urban-industrial landscape and claimed a substantial minority of the city's male workforce. Some urban iron communities specialized, such as Wheeling, Virginia, called "nail city" after its primary product, and the towns along the lower Schuylkill River outside Philadelphia that were known for sheet iron and boilerplate. Unlike frontier company towns, urban ironmaking communities usually had numerous ironworks, mostly manufacturing concerns, offering more choice for workers. Very few blast furnaces were located in American cities in the antebellum period, chiefly because of the prohibitive cost of shipping raw materials. To highlight the characteristics of this kind of community, most of my examples are towns where iron was the major, if not the only, heavy industry.

The first examples are iron towns that developed along the Ausable and Little Ausable Rivers in Essex and Clinton Counties in upstate New York. Chasms that these rivers carved on their descent from the Adirondack Mountains to Lake Champlain are now tourist attractions. In the early nineteenth century, the region's picturesque waterfalls and rapids provided excellent waterpower sites for mills and ironworks (fig. 37). The proximity of waterpower to outcrops of some of the country's richest iron ore deposits began to draw investors in the 1820s.[90] The Peru Iron Company, organized in 1822–24, anchored the village of Clintonville. This settlement was created de novo to make iron, as were almost sixty other settlements in the Adirondack iron region between 1800 and 1860. By about 1830, the Peru Iron Works included bloomeries, a rolling mill, a nailery, and a cable and anchor factory, which collectively employed between 400 and

Figure 37. Joseph Lesley's sketch of Keeseville, New York, 1857. J. Peter Lesley Papers, American Philosophical Society. The Au Sable, or Ausable, River runs in rapids over sandstone slabs at Keeseville, originally named Anderson Falls. Along with coal, water diverted from the river provided power for the rolling mill and nail factory on the right bank. The Ausable Iron Works appears in the background of figure 23.

500 men. A government report noted 284 men at the rolling mill alone.[91] Forty years later, Clintonville had a greater array of manufacturing operations, but all were still part of the Peru Iron Works.[92] Upstream, the Sable Iron Works similarly dominated the town of Ausable Forks and three smaller settlements nearby. The Sable Iron Company owned ore mines and bloomeries, tracts of forest for charcoal, a sawmill, and a factory that made barrels to ship Sable nails down Lake Champlain, the Champlain Canal, and the Hudson River to markets in the New York City area. In 1860, when the town of Ausable had 3,227 people, 85 percent of the working population was employed at the ironworks or the iron company's boardinghouses, hotels, stores, or farms. Nails, bar iron, and other iron products accounted for 94 percent of the town's manufactured goods by value.[93]

Catasauqua, Pennsylvania, was also dominated by one iron company, in this case a smelting operation. The Lehigh Crane Iron Company, established on the banks of the Lehigh River in 1839, transformed a small farming settlement into a center of anthracite pig iron production. When L. F. Henning painted its portrait about 1852 (fig. 38), Lehigh Crane had five blast furnaces "in one pile," as Lesley put it. Standing up to forty-seven feet high, they were the largest rank of furnaces in antebellum America at the time. By 1856, the company employed 650 men, more than 60 percent of the town's adult male population. Welsh immigrants dominated the top skilled and managerial positions at the ironworks.[94]

Danville, an iron town in Montour County, Pennsylvania, was one of the few urban antebellum communities with both mineral-fuel blast furnaces and rolling mills. Like Mount Savage in western Maryland and the Cambria Iron Works in Johnstown, Pennsylvania, Danville was an early producer of railroad rails. The Montour Iron Company built the first anthracite furnace in Danville in 1839. By 1846–47, the town had two anthracite furnaces and two large rolling mills. Five more anthracite furnaces and a hot-blast charcoal furnace were within twelve miles of the Montour mill.[95] In 1850, the town's 3,302 residents included hundreds of English, Welsh, and American puddlers, rollers, heaters, molders, machinists, and miners, as well as a smaller numbers of Irish, Scottish, German, French, and Nova Scotian ironworkers. The Irish were most prominent among laborers in Danville. Three of the town's six ironmasters were Americans; the others were from England or Wales.[96]

Figure 38. *West View of Catasauqua*, ca. 1852, by L. F. Henning. Photograph courtesy of the Historic Catasauqua Preservation Association. This view emphasizes the centrality of the Lehigh Crane Iron Works to the market town turned industrial center. David Thomas's blue-slate-roofed home stands apart on the hill (*upper left*).

These examples point up several characteristics that distinguished urban ironmaking communities from iron villages and frontier company towns. In addition to having many more skilled ironworkers, their population was more heterogeneous. Pittsburgh was a magnet for Welsh, English, and Irish ironworkers by the mid-1830s. Established immigrants from the British Isles opened doors for their friends and relatives, giving the top rungs of labor at Pittsburgh rolling mills a decidedly ethnic cast.[97] Chain migration also drew nonindustrial immigrants from British and continental iron centers, contributing further to the formation of communities with strong ethnic institutions. In Danville and Pittsburgh, as in coal-mining towns that became destinations for Welsh immigrants, Welsh-language chapels and choir competitions marked community life. Danville was known to Welsh Americans as the home of Joseph Parry, an immigrant puddler from Merthyr Tydfil who composed some of the finest music in the Welsh choral repertoire.[98] The mixing of ethnic groups (and of races in the South) heightened workers' sense of identity, which among British immigrant artisans was already a potent combination of class consciousness, occupational pride, and, for some workers, linguistic identity. These places were also hardest hit by economic downturns, and urban workers were most likely to respond to wage cuts by going on strike.

As key nodes in the networks of skilled industrial migration, urban iron communities played a particularly important role in the transfer and implementation of British ironmaking technologies. Bruno Latour provides a useful framework for assessing this relationship. He describes science as consisting of the networks that connect scientists and engineers to sources of information and then link that information back to "centers of calculation" where disparate information is integrated and used for making decisions.[99] In his discussion of how decisions are implemented in locations distant from the center of calculation, Latour asks, "So how is it that in some cases science's predictions are fulfilled and in some other cases pitifully fail? The rule of method to apply here is rather straightforward: every time you hear about a successful application of a science, look for the progressive extension of a network. Every time you hear about a failure of science, look for what part of which network has been punctured."[100] Substitute "industrial model of production" for "science," and one has a working hypothesis for examining the role of networks of labor skill and technical knowledge in the modernization of the American iron industry.

By "centers of calculation," Latour meant institutions at the core of powerful nations (such as government agencies in capital cities, universities, scientific laboratories, and mapping organizations) where remotely gathered information has historically been sent to be transcribed and transformed into more useful forms. Before calculation is possible, locations have to become centers of accumulation—places where information can be collected, organized, refined, and fitted into larger networks. One could interpret Lesley's gathering and disseminating of information on the iron industry and the American Iron Association's lobbying of Congress as efforts to establish Philadelphia, and the network of large-scale East Coast producers who identified themselves as leaders in the industry, as the center of accumulation and calculation for the iron industry. Latour's idea is even more useful, however, when linked to historical geographer David R. Meyer's notion of "networked machinists" in early nineteenth-century America.[101] Meyer found that both places and individuals could be centers of accumulation and calculation, though at different scales and with somewhat different functions. Within networks of machinists, gifted inventors and production supervisors, and the companies with which they were associated, emerged as "hubs" (nodes within the network) that attracted other machinists who wanted to learn certain skills, work on challenging projects, or simply hold a steady job at a successful machine shop. "Nests" of machinists in manufacturing centers such as New York City, Philadelphia, and Fall River, Massachusetts, became rookeries of innovation as the exchange of ideas and methods among fellow machinists spread best practice within and between those centers. Meyer predicts that hubs (whether firms or individuals) with more numerous and diverse contacts would become more competitive and innovative over time, while those with fewer, less diverse contacts would fall behind. In the iron industry, this prediction proved true in the Mid-Atlantic region generally and several urban iron communities in particular, as we shall see.

Shared ethnic identity was often a unifying force that engendered trust, a vital quality for teams of men working in occupations that were physically demanding and sometimes dangerous. Cultural differences, on the other hand, could fuel misunderstanding, distrust, and hostility between groups of workers or between workers and managers. This problem was most pronounced at urban ironworks in the South where managers tried to compel British immigrant artisans to work alongside slaves or to train slaves in their craft. The best-documented example of this was the strike organized by British and American puddlers and rollers at the Tredegar Iron Works in Richmond, Virginia, in 1847. When the works superintendent, Joseph Reid Anderson, pressed skilled white workers to honor their contracts and fully train their slave assistants, a number of artisans walked out. Anderson fired the strikers, evicted them from their company houses, and replaced them with slave workers. With "new and inexperienced hands" at the puddling furnaces and roll trains, mill production dropped by more than 25 percent. Although he had sworn before the strike that he would never hire another "Pittsburgh Puddler," by 1850 Anderson was obliged once again to recruit skilled immigrants and northerners to fill orders for rails, locomotives, and heavy ordnance.[102]

Anderson may also have had economic motivation to confront white mill workers in the summer of 1847, because at that time Tredegar, like many other ironworks, was suffering from falling iron prices. Pennsylvania ironmasters seriously discussed fixing wage rates that year, but agreement came too late to prevent strikes at several rolling mills. As the credit crisis deepened and became a general economic depression in 1848 and 1849, rolling mill owners in many parts of the country cut wages and reduced production.[103] In January 1850, rolling mills in Pittsburgh cut puddlers' wages by 25 percent, prompting a series of strikes led by English and Welsh artisans. When mill owners recruited American strikebreakers from eastern Pennsylvania, workers in Pittsburgh turned violent. Mill workers' wives and daughters joined the protest by pushing their way onto mill grounds and throwing dirt and waste coal into puddling and heating furnaces.[104]

Women shared the working-class consciousness of their male relatives, and they suffered as much or more from the deprivations caused by cyclical unemployment in the iron industry. Mass unemployment was new in antebellum America. As the old reciprocal relationships between agricultural and industrial labor changed, industrial workers grew increasingly dependent on wage labor in cities. The advent of truly free labor came at the cost of greater vulnerability to hunger, homelessness, and disease, for antebellum employment contracts generally allowed employers to release workers without warning, with no compensation, for whatever reason they saw fit. Historian Alexander Keyssar, writing of labor in late nineteenth-century Massachusetts, notes that unemployment was "an important precipitant of geographic mobility . . . because jobs were chronically unsteady and because unemployment levels varied considerably among the hundreds of cities and towns within the commonwealth's borders."[105] This was true on the national scale for the antebellum iron industry. In boom times, companies' growth and the geographical expansion of the industry generated more jobs than there were workers to fill them, prompting national and international migration. When the industry entered one of its periodic contractions, large ironworks shed nonessential workers. While unskilled and semiskilled workers at rural ironworks could switch to agricultural labor, there were no easy alternatives for rollers, puddlers, and coal miners. Unless they were willing to accept reduced wages and cross picket lines, skilled ironworkers could be unemployed for months at a time.

That was what happened in 1850 in Pittsburgh. A puddlers' strike began on December 20, 1849, and was not finally concluded until May 12, 1850. Among those who stayed in the city, repeatedly marching in protest against mill owners, the experience of prolonged unemployment and resistance forged stronger class consciousness and laid the groundwork for more effective labor organization in the late 1850s and 1860s. Some eastern Pennsylvania workers "stood out" in support of the Pittsburgh strikers, swelling the numbers of marchers in one protest to over one thousand.[106]

The social differences that contributed to labor conflicts also made urban iron communities vibrant and alluring places, full of interest and promise for young workers. Rapidly growing industrial towns like Scranton, Pennsylvania, sported gentlemen's smoking rooms in fancy hotels as well as ethnic taverns where workers could share a drink among friends without fearing censure from a company manager. Such places inspired ambition, even in a cautious man like Peter Lesley, who visited Scranton during his tour of the anthracite iron district in November 1856:

> Here I am in this phantasmagorical city, the creation of a day, full of grand houses, furnaces of the largest size, rolling mills, depots, railroads crossing & recrossing, rows of operative houses, broad streets, hills covered with coal operations, and a universal swarm & tumult of busy life,—here in one of the largest and best hotels in the Union, after an evening spent in the Scranton Store office with Platt & the Scrantons & Mitchell, discussing iron, getting statistics, & renewing remeniscences [*sic*] of Princeton. What an evening it has been.[107]

Disciplining Labor

Ironmaking communities differed in size, morphology, geographic setting, demography, technology, and labor relations. One thing they all experienced was the periodic shortage of skilled labor. The problem of labor scarcity was endemic in the young United States; it has often been singled out as one of the basic differences between American and European society in the late eighteenth and early nineteenth centuries. Labor scarcity was particularly acute in the iron industry. An insight from the work of Karl Marx, highlighted in the work of geographer David Harvey, helps one grasp the geographical dimensions of the problem. Marx observed that industrial capitalism demanded both the geographical mobility of labor and the constraint of labor mobility. The paradox arose from the contradictory needs and cycles of nineteenth-century manufacturing. On one hand, the period's archetypal industries—textiles, iron, and railroads—produced massive manufacturing plants and transport infrastructure that were literally bolted to the ground. Increasing the return on investment required sinking more and more capital into the immovable built environment of factories, roads, track, and ever-larger agglomerations of productive capacity. Each of these required large infusions of labor, with particularly heavy demand for skilled labor during the early phases of construction and the implementation of new technologies. In the United States, industry was also expanding geographically with unprecedented speed during the early nineteenth century. The spread of heavy manufacturing to the American Midwest and, to a lesser extent, the South created new loci of demand for labor. Both unemployment and the multiplication of employment destinations set industrial workers in motion locally, regionally, and internationally.[108]

The antebellum iron industry embodied all these contradictions. Skilled ironworkers enjoyed the advantages of being in high demand in many places at once. They also saw the many faces of persuasion and coercion as ironworks managers tried to secure their skills and bend their labor to managerial control. The American legal system shaped labor relations as well, as the antebellum period ushered in the new legal framework of employment at will. Under the previous doctrine of master and servant law, employers could invoke criminal sanctions against workers who quit or failed to fulfill their employment contracts. Imprisonment for breach of contract became illegal in the United States in the early 1820s (it remained a tool of labor discipline in Great Britain until the 1870s). Denied the means of physically restraining workers to ensure fulfillment of their contracts, employers developed what historian David Montgomery calls new forms of discipline, notably pecuniary sanctions that threatened working-class families with starvation if their breadwinners quit or failed to perform their contractual obligations.[109] Positive incentives such as relatively high wages, promo-

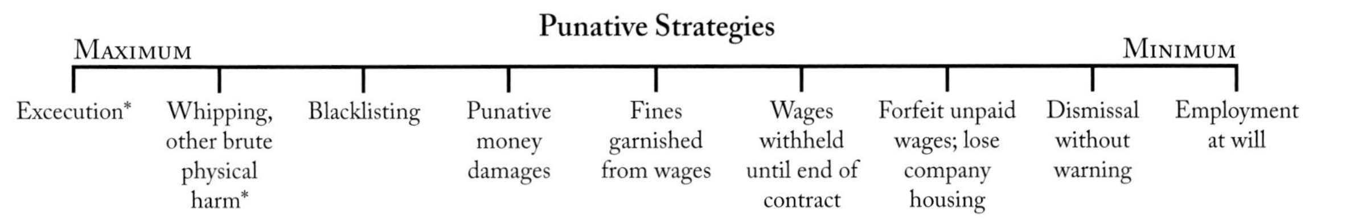

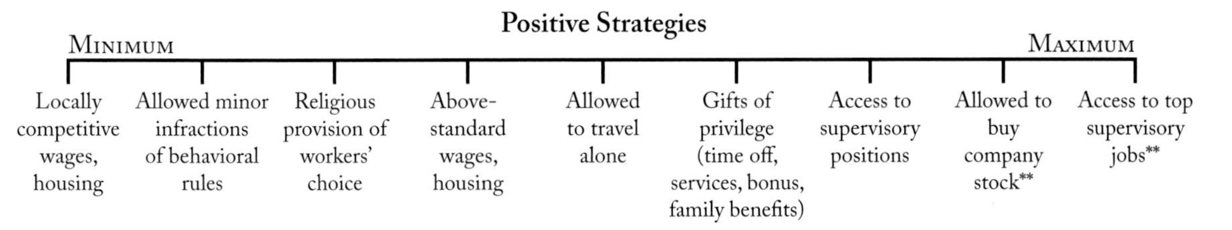

Figure 39. Continuum of management strategies used to discipline ironworkers.

tions, and the provision of schools and churches aimed to achieve the same basic goal of keeping workers in place, sustaining the "reserve army" of labor by inhibiting workers' movement.[110] Legal historian Robert J. Steinfeld argues that the conventional distinctions between indentured servitude, slave bondage, and free labor do not accurately describe nineteenth-century employment relations in Great Britain or the United States. He argues that close examination of employers' degree of control over labor shows a continuum of coercion within all three putative conditions of labor, particularly before the American Civil War.[111] I see a wider continuum in the managerial policies applied to skilled ironworkers, running from harsh coercion to generous incentives and allowance for considerable mobility (fig. 39). While in some places even the most privileged and free-ranging members of this "aristocracy of labor" were not entirely free, in other places men bound by slavery enjoyed significant degrees of freedom.[112]

Managerial policies on the positive side of the spectrum were most common, for applying discipline to ironworkers was tricky. If skilled workers broke their labor contracts by quitting, absenting themselves from work for long periods of time, going on strike, or refusing to meet contractual production quotas, employers had few options for replacing them, particularly during periods of economic growth. Antebellum employers faced self-confident opposition, particularly among their immigrant workforce. Many British artisans who emigrated in the 1830s had been politicized by movements for political reform that culminated in Chartist uprisings such as that of 1839 in Newport, Wales. The first trade union lodges associated with national industrial unions at South Wales ironworks formed in the early 1830s. Ironmasters responded by locking out trade union members and hiring militias to suppress protests. Similar experiences across Britain stiffened immigrant artisans' resistance to perceived injustice against their class.[113]

American managers commonly described skilled ironworkers as independent-minded, clannish, lazy, and fond of drink. As Christmas approached at Lonaconing Furnace in western Maryland, works manager and co-owner John Henry Alexander noted, "The Welshmen are drinking very hard. Steele thinks it cannot be helped."[114] Daniel Tyler, superintendent of Farrandsville Furnace in central Pennsylvania, complained that the Welshman in charge of building the furnace was not "used to work." In the 1850s, federal ordnance inspectors visited foundries with the aim of standardizing heavy weapons and improving safety. Their questions about production methods met stubborn silence from artisans who would not divulge their craft secrets.[115] William Weaver wrote in 1825 that "no reliance could be placed in the free White laborers who are employed about Iron Works" in Virginia, for in "moments of the greatest pressure & necessity, the proprietor must either make them advances which they will never repay, or they leave his service to the ruin of his business."[116] Thirty-five years later, one of Weaver's furnace superintendents tersely summed up the labor problem: " 'the white hands 'damn them . . . won't stick.' "[117]

Historian Charles B. Dew argues that ironmasters overseeing slaves had to strike a "delicate balance" between coercion and reward in order to keep the peace and guarantee slaves' productivity.[118] I see a similar balancing act going on at many antebellum ironworks that employed free white labor. Managers used strategies ranging from corporal punishment to generous incentives. Execution and whipping, the two most extreme punitive strategies, were used only against slaves, while only free white workers were allowed to buy company stock or become superintendent of a large ironworks. Otherwise, one can find examples of all strategies being used with free and slave labor.

The most extreme case of physical coercion in the antebellum iron industry played out in western Tennessee in the late 1850s. The Cumberland Rolling Mill, on the Cumberland River in Stewart County, Tennessee, was one of the largest rolling mills in the antebellum South. The mill, its coal mines, and the dozen or so charcoal blast furnaces that supplied pig iron to the mill had a combined full-time workforce of perhaps six hundred men, the vast majority of whom were slaves, including many skilled slaves at the rolling mill. According to the 1850 census, the mill had no more than fifteen white employees, all natives of Appalachia except for one Scottish engineer.[119] In 1856, the Cumberland mill was the center of a slave escape that whites believed was a violent insurrection. The details of this event, published in local newspapers, amplified white hysteria. Posses captured fifty-six of the escaped slaves and executed at least nineteen. The lack of company records makes it impossible to know for certain what made the slaves try to escape and whether the executed slaves were in fact skilled ironworkers and coal miners, as the newspaper reports and posses claimed, though it is likely that they were.[120] The violent events in Tennessee's western iron district proved to many southerners that gathering hundreds of male slaves at industrial works and permitting them to hold skilled, semiautonomous positions invited conspiracy and revolt. A contemporary observer judged such slaves " 'the most corrupt and turbulent' members of their race"[121]—as unreliable as white artisans but far more dangerous.

Although ironworks managers in the North did not hold such extreme views of their employees, some attempted to exercise milder forms of physical restraint that harked back to the privileges masters had possessed under master and servant law. When miners working for Farrandsville Furnace left the company's coal diggings, where they had been waiting more than a month to be paid, the superintendent had them arrested and held in jail for weeks. He meant to make them an object lesson to prevent other "desertions." At Lonaconing Furnace, discipline was carried out more in the spirit of the new laws. If workers misbehaved, left the works without permission, or simply were no longer needed, they were fired. If they refused to leave, the manager had them escorted off company property.[122]

For part-time or seasonal workers at charcoal iron furnaces, employment contracts specified particular tasks, such as building a boardinghouse or charcoaling so many cords of wood. Workers at mineral-fuel furnaces and rolling mills were typically employed for a term of years at fixed rates of pay. These operations had to retain their skilled workforce through the winter or face the ordeal of hiring workers shortly before waterways opened for navigation, when their competitors were also scrambling for labor. Retaining workers was expensive, however. Farrandsville Furnace was not alone in leaving workers unpaid for lengthy periods when bad weather, broken machinery, or low demand for iron lessened company liquidity. Going without pay became a chronic complaint among ironworkers. As one of the miners who remained at the Farrandsville coal diggings wrote to the mine manager, "no one likes to work no more than a month without is monney [*sic*]."[123]

Competitive wages were the most common positive incentive used to attract and keep skilled workers but by no means the only one. American iron companies also offered immigrant artisans exceptional opportunities for advancement that appealed to their ambition and pride. The strongest incentives were access to salaried positions with supervisory responsibility and the privilege of buying shares of company stock. Other workers were offered less costly but still meaningful rewards. Gouvernour Kemble, owner of the West Point Foundry in Cold Spring, New York, acknowledged his immigrant Irish workers' religious preferences by building them a Catholic chapel once he saw their lack of interest in

Protestant services. Lonaconing's Alexander, a devout Episcopalian, hired clergy to minister to his employees, but he also allowed the Welsh to attend their own Nonconformist services led by their chosen bilingual preacher.[124] Workers at many rural ironworks were allowed minor infractions of the rules. Overtime pay, usually in the form of credit at the company store, was commonly offered for work done after dark, on a holiday, or beyond the normal ten- to twelve-hour workday. During the Civil War, when skilled ironworkers were particularly scarce in the South, Alabama's Shelby Iron Company treated its white rolling mill workers to magazine subscriptions and the occasional use of slaves' labor. More important, Shelby's manager maintained a racial division of labor that preserved white immigrants' status. Forcing white workers to accept slave apprentices, as Anderson had attempted to do at Tredegar, or putting crews of mixed ethnicities to work in the same coal mine, as was done at Lonaconing, increased the chances of conflict and lower productivity.

If southern ironmasters had to strike a delicate balance between coercion and reward with slave labor, one would expect that managers of free white ironworkers had to err even more on the side of liberal incentives. It is therefore surprising to find that managers used so many different strategies to make workers "stick," from imprisonment to exceptional privileges that blurred the line between labor and management. The full continuum of strategies ranged from involuntary labor under terms strikingly similar to indentured servitude to true employment at will. There is no clear evidence of a progression from punitive to positive incentives between 1830 and 1860. Managerial practices throughout the period have an ad hoc quality, reflecting managers' uncertainty about their role in relation to changing systems of industrial production, as well as workers' testing of limits during a time when metalworking skills were crucial for most kinds of ironmaking.

Another surprising discovery was that variations in managers' responses to the scarcity of skilled labor did not follow the regional divides of North and South. In both regions, some managers allowed skilled workers to practice their craft as they saw fit and provided good compensation and housing, while other managers stinted workers' pay, failed to provide adequate housing, and tried to impose strict control over workers' behavior at work and in their private lives. Generally speaking, coercive measures were most commonly applied by managers who lacked experience and technical knowledge or who were most anxious about the loss of skilled workers, either because of their own inability to supervise operations or because of pressure from company owners to generate profits. Managers who were familiar with the iron business or who had come up through the ranks of skilled labor usually favored positive strategies and were far less likely to impose strict spatial discipline on workers. Philip Scranton found similarly favorable attitudes toward workers among proprietors of textile mills who came from mechanical backgrounds.[125]

These practices had economic consequences. The more forcefully managers at technologically advanced ironworks tried to constrain workers' geographical mobility, the more likely they were to lose them or to precipitate a crisis.[126] In an industry that was undergoing rapid technological change and geographical expansion, positive strategies offered the best means of holding essential workers in place. Ironworks managers who tried to restrict workers' mobility and power did not necessarily "curb the aggregate mobility of labor power in ways . . . inimical to the reproduction of the capitalist system as a whole,"[127] as Harvey claims, but they did impair their companies' ability to compete in the American marketplace. That managers tried so many strategies to get and keep skilled free labor, troublesome as it was, indicates how vital skill was to the antebellum iron industry.

Where workers' energy, skill, and cooperative work culture were incorporated into American ironmaking, labor was a positive force in production and an important factor contributing to the industry's growth. But as this chapter has shown, relations were not always good between ironworkers and their American employers. Labor relations were subject to the strains of personal conflict at the work site as well as problems caused by national and international forces, such as credit crises and competition from foreign imports. Issues at every scale affected ironworks to varying degrees, depending on the many dimensions of industrial production and work life that collectively constituted the circumstances in which a given company operated.

Up to this point, I have looked at the iron industry from a fairly broad perspective, with the aim of identifying key differences that distinguished the nature of ironmaking at the regional scale and the general characteristics of workplace environments and labor relations. One needs more than typologies, however, to explain the regional variations of technology and change in the American industry. The next two chapters examine that question in situ, looking closely at four iron companies whose owners set out to replicate part, or all, of the British model as closely as possible. One company failed badly. Another struggled to make a profit and never fulfilled its owners' hopes. The other two not only succeeded but became models of best practice, though both differed in significant ways from the British works that inspired them. The stories of these companies illuminate the role individuals played in industrial change, for good and ill. They also reveal how individuals' decisions were shaped by historical and geographical circumstances. In each case, exceptionally detailed company records allow us to see how impersonal economic forces and very human hopes played out in places that inflamed industrial ambition.

It is the commencement of a new Era, the making of Iron from pit coal in America—they are ambitious of the fame of first bringing it to bear for the benefit of the States. . . . [I]n a few years, I shall blow up such a blaze in America that the influence of all the Tories in England and Wales will not readily extinguish.

EDWARD THOMAS, January 11, 1836

3 High Hopes and Failure

The 1820s and early 1830s were a period of high hopes for industrial development in the United States. The success of large textile manufacturing firms in New England, beginning with the pathbreaking cotton mills at Waltham and Lowell, Massachusetts, had proved that the United States could compete with British industry. The Erie Canal, completed in 1824, was one of many ambitious "internal improvements" built to stimulate commerce between seaboard manufacturing cities and western agricultural settlements. America's first railroad, the Quincy Granite Railway, was built in 1826 to enable construction of the Bunker Hill Monument in Charlestown, Massachusetts. Although it was only two miles long, the Quincy Railway's ironclad rails became an early model for heavy-duty freight lines. The Mauch Chunk Railroad's inclined plane was built in 1827, and construction of the Baltimore and Ohio Railroad began in 1828. By 1833, three railroads were under construction in Massachusetts to haul passengers and freight on solid iron rails.[1] The exciting notion that railroads might span the nation gave fresh impetus to the older idea that the United States should achieve economic independence by developing its domestic manufactures. If American companies could replicate the British model of mass-producing iron, they could supply all the rails and locomotives the country needed. There would be no need to import rails and sheet iron from Britain. Given the country's vast resources, American firms ought to be able to manufacture all the steam engines, textile machinery, and other iron wares required for industry and agriculture. In short, the next leap forward in America's economic development required large-scale iron production, which meant making iron with coal. Beginning in 1825, the Franklin Institute in Philadelphia offered a gold medal to the individual who could produce the most iron smelted with anthracite. The medal went unclaimed for nearly fifteen years, despite the additional incentive of $5,000 that Philadelphia businessman Nicholas Biddle and his associates offered to the first person who could keep an anthracite furnace in production for three months.[2]

The coal-fired ironworks that many would-be American investors looked to as models were the large operations in South Wales and along the river Clyde in Scotland. Dufrénoy's widely circulated technical manual touted Cyfarthfa and Dowlais as the epitome of modern ironmaking because of their scale of operations, integrated production, and low costs per ton.[3] By the 1830s, these two Welsh companies had an international reputation as leaders in applying new technologies. They were among the first British iron companies to embrace the coke-smelting technology developed in the early eighteenth century by Abraham Darby at Coalbrookdale, Shropshire. Dowlais probably used coke in its furnaces from its establishment in 1759, while Cyfarthfa, chartered in 1765, built all its furnaces to burn coke. In 1787, Cyfarthfa owner Richard Crawshay was the second ironmaster to take out a license to employ the puddling and rolling process developed by Henry Cort. In the 1790s, Cyfarthfa was the most productive ironworks in the world.[4]

The French were the first Europeans to replicate the Welsh model of integrated iron production (which they and others often called "the English method"). Government-sponsored missions to Merthyr Tydfil in the 1780s laid the groundwork for building a state-of-the-art ironworks at Le Creusôt in Burgundy. After the Napoleonic Wars, renewed contact across the English Channel informed the construction of a number of "English-type" forges and integrated ironworks in other parts of France where coal and iron ore deposits closely matched the minerals in South Wales. Georges Dufaud, who owned an iron trading company in Paris, established personal connections with the Crawshays that facilitated the construction of furnaces and a forge at Fourchambault, also in Burgundy. Engineer François Cabrol's visit to Merthyr in 1826–27 similarly enabled the transfer of Welsh methods to the Aveyron region of south-central France.[5]

British labor was part and parcel of the transfer of British technology to France. An ironworks built by two English entrepreneurs in the early 1820s at

Charenton employed 135 English and Welsh workers. In 1827, all the foremen in the rolling mill and forge at Le Creusôt were British immigrants. Fourchambault's skilled workforce was dominated by English and Welsh artisans. At Decazeville and the other company towns built by Cabrol's firm, virtually all of the engineers, puddlers, heaters, rollers, and polishers during the early years of production were immigrants from Wales. Some went home after a few years or emigrated to the United States. Others married French women and settled in as permanent residents.[6] In these places, importing large numbers of experienced engineers, foremen, and artisans facilitated implementing imported methods of iron smelting and refining. This accomplished two goals in one stroke: it enabled firms to recoup their heavy start-up costs relatively quickly, and it implanted an industrial work culture that otherwise would have been slow to develop in the rural locations where the new works were situated. The board of directors of the Decazeville works claimed, "It is perhaps without parallel in France that a single enterprise has been founded so quickly." Local people may not have liked the *ouvriers cosmopolytes* who worked in their midst, but the production and sale of marketable iron, and the profits it yielded, were exactly what investors wanted.[7]

Several American companies took the same approach to replicating the Welsh model in the United States in the 1830s and early 1840s. This chapter tells the story of two such attempts that failed, or almost failed. The next chapter looks at companies that were much more successful. These tales of failure and success reveal that while American entrepreneurs grasped some of the key features of large-scale ironmaking in Great Britain, they missed others. French and American reports on the British iron industry throughout the antebellum period emphasized the low unit costs of Welsh bar iron and Scottish pig iron, which they attributed chiefly to the efficiency of new technologies, particularly coke furnaces and rolling mills, and the low cost of British labor. Observers also recognized the economic advantages of proximate mineral resources, integrated production, and transportation by canal and rail.[8] All these factors required heavy initial capital investment but promised handsome returns. Aside from mineral deposits, they all could be transferred physically to new sites of production or replicated from plans. A labor recruiting agent for a Pennsylvania iron company told his employer in 1835 that to replicate Dowlais, one simply had "to cut down the Trees, build an air furnace, mine the Coal and Iron Stone, and smelt the ore and Roll the Iron into merchantable Bars and Bolts." So long as a company had knowledgeable workers to supervise construction and man the machines, the process should not be too difficult.[9]

This mechanistic view of technology transfer failed to take into account a number of crucial factors. The most fundamental one was the chemical content of the raw materials used to smelt iron. Although American iron entrepreneurs were aware that the quality of mineral resources affected the quality of the final product, almost none of them knew how to assay iron ore and coal. They hired geologists to do that, but chemical assays were not always reliable. Geology was a very new science in the early nineteenth century, the chemistry of ores and coals only partly understood. The only way to determine the suitability of local resources for a given method of smelting or refining was to conduct physical trials on specimens, such as breaking bars of iron to examine the metal's crystalline structure or bending a sample to test its tensile strength. In the absence of precise knowledge, iron companies relied on tradition and reputation.[10] The shift from charcoal to coal introduced new uncertainty into ironmaking formulas because trace amounts of impurities in coal, such as sulfur, phosphorus, and manganese, could make iron that was either ideal or completely unusable for a given product. This problem was not peculiar to the United States. The coke smelting process that Darby developed using Shropshire coals only gradually diffused to other parts of Britain. It took six decades for firms across Britain to adjust mineral-fuel technologies to the resources available to them. In the meantime, British inventors such as Henry Cort were developing other techniques to remove impurities from coke iron so that it could be used in a wider range of products than the castings that had become Darby's specialty.[11]

The second factor that slowed adoption of British coal-fired technologies in the United States was the distance between resource deposits. Each of the locations in England, Scotland, and Wales where iron was smelted with coal in the early nineteenth century (fig. 40) had readily accessible, proximate deposits of iron ore, limestone, and metallurgical coal (coal that coked well or could be used directly in a furnace). The geographic circumstances of Merthyr Tydfil in this regard were particularly advantageous. As historian Chris Evans notes in his study of Merthyr's early iron industry, "Works like Penydarren and Cyfarthfa were constructed with integration in mind, laid out so that each successive operation followed its predecessor in a continuous flow downhill, following the contours of the valley location."[12] Figure 41 highlights the steps of production at the Dowlais Iron Company, as revealed in a manuscript map produced by the British Ordnance Survey in 1851.

The layout of Dowlais, like other Merthyr ironworks, followed a linear sequence from upland mines to shipment down the Taff Valley to the Cardiff Docks on the Bristol Channel. Coal, ore, and limestone were mined and quarried from the hills immediately above the works (off the right edge of the map). Tram cars carried the raw materials downhill to the blast furnaces. Clay from nearby pits was also carted down to be milled and manufactured into firebrick to line the furnaces. Next to the brick factory at the upper end of the Dowlais grounds were calcinating kilns where iron ore was heated to drive off impurities. The blast furnaces, thirteen of them in 1851, stood in rows with their backs to a hill so that they could be loaded easily at the top. Pig iron produced by the furnaces was

Figure 40. British iron regions and resources, ca. 1848. This map, based on Robert Hunt's map of coal, iron, and furnaces in blast, highlights the clustering of British ironworks in regions where coal and iron ore were found close together.

shuttled by rail farther downhill to the rolling and puddling sheds or the forge. Just below the forge stood the Dowlais railroad station and (off the left side of figure 41) a turning basin for the Glamorganshire Canal. Trains and canal boats carried pig iron, rails, forgings, and cast-iron products to the docks at Cardiff,

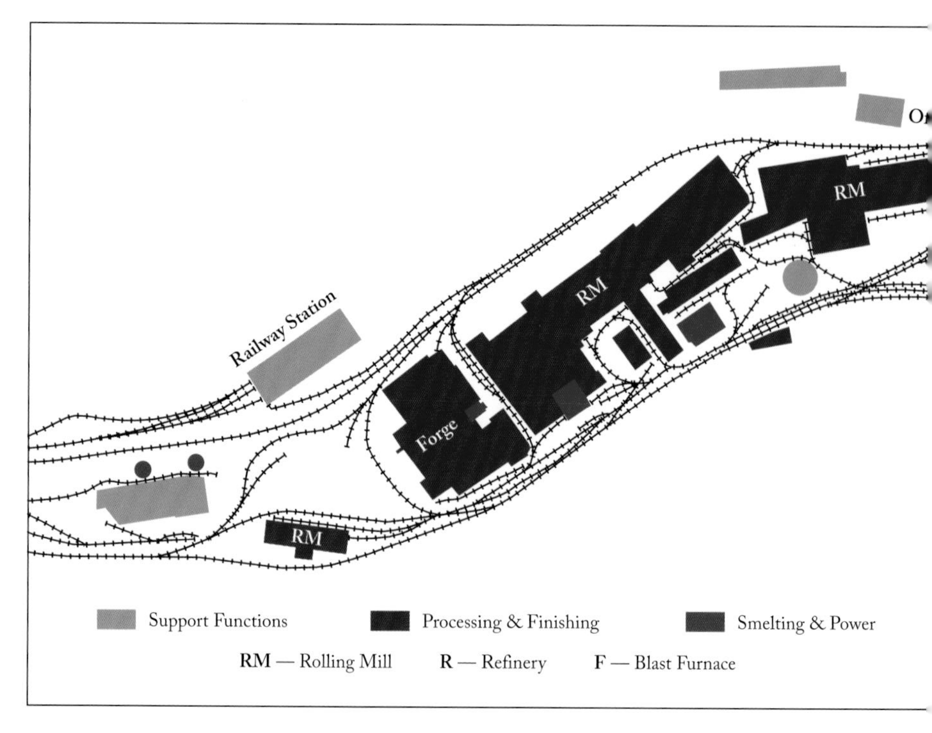

Figure 41. Dowlais Iron Works, ca. 1852. This diagram is based on an Ordnance Survey map that shows the size and function of virtually every building at Dowlais and other ironworks in Merthyr Tydfil. The sequence of production ran from right to left, beginning with smelting iron from ore at the furnaces, then refining pig iron at nearby refineries, then transferring bar iron downhill to rolling mills, forges, and foundries where final goods were produced. A journey of about thirty miles by rail or along the Glamorgan Canal delivered finished iron to the docks at Cardiff. This integrated system became a model for American iron manufacturers.

which served ships of all size. With just thirty miles from Merthyr to Cardiff, shipping costs were modest. These geographical and geological circumstances minimized the costs of transportation and production for Dowlais as for many British producers.

British ironworks also benefited from relatively low-cost labor, thanks to rapid population increase and improvements in agricultural productivity.[13] British wages at most levels of skill were well below American wages, which enabled US companies to entice British artisans by offering them handsome pay and often paying for their passage. What American employers did not factor into their calculations regarding imported skilled labor was the importance of the culture of work, which related both to workers' ethnic and occupational identities and to their social lives in their home countries. Better pay did not necessarily make immigrant workers productive or obedient. Some well-paid British artisans turned fractious and unmanageable in the United States. As the previous chapter suggested, the nature of labor at large ironworks cultivated pride among skilled workers. In Britain, artisans enjoyed considerable independence in doing their work as

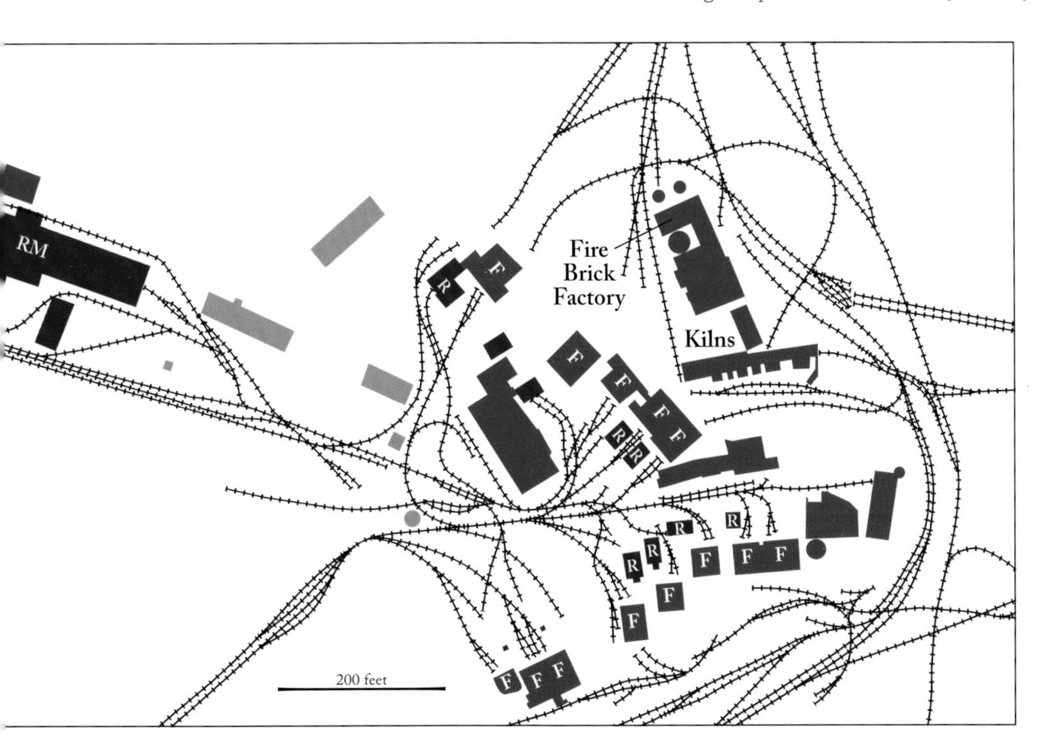

they saw fit. Strong regional identities reinforced loyalties rooted in class, occupational, and ethnoreligious identities. Ethnic identity may have been particularly important in South Wales. The migration of Welsh speakers from the countryside to iron and coal towns such as Merthyr and Tredegar created a distinctly Welsh industrial culture. It was so strong by the 1830s that it behooved English migrants to learn the Welsh language if they wanted to climb the occupational ladder.[14] On the shop floor of most major British ironworks, as in the mines, men worked in small crews with kin and ethnic brethren. Ethnic and occupational solidarity fostered the trust that helped rollers and catchers, puddlers and their assistants, mine bosses and their crews work smoothly together. The same qualities seemed obdurate and "clannish" to many American employers, as they had to colonial ironmasters and would to the next generation of iron and steel makers.[15]

The work culture of large-scale British iron manufacturing was also very urban. In Merthyr Tydfil, Dowlais artisans' homes on "Forge Row" and "Finery Row" were next door to workingmen's taverns: the Bush Inn, Black Bell, Owain Glyndwr, Prince Albert, King's Head, Rolling Mill, and Forge Hammer. Pub culture was well established in industrial South Wales, as were the sometimes contradictory influences of Old Testament, Nonconformist chapels and political movements for workers' rights. By 1851, Merthyr had a population of over 46,000. At least half the working population was employed in the ironworks and coal mines.[16] These numbers help explain how the ironworks of South Wales could afford to lose hundreds of skilled ironworkers a year without suffering labor shortages. Overseas demand for skilled labor may even have served as a safety valve for the volatile town. At the same time, the concentration of large ironworks provided alternative employment for skilled and unskilled workers within their home region. Migration to the United States gave industrial workers even more opportunities for work and for advancement. American iron-manufacturing cities such as Wheeling, Pittsburgh, and Boston were certainly urban. In remote mountainous iron districts, however, European immigrants found few of the amenities or employment options to which they were accustomed. The American industrial frontier was not a hospitable environment for skilled immigrant workers and their families, nor was it conducive to peaceful labor relations.

The last two factors that Americans did not fully appreciate were the most geographic of all: distance and weather. Merthyr Tydfil's ironworks were just thirty miles from Cardiff. Ironworks along the river Clyde were even closer to the sea. Coalbrookdale and other iron towns in the upper Severn Valley were up to one hundred miles from the coast, but the Severn River was navigable for that distance. Canals connected the Midlands iron district to London and Liverpool. In short, distances to major domestic markets added little cost to British iron, and proximity to the sea provided access to international markets by the least expensive mode of shipping—ocean transport.[17] In the United States, only 33 percent of antebellum blast furnaces and 52 percent of rolling mills lay within one hundred miles of the Atlantic coast, and some of them lay behind significant topographic obstacles.[18] As for weather, British winters are chilly and damp, but the island's maritime climate rarely produces snow or freezing cold in the iron districts.[19] British canals usually operated year-round, as did trains and ironworks. The northeastern United States has a continental climate. During the coldest winters of the 1830s and 1840s, the upper branches of the Susquehanna River in central Pennsylvania froze over. Canals were usually drained to avoid their freezing, though of course a drained canal was useless for transportation.[20] Snowstorms in the Allegheny Mountains stopped overland traffic for days at a time. The annual spring thaw made muddy roads impassable to heavy freight, while freshets (a mild word for often destructive spring floods) made downstream passage dangerous and sometimes impossible.

Taken all together, these environmental and cultural conditions constituted factors of production that were very different in many parts of the United States than in Britain. American iron manufacturers could not match the productivity of the British industry until they had solved the frustrating and expensive problems posed by their own country's physical geography, mineral geology, labor supply, labor relations, and the vastness of American distances. The weather just made the other problems worse.

Farrandsville

The most audacious attempt to replicate the Welsh model of ironmaking on American soil in the 1830s happened at least partly by accident in a remote central Pennsylvania valley at a place called Farrandsville. It was named after William Powell Farrand (1777–1839), a Philadelphia bookseller and publisher who went bankrupt and then remade himself as a commission merchant sometime around 1818. His involvement in literary circles, particularly as publisher of the short-lived *American Review of History and Politics* (1811–12), put him in contact with the cultured businessmen of Philadelphia and other East Coast cities.[21] It seems likely that he knew the family of Robert Ralston, a leading Philadelphia merchant, whose son Gerard Ralston was a literary man involved with the Historical Society of Pennsylvania. Gerard's older brother, Matthew Clarkson Ralston, a merchant like their father, was an early promoter of railroads in Pennsylvania.[22] Matthew Ralston was also one of the proprietors of the Lycoming Coal Company, for whom Farrand worked as agent. Chartered in 1828 to mine bituminous coal, the Lycoming Company owned three thousand acres of land along Lycoming Creek, a tributary to the West Branch of the Susquehanna River in central Pennsylvania. The company's prospectus in 1828 claimed that its coal was the equal of Newcastle's—or better, because the coal lay "in horizontal veins, elevated considerably above the ordinary level of the adjacent country," so that it could be "mined without encountering any of [the] difficulties" associated with deep-pit mining. Simple drift mines would be adequate to extract the estimated 13 million tons of coal on the property. The company's holdings further included iron ore and clay suitable for firebrick. All that was required to make the property yield its wealth was "the completion of the Pennsylvania canal to the mouth of Lycoming creek, . . . an event of most auspicious promise to [the company's] future operations, for it will open for their coal a direct canal communication to all the towns on the Susquehanna, to the capital of the state, and to her great commercial metropolis," Philadelphia.[23]

In the spring of 1832, four "men of Boston," as they described themselves, took the bait and purchased the Lycoming Coal Company. The new partners of the reconstituted company were among the wealthiest and most experienced businessmen on the eastern seaboard. Thomas Handasyd Perkins, Patrick Tracy Jackson, Edmund Dwight, and George W. Lyman were leading members of Boston's "enterprising elite," the merchants who built the country's first large-scale textile mill at Waltham, Massachusetts, in 1813–14 and went on to finance the textile mills at Lowell.[24] A brief review of their careers shows what a formidable array of experience they brought to their Pennsylvania venture.

Thomas Handasyd Perkins (1764–1854) was known as "the Colonel" after his honorary appointment as lieutenant-colonel of a military corps in Boston. He made his first fortune in the China trade, which he entered as a young merchant in 1789. In league with the North West Company, the shipping firm of James and T. H. Perkins monopolized the fur trade from the Columbia River region to Canton until 1821.[25] Perkins directed the construction of the Quincy Granite Railway in 1826; invested in the Western Railroad through central Massachusetts, chartered in 1833; and was a major investor in the Locks and Canal Company, founded at Lowell in 1824, which built mill, canal, and rail equipment and oversaw the development of the mill complexes. Perkins's other investments as a member of the Boston Associates included major textile companies at Lowell and elsewhere in Massachusetts, the Boston Insurance Company, and the Massachusetts Hospital Life Insurance Company. Perkins served as an officer in several of these companies and was elected president of the Boston branch of the Bank of the United States. He also served many terms in the Massachusetts state legislature, primarily in the senate. Perkins moved among the highest circles in his home city and Washington, DC. As a young man he was offered the post of secretary of the navy by President George Washington. Perkins declined, reportedly telling the president that "he owned a larger fleet of vessels than the United States Navy and believed he could do more good by continuing to manage his own property."[26]

In 1832, when Thomas Sully painted his portrait (fig. 42), Perkins was at the height of his wealth and influence. He was an extremely successful businessman, a political figure, and a generous philanthropist.[27] The portrait suggests the ease and confidence of a man of the world, a man accustomed to taking risks. From his warehouse office on Boston's India Wharf, Perkins kept a close eye on the fluctuating fortunes of American commerce and foreign competitors. One of the opportunities he had spied was the need for increased domestic iron manufacturing when the Napoleonic Wars lessened the flow of British imports to the United States. In 1807 Perkins, his brother James, and six other Boston merchants invested $200,000 in the Monkton Iron Company in Vergennes, Vermont, a small town six miles from Lake Champlain. The company built a charcoal blast furnace to smelt pig iron from local ores, hired molders to produce cast-iron wares for regional markets, and erected a rolling mill to manufacture rolled bar, sheet iron, and wire for sale in Boston and New York.[28] The ironworks produced almost no income until the War of 1812 created a market on its doorstep, as the American navy desperately needed ordnance to fight the British on Lake Champlain. By 1814, Perkins was charging $60 a ton for cannon shot. Local histories credit Vergennes iron with Commodore Thomas Macdonough's victory in the Battle of

Figure 42. Thomas Handasyd Perkins, 1831–32, by Thomas Sully (1783–1872), oil on canvas, 287 × 195.6 cm, Boston Athenaeum.

Plattsburgh in September 1814.[29] Although the Monkton Iron Company closed in 1816, the experience had shown Perkins what could be achieved in the iron industry if one was ready to invest in the right place at the right time.[30]

Patrick Tracy Jackson (1780–1847) apprenticed with a New England merchant and worked in the East Indies trade as a young man. His first brush with bankruptcy in 1811 worried his brother-in-law, Francis Cabot Lowell, but with what became his trademark resiliency, Jackson rebounded from his losses and became one of the largest shareholders in the Boston Manufacturing Company, which built the cotton textile mills at Waltham.[31] He was superintendent of the mills in 1816. "The same driving energy that had earlier led Jackson into business ventures of dubious soundness," Robert Dalzell writes, "made him invaluable as field commander of the company's operations. His skill at supervising construction and managing labor was unsurpassed. He also had a talent for tapping available sources of credit in Boston and so was given the task of raising the short-term loans the company occasionally required, though always with the specific approval of the directors."[32] Jackson had a sharp eye for business opportunities. He and Nathan Appleton had the idea of controlling the entire waterpower of the Merrimac Falls at East Chelmsford, which became the engine for driving the mills at Lowell. Jackson went on to invest in nine textile companies, was an officer in several other Boston Associates ventures, and was a guiding force in the Locks and Canals Company.[33] He managed the Appleton mills at Lowell from 1828 to 1837 and supervised construction of the Boston and Lowell Railroad, which opened to much fanfare in 1835:

> To inaugurate it Jackson, using his largest locomotive, took a party of two dozen directors and stockholders up to Lowell. Making the trip in an hour and seventeen minutes, they dined on salmon brought along for the occasion and then left for Boston again, arriving an hour and twenty-three minutes later. As the train sped by "the country people" gathered "on the bridges and banks," one of the travelers wrote afterward, "the Track labourers swung [their] hats with huzzahs."

By 1835, Jackson was worth almost $400,000.[34] In his history of the Boston Associates, Dalzell notes that Jackson "remained to the end the plunger he had been as a young man. In the mid-1830s—evidently restless with the returns he was receiving—he liquidated almost his entire interest in the mills. The proceeds went into a series of real estate ventures that turned out disastrously, leaving him with less than a quarter of his initial stake."[35] One of those ventures was the Lycoming Coal Company. Neither Jackson nor Perkins fit the mold of the Boston Associates as Dalzell portrays them. Far from being cautious merchants weary of the labor and risk of oceanic trade, both men seem to have loved risk. Both had a mechanical bent, taking a keen interest in the details of industrial processes and machinery. And they both thrived on being the first to attempt something big and new.

The other two founding partners in the Lycoming Coal Company, Edmund Dwight and George W. Lyman, lacked Jackson's energy and Perkins's breadth of experience, but both had accumulated substantial personal fortunes through their investments in textile firms and related ventures. Dwight sank much of his capital in the textile mills in Chicopee, whose location in western Massachusetts inspired his campaign on behalf of the Western Railroad in the 1830s and 1840s. Lyman was the most experienced corporate manager in the group. He was a principal stockholder or officer in eleven textile companies, a director of the Columbian Bank and the Boston and Lowell Railroad, and vice president, then president, of the Massachusetts Hospital Life Insurance Company.[36] Dwight took on the role of managing director and agent of the Lycoming Coal Company in 1833. Lyman later filled those roles.[37]

These men had good reason to be interested in coal lands in Pennsylvania. New England had no workable coal deposits. Railroads ran on coal. Securing a regular supply of bituminous coal on their own terms would clearly benefit the owners of the Western Railroad and the Boston and Lowell Railroad. Coal was also needed for fabricating locomotives, a new branch of manufacturing that the Locks and Canals Company in Lowell was preparing to undertake about the time the four Boston investors were contemplating their purchase of the Lycoming Coal Company.[38] More broadly, coal mines represented a promising investment frontier. The four partners had years of experience as what we now call venture capitalists. Like the Ralstons and many other East Coast businessmen, they saw potential for high return on investment in the coal provisioning trade. If they could move swiftly to gain control of eastern markets, they could profit handsomely as suppliers of coal for industry and home heating. The Lycoming property also had iron ore and clay suitable for making firebrick. It appeared to have all the ingredients for making iron with coal—or rather, with coke, for the partners considered building coke ovens at Farrandsville as early as July 1832.[39]

The geographic location of the Lycoming property made it particularly enticing. It lay along the northeastern fringe of Pennsylvania's great bituminous coalfield, the next bonanza mineral district west of the anthracite coalfields (fig. 43). In the late 1820s and early 1830s, prospectors and geologists had begun to map the bituminous district's extent and assess its quality. The Lycoming lands stood at the gateway to the bituminous district along the West Branch of the Susquehanna River. Channel improvements since 1818 had made navigation on the West Branch somewhat more reliable. The gradual extension of the Susquehanna and Tide Water Canal toward the West Branch of the Susquehanna River gave reason to hope that Lycoming coal could be shipped by canal to industrial works along the Lehigh, Schuylkill, and Union Canals. Furthermore, freight descending the Susquehanna all the way to Port Deposit near the head of Chesapeake Bay could easily reach Baltimore and New York. New railroads extended the potential reach of Lycoming coal to industrial towns throughout southeastern Pennsylvania and promised excellent connections to Philadelphia, the most important market in the region.[40]

The Boston investors' acquisition of the Lycoming lands was also a step toward developing western markets for coal and iron. From 1825 to 1839, the

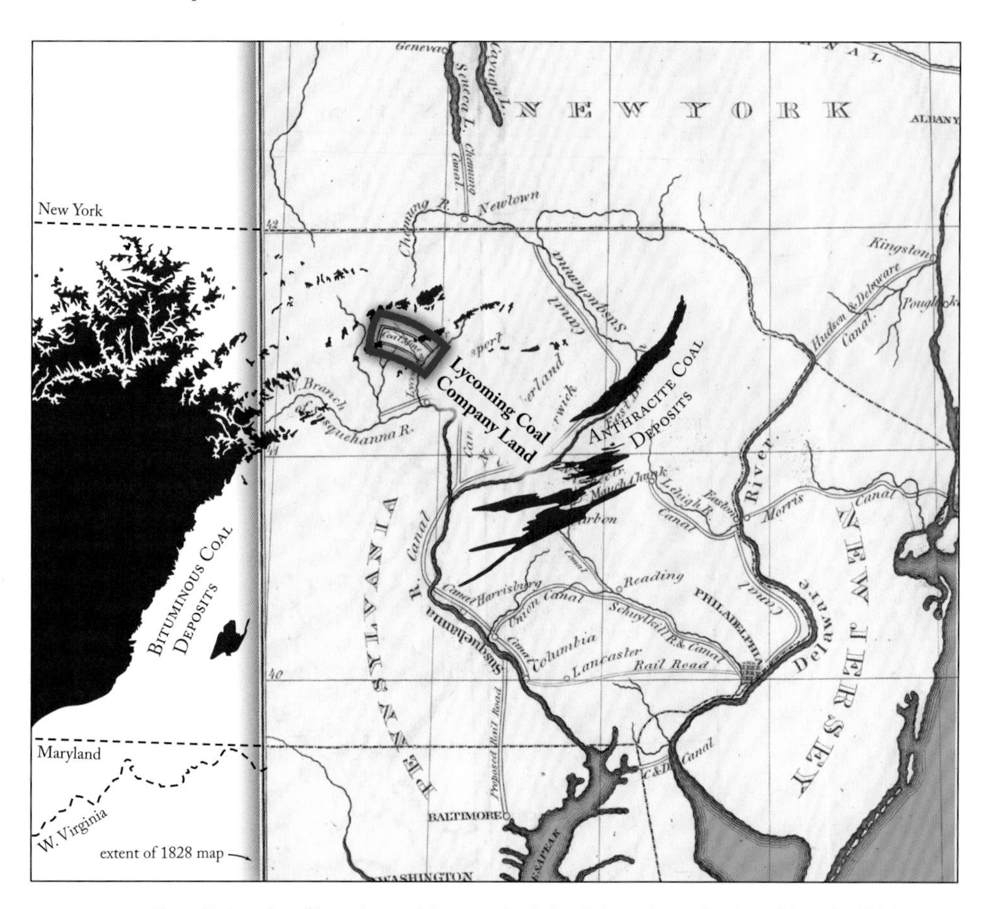

Figure 43. Location of Lycoming Coal Company lands in relation to Pennsylvania coal deposits. This figure combines a map from the 1828 company prospectus (*colored portion on the right*) with present-day geological data on bituminous and anthracite coal deposits. The company's property appeared to be ideally situated on the eastern fringe of the bituminous coal district, within range of the west branch of the Susquehanna River and canals connecting to Philadelphia.

Pennsylvania legislature invested significant sums to survey routes to connect the West Branch of the Susquehanna over the mountains to the Allegheny River. This "northern route" would traverse a shorter distance from Philadelphia to Pittsburgh, and Lake Erie, than the more winding southerly route along the Juniata River. The mountains beyond Farrandsville ultimately defeated canals, chiefly because the headwaters of their rivers lacked sufficient water, but in the 1830s the prospect of a water route to the Great Lakes drew much speculative interest to the upper reaches of the Susquehanna.[41] To entrepreneurs eying mineral resources in Pennsylvania in the early 1830s, the region around Farrandsville seemed ideally situated between the booming industrial East and all that the West might become. It connected two vast regions that possessed limitless potential for generating wealth. Blast furnaces and forges in the Juniata iron district already did good business with rolling mills, foundries, and machine shops in Philadelphia and Pittsburgh. A new ironmaking center in Lycoming County could be expected to challenge the Curtin Ironworks and other well-established ironmakers in central Pennsylvania.[42]

Perkins, Jackson, Dwight, and Lyman meant to make a quick start with their new venture. They hoped to ship 5,000 tons of coal in 1832 and 50,000 tons the following year, which Perkins estimated could net profits of $100,000.[43] Attaining that goal would require careful planning and speedy execution. An associate had told Farrand when the company first formed that "the new Proprietors are Gentlemen habitually and scrupulously exact in all manners of accounts." This was borne out in Perkins's long letter of advice to Edmund Dwight, who took first watch as manager of the company. Perkins wrote that the company should build houses for the miners they hoped to attract—fifteen or twenty to start, up to one hundred men eventually. The settlement, called "Minersville," should be near the mine openings for efficiency's sake, but not too far from the river. "Miners are gregarious, and would not like to be isolated," Perkins noted. He advised that the houses be built in a row, with quarter-acre gardens. Perkins calculated how many riverboats would be needed to float the coal downriver and how many oceangoing vessels would be required to get it to Boston and New York. The crucial calculus was that Lycoming coal must undersell Virginia bituminous coal. Steamboats would be advantageous, Perkins thought, if the company could ascertain water depth and hazards such as rapids along the whole length of the Susquehanna. He recommended selling beef, pork, bacon, flour, molasses, and other foods at a company store, as was done at Merthyr Tydfil, "as it would prevent the men being taken from their work, and furnish the articles at a lower price than they could buy them for with money." He also thought workers should be paid in cash, since miners preferred it and would accept lower wages if paid money rather than scrip. The key was to get the coal to Philadelphia as soon as possible, so that its quality could be tested against other coals. "Sugar bakers, workers in iron generally and twenty other crafts, use [bituminous coal]," Perkins concluded. "If two or three practical men, pronounce it to be superior to Richmond [coal], and equal to British coal, there is no doubt the use would be very great."[44]

The proprietors' plans for reaping a quick profit from their coal almost immediately hit problems. Miners were scarce in central Pennsylvania, as were the manual laborers and carpenters needed to build the one thousand river arks that Dwight estimated would be necessary to transport the company's coal to Port Deposit in the spring.[45] When Farrand finally managed to ship a few tons of coal to Boston in March 1833, Dwight was "very glad, as it will conquer incredulity." "The *whole secret of the Business now,*" he urged Farrand, ". . . is, to get the *greatest quantity to Markett [sic] at the lowest possible freight.*" As costs mounted to $100,000 after a year of paltry returns, Dwight changed strategy. He recommended that Farrand sell coal to towns and blast furnaces along the Susquehanna River, to save the cost of transport to coastal cities and pocket whatever income the fuel might bring from "interior markets."[46] But deliveries continued to lag far behind expectations. The last straw came when Dwight and the other proprietors visited the mines in July 1833. "We see a number of your Arks on the Bank of the River, as we pass along the Road," he wrote from Williamsport. "Unless *attended to* thoroughly, the property will be *wholly lost*." Which indeed it was. The arks, poorly built by Dwight's standards, had been washed ashore, damaged, or completely destroyed either by the force of the spring freshets or because of boatmen's carelessness.[47] A few weeks later, Dwight gave Farrand a sharp warning.

> Hitherto, your experiments have been very disastrous, as regards the *transportation of coal*, & so far from encouraging further trials have nearly led our Proprietors to the determination, to stop all further proceedings—a heavy expenditure has been incurred for mining & transporting coal & not more than 700 Tons, of 7,000 mined, have reached a markett [*sic*]—the rest is either in the River, or on its Banks, or stranded in arks along the shore. Of course unless the Business can be pursued with more certainty of success, & profit, & less hazard & loss than it has hitherto been pursued, it will soon come to an end.[48]

Fed up with the business, Dwight resigned as "active manager" of the mines in the fall of 1833. In one of his last letters Dwight told Farrand, "It would now seem that the difficulties of getting the Coal to Phila[delphia] are nearly insuperable, at a rate that will pay a profit—and that when *there*, Phila. is no markett [*sic*]." Anthracite coal merchants had been circulating bad reports about bituminous coal, and Lycoming coal had persuaded no one to think better of it. Dwight concluded that some of the company's customers thought Farrand's representations regarding Lycoming coal were "nearly *all moonshine*. Your character, therefor [*sic*] is at stake, *to make good*, your statements."[49]

The disappointments and expense of the first eighteen months were repeated again and again over the next six years. Had the proprietors abandoned the coal business in 1833, they would have lost an investment roughly equal to the cost of one good-sized textile mill. Instead, they replaced Farrand with a man whose character and experience seemed much better suited to the job. Daniel Tyler (1799–1882) was a West Point graduate, class of 1819. After serving as adjutant of an artillery school, he was sent to France in 1828 to study that country's much-admired artillery methods, with which he was already familiar as translator of a French military treatise. The superiority of French military practice and arms manufacture made a deep impression on the young lieutenant. When he returned to the United States in 1830, Tyler was hired as an arms inspector for the United States government (fig. 44). While inspecting Springfield Armory, he concluded that workers' inefficiency was inflating costs. His proposal to reorganize the armory under a military superintendent raised vehement protest from workers and the armory's acting superintendent. Tyler's proposal was not adopted, but he was promoted to superintendent of inspection of contract arms. His efforts to raise standards continued to meet stiff opposition from federal contractors. In his memoirs, Tyler claimed to have improved the quality of private arms manufacture in just six months, "but my fidelity to the Government in this instance," he noted, "as well as in that connected with the Springfield Armory proved a bar to my professional advancement." President Andrew Jackson refused to approve Tyler's appointment as captain. He resigned from the army on May 31, 1834.[50]

Figure 44. Lieutenant Daniel Tyler, 1830. From *Daniel Tyler: A Memorial Volume* (1883). This handsome portrait shows Tyler as a confident young military officer. Recently returned from France, where he studied the latest in artillery technology and French military theory, he was eager to apply his knowledge to improve the quality of American ordnance manufacturing.

By that time Tyler had been serving as general manager and superintendent of the Lycoming Coal Company's mines for several months. His six-year contract paid an annual salary of $3,000—a very high rate for the time—plus a house at Farrandsville, with a bonus of up to $2,000 if he contained mining costs. Tyler's military background, knowledge of ordnance manufacturing, and demonstrated interest in industrial efficiency all recommended him as a man who would take charge more effectively than the former bookseller from Philadelphia. He was, furthermore, the son of a well-educated, well-connected Connecticut family. Historian William C. Davis, who chronicled Tyler's later career as a Union general in the Civil War, described him as possessing a commanding presence. "His bearing was markedly distinguished, his carriage erect . . . , and his eyes deep set and piercing. One might expect much of him."[51] The Lycoming proprietors certainly did. Partly based on his assessment of their Pennsylvania holdings, they decided to petition the Pennsylvania state legislature to amend their charter to allow the company to purchase more land and to manufacture iron. Expecting that Pennsylvania would support them as Massachusetts always had, the proprietors did not wait for the legislature's approval. With Tyler as general manager and Charles Russell Lowell, Patrick Tracy Jackson's son-in-law, as the company's new agent, the Lycoming Coal Company set its sights on becoming the first American firm to produce iron with coke.[52]

There was much to be done. In addition to continuing to mine coal for shipment downriver, the company had to build coke ovens and a large blast furnace. The latter required British expertise, not only because coke furnaces were built to different specifications than charcoal furnaces but because they required more specialized equipment, including a large blowing engine and hot-blast apparatus that heated the air before it entered the furnace stack. Construction of the furnace, ovens, additional worker housing, and all the necessary outbuildings involved marshaling materials from across eastern Pennsylvania and hiring skilled workers ranging from carpenters and plasterers to metalworkers, molders, and masons.

While Tyler supervised the mines, housing construction, and the search for general labor from the local area, the owners debated how to secure skilled workers, beginning with a furnace builder. Some argued for hiring a Scotsman. Charles Russell Lowell's father had been a student at Edinburgh University, and his uncle Francis Cabot Lowell's plans for Waltham had been inspired by Scottish industrial communities. Lowell agreed with William Lyman's recommendation that "a Scotchman from [the] Muirkirk . . . works would be preferable to a Welchman [*sic*]." The Muirkirk Iron Company, founded south of Glasgow in 1787, was one of Scotland's largest pig iron producers in the 1830s. Lowell thought that the "rational character" of the Scots "is certainly better & more resembles that of our own people with whom [the builder] would be most in contact."[53] The firm decided, however, to focus its recruiting on South Wales. They may have been swayed by Gerard Ralston, who had moved to London to set up as a commission merchant with his brother Ashbel. They became major importers of British iron to Philadelphia, including sales of Welsh rails to American railroad companies. Gerard's personal connections with ironmasters led to his marriage in 1838 to Isabel Crawshay, daughter of Cyfarthfa Iron Works owner William Crawshay.[54] Ralston told Tyler that he hoped to persuade Thomas Evans, superintendent of the Dowlais Iron Company, to emigrate to Pennsylvania. When letters from his London office got no response, Ralston dispatched a surrogate to seek out Evans.[55] Antes Snyder, a civil engineer, told Ralston in April 1835 that his visit to Dowlais and Cardiff had "*entirely failed* in finding the man we want! In fact the requirements are of such a nature that few persons possessing them are to be found and they seem too highly appreciated to be obtained at a moderate cost."[56]

Not until late that summer did Ralston secure the services of an experienced Welshman. Edward Thomas (b. 1790) worked in some skilled capacity, most likely as an engineer, at the Plymouth Iron Works, one of the smaller iron companies in Merthyr. In addition to his salary and passage to America, Thomas asked for a house and home fuel, standard compensation for agents and managers at

British ironworks but terms that were not part of his original contract with the Lycoming Coal Company.[57] He had no way of knowing that the absence of those provisions was the first sign that his new situation might fall short of his own high hopes. Thomas's private ambition was to obtain the American patent for making iron with anthracite. He thought the patent would reap a fortune in royalties. His letters home to relatives in Wales do not explain how he aimed to perfect anthracite ironmaking while working for a firm bent on making iron with coke.[58]

When Thomas arrived at Farrandsville in November 1835, he encountered "a busy place" whose promise was already attracting interest from workers around the region.[59] William Lyman had built the Franklin Nail Works, a midsize rolling mill on Lick Run, where he intended to manufacture nails using Farrandsville coke pig iron as his crude iron supply (fig. 45). Thomas noted that the puddlers and rollers at the mill came from Wales and Staffordshire.[60] The industrial hamlet had a boardinghouse, a hotel, and six "double houses" (duplexes in today's parlance), one of which may have served as temporary accommodation for Thomas while he awaited construction of the house he thought had been promised him.[61] Several other Welsh and English immigrants reached Farrandsville in November as well, including the families of several miners from Newcastle.[62] Housing them all became a problem, especially with winter drawing in.

Figure 45. Farrandsville, ca. 1839. Based on an original sketch map by Michael P. Conzen. The contour lines suggest the narrow valley and steep hillsides running down to the Susquehanna River. The industrial hamlet included company housing for miners in "Minersville" as well as housing for furnace and rolling mill workers. The hotel on the riverbank was built in anticipation of visitors. The canal lock was begun by 1839 but was not yet complete.

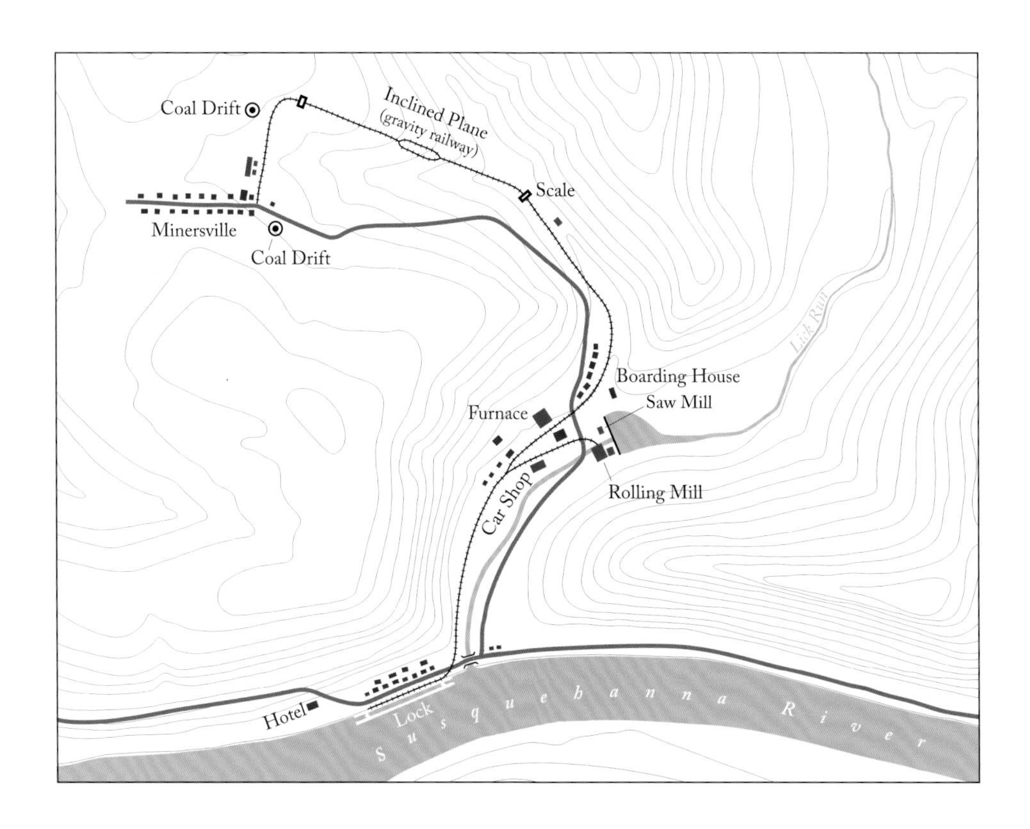

Thomas was warmly welcomed by Thomas Perkins and Daniel Tyler. His initial impression of Tyler, as he told his sister Bess, was mixed. "Mr. Tyler is, I believe, a man of good family, has been in the army and for some time engaged by the government—an inspector of arms to be delivered in their arsenal or armory—he knows nothing whatever of the business he has to command here—he is like other men accustomed to the army—authoritative and hasty, but to me he has behaved very well hitherto." Perkins "expressed his hope that I would have no desire to return" to Wales. He assured Thomas "that salary was no object with them." "It appears to be more a matter of pride than profit which induced them to proceed with the Furnace here," Thomas wrote. "It is the commencement of a new Era, the making of Iron from pit coal in America. They are ambitious of the fame of first bringing it to bear for the benefit of the States."[63] Thomas was dubious about the company's resources, however. "Mr. Treadwell's representation to me that their mineral deposits here are similar to those in Wales, having bituminous coal from stone mines and limestone vein, is a complete Yankee fudge—it so happens they only have some ballmine [clumps of low-grade iron ore], a small quantity laying beneath a bed of superior fine clay . . . and ores must be procured from a distance." The company's coal lands had veins of good bituminous coal, but it contained "more earthy matter than [any] coal I have ever seen." Furthermore, the coalfield being mined when he arrived was "very limited"—just 150 acres, a fraction of the area advertised in the 1828 Lycoming prospectus.[64]

William Farrand confirmed Thomas's assessment of local iron ore. He had hunted all summer for good ore on the Lycoming property and, surreptitiously, on adjoining land. "I fear your lower iron ore is a sandy lean iron stone," he wrote Tyler in November 1835. "Your specimens on Larys creek," part of the property lying north and east of the furnace site, "are a lean ore, so lean as generally to be poor."[65] A year later, the company bought land with higher-grade ore deposits on Hemlock Creek in Columbia County, near the North Branch of the Susquehanna about one hundred miles by water from Farrandsville. The Hemlock mine provided most of the ore smelted at Farrandsville for the rest of the decade.[66]

As ore prospecting continued, construction of the blast furnace began. The one document that clearly attests to Edward Thomas's skill is a letter he wrote to the Lycoming Coal Company in November 1835, attempting to explain the relation between the proportions of a blast furnace and the diameter of cylinders in blowing engines of various sizes. The letter bristles with technical language and mathematical calculations. Thomas may have been trying to impress his employers with his expertise. The more important subtext was Thomas's concern that the large furnace they intended to build would not work properly unless it had a proportionately large and powerful blowing engine. (Perkins had settled on a width of sixteen feet at the boshes [the widest internal dimension of a blast furnace], larger than any American charcoal furnace at the time but somewhat smaller than the coke furnaces at Dowlais and Cyfarthfa.) Thomas noted that ore quality, the percentage of iron in the ore, also affected furnace production, though he did not spell out the likelihood of poor yield from the company's low-grade ore. Diplomatically, he merely concluded that "to make a large quantity of iron, the necessary means must be at command for making it—a furnace of sufficient size—minerals of good quality—and blast in sufficient quantity, [for] without such means it cannot be done in any place."[67]

In his letters home, Thomas more frankly painted Farrandsville and other central Pennsylvania ironworks as technologically backward operations in a wilderness of "panthers, bears, wolves, . . . [w]ild turkeys" and "uncivilized" people. Expecting the worst, he had bought several dozen bottles of sherry and "the best French brandy" from a fellow Welshman in Philadelphia before ascending the Susquehanna. But Thomas was also excited to be working for a company that regarded Farrandsville as the testing ground for a grand plan. He wrote Bess that "if it succeeds, they [will] enter upon it elsewhere upon a large scale in fact upon a larger scale than any one work now existing in Wales."[68] The furnace he had been hired to build was to be the centerpiece of an ambitious experiment that might ultimately become the Lowell of iron manufacturing.

Construction went much more slowly than expected. The furnace was not finished in time to ship pig iron downriver on the spring freshet in 1836 as the company had hoped, or by the summer. Meantime, serious problems arose between Thomas and Tyler. By the end of June, they were barely speaking. Thomas had been fired, or knew he soon would be, when he wrote to Tyler, "I feel the most ardent wish and interest to render the Lycoming Coal Company the best services [I] am capable of during the time I remain here and will endeavor to prevent any thing from being done wrong as to the preparations for erecting the blast furnace, unless you order otherwise, which hitherto has been too much the case."[69] Tyler left the hearthstone for the furnace untended by the riverbank, risking damage that would make the massive piece useless. More seriously, Tyler had not been giving all his attention to company business as his contract required. He hired a relative to manage the coal mines so that he could spend time at a nearby foundry that was casting parts for the rolling mill and furnace. Thomas wrote to his sister that Tyler was pocketing profits from the foundry while attempting to hide his involvement. "He is now represented to be a very bad character," Thomas sniped. Tyler's misbehavior had caused "so great a coolness" between him and William Lyman that Lyman would "have nothing to do with him."

Not only was Tyler violating his employment contract, he was putting the whole venture at risk by contracting for second-rate parts for the furnace's most vital piece of equipment. Among US steam engine manufacturers in the 1830s, only the West Point Foundry in Cold Spring, New York, made engines of sufficient size and precision to blow a coal-fired furnace. Tyler should have realized—and Thomas doubtless knew—that a good blowing engine could not be assembled piecemeal from parts manufactured by an inexperienced founder in central Pennsylvania. The order to the local foundry was canceled. Meanwhile the nearly finished furnace sat idle.[70]

The proprietors in Boston, however, consistently sided with Tyler. In August 1836 they canceled Thomas's contract. When Lyman warned the board of Tyler's malfeasance, they voted that Lyman's complaint "does not in the slightest degree impair the confidence which the Co. has always reposed in the 'integrity, veracity & sense of justice' of Mr. Tyler & that his conduct of their affairs as agent at their mines fully sustains the character for talent & integrity which he has previously acquired." They even sent Tyler a copy of Lyman's correspondence "to place him on his guard against the efforts now making to injure his reputation."[71] Tyler's strongest advocate may have been Charles Russell Lowell, whose warm, personal letters to Tyler suggest that the two men had become fast friends. They shared a love of travel, a yen for European sophistication, and a lack of concern for technical details. The senior partners chose to entrust Farrandsville to the next generation—a decision they would soon regret.

In October 1836, the board sent Tyler to Britain to find Thomas's replacement. This time Tyler went straight to Glasgow, where he hired James Ralston (apparently no relation to the Ralstons of Philadelphia) as furnace keeper and manager, and another Scot, Robert Graham, as his assistant. Tyler also ordered a blowing engine to be made by Camlachie Foundry, a leading marine engine manufacturer on the river Clyde.[72] As he prepared to sail back to Philadelphia, Tyler crowed to the acting manager at Farrandsville that his trip had been a great success—in fact, a revelation.

> I have made up my mind clearly that we shall be able to make iron with tolerable good advantage. I have also, almost concluded, that we shall be able to use our coal raw with hot blast. . . .
>
> The more I see of this furnace business the more I am convinced there is no mystery in it. I feel confident we shall make iron, & from the start. I have secured the services of a furnaceman, who not only understands the business but is used to work. He comes out to work not to overtake.[73]

Edward Thomas's high opinion of himself may well have seemed overbearing, particularly if the Welshman's sense of technical superiority exceeded his actual ability to build a coke furnace or to adapt Welsh smelting technology to conditions at Farrandsville. But Tyler's handpicked Scottish successor did no better. By February 1839, Tyler and Ralston had submitted to arbitration to settle their differences. The arbitration agreement stated that Ralston had "entirely failed in benefitting . . . Tyler as furnace manager." Ralston lasted several months longer at Farrandsville than Edward Thomas had, with no more iron to show for it.[74] At some point during his tenure the furnace was finally completed, but Ralston was unable to put it in blast with Lycoming coal. That was not accomplished until December 1838 under the supervision of another Welsh immigrant named Benjamin Perry, an illiterate but highly skilled artisan who had formerly been a manager at the Pentwyn Iron Works in Abersychan, Monmouthshire. Perry went on to make a specialty of blowing in American coal-fired blast furnaces. He joined William Lyman at Pottsville Furnace in 1840 and is credited with blowing in that furnace as one of the first in the United States to achieve sustained iron production with anthracite. Later that year Perry blew in two other anthracite furnaces near Danville.[75] Meanwhile, Farrandsville Furnace remained in blast but produced only a few hundred tons of iron in 1839 and early 1840, none of which found a ready market. One forge master described its product as completely unsuitable for bar iron or shapes. Wrought iron made from Farrandsville pig iron was so brittle when "cherry red," he complained, that it "would brake [*sic*] under the hammer"—a classic description of "red short" iron that was suitable for cut nails or cast iron but not for foundry iron or rolled shapes, such as railroad rails. Far from making iron with raw coal as Tyler had predicted, Farrandsville Furnace seems to have managed to produce marketable iron only when charcoal was added to the company's coal.[76]

Tyler's mismanagement extended to the lower ranks of labor as well. Shortly after he returned from Scotland, he dispatched some of the English and Welsh coal miners the company had recruited at considerable expense to mine iron ore along Hemlock Creek. They found the site bereft of housing, just as Edward Thomas had found no accommodation ready for him at Farrandsville. As the spring of 1837 wore on, financial conditions worsened and Tyler began withholding the miners' pay. By May 10, when the bank panic began in New York City, the Hemlock miners were fed up with physical discomfort, late pay, and the inadequate housing provided for their wives and children back in Farrandsville. They may also have been discontented working shallow ore mines that required little of their coal-mining expertise. That summer, most of the ore miners "stood out" (went on strike), and several of the unmarried Welshmen ran away.[77]

The deserters, as former lieutenant Tyler called them, had gone to Pottsville, a rapidly growing coal town in the anthracite region of southeastern Pennsylvania and a center of Welsh industrial immigration in the 1830s.[78] Tyler's response was to prosecute the miners for violating their contracts, which bound them to work for one year and to pay back the cost of their passage through deductions from their weekly pay.[79] He enlisted the local sheriff to hunt down the men in Pottsville, despite his mining supervisor's advice that he drop the case and recruit new workers from Pottsville's growing pool of unemployed miners. Tyler wrote to his legal counsel, "I am desirous of making an example of these men, to prevent other desertions, as well as to do justice to our Company, & I have to request that you will immediately arrest them in a suit of damages say $500 for violation of contract, & unless they can get bail, they will have to be committed [to jail], an affair, which I fancy they will not relish."[80] The men were caught and jailed in Columbia County. For weeks they remained there, content to let the company pay for their bed and board. "As for paying the debts," one of Tyler's representatives informed him, "they say it is utterly impossible, for they have not $5 among the whole of them." Besides, the men claimed "that they had permission to leave the works, that is to annul their contracts for one years [*sic*] services,—and that there was no work for them . . . to do."[81] To Tyler, as to British employers of contract labor in this period, lack of work did not give workers the right to leave and seek paid employment elsewhere.[82]

Tyler's charges against the absconders never went to trial. The jailed miners appealed to the insolvency laws, which exempted them from bail. They eventually apologized to Tyler and begged him to take them back. With the economic depression biting deep in Pennsylvania mining towns, they had few alternatives. "They are now satisfied," the company's lawyer wrote, that at Hemlock Creek "they had better wages, better living, & more comfortable houses, than the miners at Pottsville or any other place in the State."[83] Tyler had won the battle, but at the cost of nearly seven months of disruption and delay in developing the Hemlock mines.

Technically, the Farrandsville miners were imprisoned for debt. Had they been able to pay their bail, they would have been free to go. Tyler could not have had them tried and imprisoned for violating the terms of their employment contracts, for US courts had decisively ruled in 1821 that criminal sanctions could not be applied to workers who breached their labor contracts.[84] Tyler's angry letters, however, show that he was determined to punish the men as severely as the law allowed, and that he meant their confinement to apply pressure on other workers at Farrandsville. By training or temperament, Tyler was disposed to apply harsh discipline to disobedient underlings. Several years as agent and sometime manager of coal and ore mines had taught him little about industrial labor relations. Tyler remained as peremptory and convinced of his superiority as he had been as a young arms inspector for the US military.

I will return to the issue of labor mobility later in this chapter, since it was a chronic problem at frontier sites of iron production. In the case of Farrandsville Furnace, poor labor relations were part of a larger set of problems and miscalculations that reflected the proprietors' ignorance of the region and the business of making iron with coal. The company's dirty coal was unsuited to coke smelting. The proprietors had counted on canal transportation to get their coal and iron to market, but construction of the canal was repeatedly delayed. The Susquehanna and Tide Water Canal was not completed until 1840, and the improvements planned between Lock Haven and Farrandsville were never entirely finished.[85] Much coal was lost in spring freshets that dashed the company's arks to bits. Perkins's notion of using steamboats was a pipedream, given the rapids and falls along the upper Susquehanna. The venture was also a social failure. Farrandsville and Hemlock Creek were rough frontier settlements with none of the attractions of urban life that Welsh and English miners and ironworkers enjoyed in industrial towns like Sheffield and Merthyr Tydfil. Although some of the immigrants disobeyed company rules against drink (corn liquor was readily available from local stills),[86] drinking can't have been as pleasurable in the woodlands of Pennsylvania as it was in workingmen's taverns back home. Community life was dreary for Farrandsville's women and children as well, with no school for workers' children, no church or chapel, and no shops but the company store.

The proprietors in Boston weathered the first blows of the Panic of 1837 and continued to invest in Farrandsville until early the next year, when Patrick T. Jackson visited the works to see for himself why no iron was being produced. He returned to Boston guardedly optimistic that the furnace might yet succeed, despite evidence that the company's coal "appears to soften or melt so as to choke up" in the furnace "& thereby prevent the blast from producing its proper effect." He told Tyler to begin making charcoal in case coal proved unworkable, and he ordered the dismissal of all but a skeleton crew to run experiments with a cupola furnace to determine whether marketable iron could in fact be made at Farrandsville. Jackson also proposed himself as agent, seeking board approval to take charge of operations.[87] The board now showed signs of internal division. They appointed Charles Russell Lowell to review his father-in-law's proposal, which he rejected. Rudderless, the company drifted through the summer and fall while Benjamin Perry slowly figured out how to make the furnace produce. Meanwhile William Lyman and his partner Robert Bennet Forbes shut down the Franklin Nail Works, in which they had invested roughly $50,000. The mill never did make nails from Farrandsville iron.[88]

Not until June 1839 did the board finally confront Daniel Tyler. He was replaced as agent by George W. Lyman, while Lowell was demoted from treasurer to company clerk. Some months later, Jackson's ire with his son-in-law exploded into public view. On Valentine's Day 1840, Boston newspapers published the news that Charles Russell Lowell had been declared bankrupt. In her biography of Lowell's son, Charles Russell Lowell Jr., Carol Bundy describes the family's humiliating disgrace. "The Lowells' house was seized and sold, as was their furniture and silver. Throughout the spring Charles appeared at Patrick Tracy Jackson's front door, looking wretched and wanting to see his wife and children." Jackson, she continues, "would have nothing to do with his son-in-law." Bundy concludes that Jackson held Lowell responsible for destroying many family fortunes through his mismanagement of the Farrandsville operation. Besides wasting his own wealth, Lowell squandered funds entrusted to him by his father, abolitionist and Congregational minister Charles Lowell, as well as money Jackson had provided to shore up the younger Lowell's failing finances.[89] Lycoming Coal Company records, however, show that Jackson also dug himself into a hole by urging his partners to continue trying to make iron at Farrandsville for six years without any return on their investment. Lowell was a callow young man who played at running an industrial concern; he never learned what the business required. His much more experienced father-in-law should have recognized the warning signs long before the company's hopes guttered and died in 1839–40.

Both men trusted Daniel Tyler too much. Seeing Tyler as one of their own kind blinded them to his personal flaws and his limited knowledge of the iron industry. Tyler's overconfidence blinded him to his own ignorance of nearly every aspect of ironmaking. Rather than seeking advice from more experienced managers or trusting the knowledge of the skilled workers he supervised, Tyler bluffed and blundered his way through other men's fortunes. Tyler is certainly not alone in the annals of American industry for emerging from a fiasco with his reputation more or less intact. He went on to build a long career in industry and finance as president and superintending engineer of a series of railroad companies, including the Dauphin and Susquehanna Railroad and Coal Company and the Schuylkill and Susquehanna Railroad Company. During the Civil War, Tyler returned to military service as colonel of the First Connecticut Infantry and then accepted an appointment as brigadier general of Connecticut volunteers (fig. 46). In the First Battle of Bull Run, outside Manassas, Virginia, Tyler made a series of serious errors of judgment. He ignored orders, marched his men away from the battle, and hung back while his troops were routed by the Confederates. Tyler remained in uniform until 1864, "achieving nothing and spouting much venom" against the commanding officer whose orders he had disobeyed. He was never again allowed a significant field command. Civil War historian Richard Sewell's tart epithet—"the incompetent Daniel Tyler"—applies to the man's early industrial career as well.[90]

From 1838 to 1840, all Lycoming Coal Company lands and movable property were sold at auction. An auction inventory for items from the Farrandsville hotel suggests the elegance to which Tyler and Lowell aspired. In addition to kitchen equipment and bedroom furnishings, the auction included cut-glass decanters, ivory-handled table knives and forks, champagne glasses, two large mahogany tables, cherry dining tables, and damask tablecloths. The auctions all together netted $31,000, which left the company's total loss at approximately $400,000.[91] Thomas H. Perkins's personal losses in the venture left him "deeply embarrassed." This phrase, a common euphemism for debt and commercial failure in the antebellum period, conveys the wounded pride Perkins must have felt from the public failure of an audacious industrial venture. The depth of his embarrassment may explain why there is not a whisper about Farrandsville or the Lycoming Coal Company in his published biography. Patrick Tracy Jackson never recovered his former wealth. Charles Russell Lowell's personal disaster left him "a broken man." He suffered poor health and failed in another business venture in Boston. He finally found stability as assistant librarian at the Boston Athenaeum, "where he helped prepare the first card catalog of its collection: a fastidious, time-consuming occupation for which he was well suited and where he could do no damage."[92]

The failed attempt to make coke iron at Farrandsville had a chilling effect on the development of the iron industry in central Pennsylvania. No other coke furnace was built in Lycoming or Clinton County before the Civil War, and no coal-burning furnace of any kind was built in the area until 1855.[93] One can sense the shadow cast by Farrandsville in the uncharacteristically cautious language of an 1843 promotional brochure touting nearby Ralston Furnace to prospective investors. Readers at the time may have recognized an allusion to the Lycoming Coal Company's failure in this passage:

Figure 46. Brigadier General Daniel Tyler, ca. 1861–65. Courtesy of the Library of Congress. Tyler had endured many difficulties by the time this formal portrait was taken. He survived the debacles at Farrandsville and the First Battle of Bull Run, but both experiences left their scars.

> It may be well to observe, that the most exaggerated ideas prevail as to the mineral wealth of wild and especially untried districts;—for bounteous as nature has been in this respect to Pennsylvania, it is not easy to find bituminous coal, rich ore and water-power in juxtaposition in a locality easy of approach. A rigorous examination generally shows some radical deficiency, and where this is not made previous to commencing works the most unfortunate results have occurred. Thus, more than half a million of dollars was expended in a situation where they have since discovered, that neither coal nor ore can be found in quantities sufficient for practical operations, and not only is this large sum lost to the stockholders, but its waste is actually the means of deterring others from embarking in the business even under favorable auspices.

Ralston's proprietors—who included Archibald McIntyre of the struggling, geographically isolated Adirondac Furnace in Tahawus, New York—tried to assure investors that they had taken all necessary precautions to ensure the success of Ralston Furnace. It never produced much, however, and a later attempt to smelt iron with charcoal and anthracite failed to produce marketable pig iron. When Lesley visited Ralston in June 1857, he described it as "a hole & no place."[94] Astonville Furnace, built before Ralston and Farrandsville Furnaces by Matthew C. Ralston and associates, produced little iron until Ralston's death in 1840. The *History of Lycoming County* describes the failure there as approaching that of Farrandsville. "Investments running into hundreds of thousands of dollars were made" in Astonville Furnace, the chronicle says, "but from all indications the experiment was not attended by a high degree of success." Matthew Ralston's fortune "dwindled to nothing, and after a few years of futile struggle, he gave up the fight and died a poor man."[95]

Hopes for extracting the mineral wealth from Farrandsville persisted despite the stories of loss. In December 1853 the newly incorporated Farrandsville Company began purchasing coal, ore, and timber lands, first the former ironworks site and mines of the Lycoming Coal Company and then large tracts along Tangascootack Creek on the south side of the Susquehanna. By June 1857 the Farrandsville Company had bought more than six thousand acres.[96] In his report to the stockholders in 1856, company president Christopher Fallon described the property in much the same terms the Ralstons had used in their prospectus of 1828. The coal land was "nearer the New York market by more than 80 miles, than any other from which a good bituminous coal can be obtained in large quantities, and besides being nearer is believed to be the only large bituminous coal field, the produce of which can be taken to the eastern markets without crossing the high grade on the Alleghany [*sic*] mountains." Fallon acknowledged that the coal beds above Quin's Run in Farrandsville lay in thin veins that were "interlaid with so much slate as to make [them] comparatively unprofitable to work for export," but he confidently reported that this coal could be converted to coke for use at ironworks. The blast furnace they had acquired was of such "superior" construction that it could be put into working order "at little expense." The company's best prospects lay on the south side of the river, where its lands contained "large, indeed, for all practical purposes, inexhaustible quantities of bituminous coal of the very best quality" as well as "valuable iron ore" that could be "mixed with the superior haematite ores from Nittany Valley" to produce "a very superior iron . . . at a small cost, using coke made from the dust or fine coal necessarily made at all mines." Once the old canal dock was improved and a bridge built across the Susquehanna to transport the Tangascootack coal to the canal, Fallon expected "to do a large business" shipping coal to iron companies down the Susquehanna as soon as canal navigation opened in the spring.[97]

Christopher Fallon and his brother John Fallon were Spanish immigrants born in Cádiz, Spain, sons of an Irish father and a Spanish mother. Christopher (1809–63), also known as Cristóbal, became a successful lawyer in Philadelphia. He served on the boards of several Pennsylvania railroads, including the Sunbury and Erie Railroad, which began construction along the West Branch of the Susquehanna in 1853 and was nearing Farrandsville in 1856. That year, he narrowly defeated Daniel Tyler to be reelected president of the West Chester and Philadelphia Railroad, a branch of the Pennsylvania Railroad. John (Juan) Fallon (1819–85) worked briefly as a railroad engineer after following Christopher to Philadelphia in 1836. He became a lawyer as well.[98]

According to a history of the Philadelphia Hibernian Society, the two brothers served for some years as American agents for the dowager queen Cristina of Spain, investing her funds in Philadelphia real estate. In Farrandsville, some accounts credited the Fallons as agents for Cristina, others as the agents for her exiled daughter, Queen Isabel II.[99] Correspondence in the Fallon brothers' papers reveals that they were in fact acting on behalf of the Duke of Riánsares, consort of Queen Cristina, and a number of other Spanish aristocrats. In 1859 the duke held stock in the Farrandsville Company valued at more than $164,000. By that time, however, the US iron industry and much of the coal industry had ground to a halt because of the economic depression precipitated by bank failures in 1857. The Fallon brothers had a serious falling-out in 1858 and dissolved their partnership in Farrandsville. John assumed much of the company debt and struggled to satisfy his foreign creditors.[100] He appears to have held on to the property, perhaps producing a little bar iron, until late in the Civil War. He finally sold the works in 1864 or 1865.[101]

When I visited Farrandsville in 2004 and heard the tale of a Spanish queen's investment in the iron industry, it seemed too far-fetched to believe. As is so often the case, however, the story has elements of truth that have left traces in the landscape. A stream called Queen's Run ("Quin's Run" in Christopher Fallon's

1856 report to stockholders) flows past former coal mines near Farrandsville. The ruined foundation of a very large rectangular structure near the former canal basin on the Susquehanna River matches the alignment of a porticoed building called the Queen's Mansion that was photographed sometime in the late nineteenth century.[102] Local lore has it that the mansion was built in anticipation of a royal visit that never came to pass. The Fallon House, a large italianate hotel that opened in 1856, still bears a placard saying it was "built by Maria Christina queen of Spain."[103]

Like the Boston capitalists who preceded them, Christopher and John Fallon and their foreign investors hoped to tap the geological treasures of Pennsylvania's rugged interior. Also like some of those involved with the Lycoming Coal Company, the Fallons knew next to nothing about the iron industry. They were seduced by the site's powerful location advantages, which they believed would translate into a competitive edge over coal and iron companies that were farther from eastern markets, but they badly misjudged the quality of the resources they had acquired. Like their predecessors, the Farrandsville Company lost virtually all their investment, between $400,000 and $600,000.

The second attempt to make coke iron at Farrandsville was the last. After the Fallons left Lycoming County, Farrandsville attracted no other coal or iron ventures in the nineteenth century. Smaller industries kept the settlement alive after the Civil War, including a firebrick factory, a sawmill, and a cigar factory.[104] The stone furnace still stands, its pharaonic stack a landmark of the speculative fever that envisioned the upper reaches of the Susquehanna as a mecca for American industry.

Lonaconing

As the ironworks at Farrandsville were being built, similar visions of industrial greatness drew another cast of characters to the western panhandle of Maryland. John Henry Alexander (1812–67), son of a merchant in Annapolis, learned of the extensive bituminous coal deposits in Allegany County, west of Cumberland, in the mid-1830s while working as the state's chief engineer and assisting with the first geological survey of Maryland. In 1836, Alexander formed a partnership with Philip T. Tyson, who had training in geology and chemistry. The two men purchased eleven thousand acres of woods and mineral land along George's Creek, about six miles south of Frostburg, then set about conducting a detailed survey of their holdings in order to attract investors.[105]

The 1836 prospectus for the George's Creek Coal and Iron Company laid out an even more ambitious plan than that afoot at Farrandsville. Alexander aimed to build four coke-fired blast furnaces powered by steam engines, with hot-blast apparatus, and a rolling mill complete with puddling furnaces, steam hammer, and foundry. The company would produce merchant bar, railroad iron, and heavy castings. It would also mine coal to sell and to fuel the ironworks. Alexander predicted that the company town, called Lonaconing, would eventually house up to one thousand people. He and Tyson hoped to be first to supply rails to the Baltimore and Ohio Railroad as its construction pushed west from Harpers Ferry. Whether the railroad was built through Frostburg or followed the Potomac River southwest of Cumberland through Westernport, it would pass within a few miles of Lonaconing. The partners also expected to capitalize on the planned extension of the Chesapeake and Ohio Canal, which was within fifty miles of Cumberland in 1836. Both transportation links would connect George's Creek to Baltimore's growing manufacturing firms and the city's port, from which ships could carry coal and iron to Philadelphia, New York, and more distant markets. The total investment would be $167,930.[106] Alexander predicted that "the time will come when . . . the Western County of Maryland shall be looked upon as the Wales of North America."[107]

In many ways the early years of the George's Creek venture mirrored those of the Lycoming Coal and Iron Company. Both were built in anticipation of transportation improvements that the companies' survival would depend on. They were at either end of the physiographic divide called the Allegheny Front (fig. 47), which in the 1830s marked the fairly sharp transition from the long-settled

Figure 47. Location of Lonaconing and Farrandsville. The map shows the extent of the Susquehanna Canal (to Farrandsville) and the Chesapeake and Ohio Canal (well short of Lonaconing) about 1838. The Allegheny Front is the curving mountain range visible in the center-left part of the map.

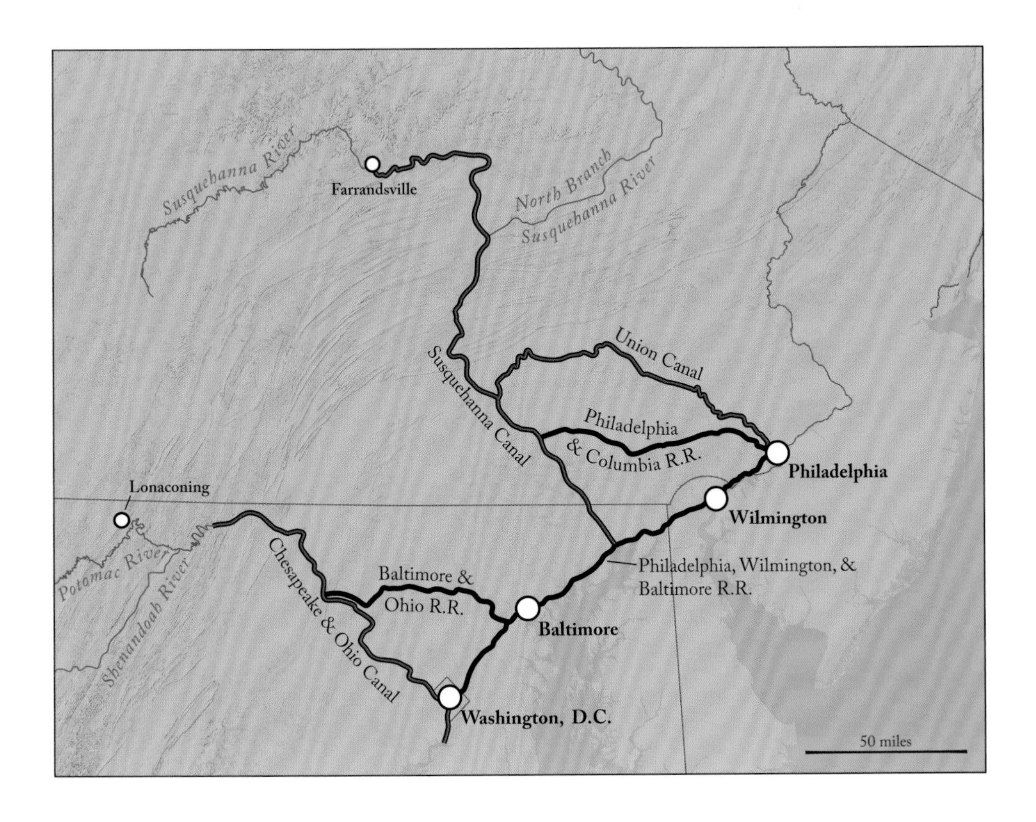

ridge and valley lands of the Mid-Atlantic to the sparsely populated uplands of the western Allegheny Mountains. Outcrops of bituminous coal had been found all along the Allegheny Front as early as the 1750s.[108] Coal mining first lured industrial prospectors to the region; ironmaking was an afterthought. Iron appealed chiefly because of its potential to return high profits if companies could produce finished iron whose added value would justify the cost of shipping over long distances. As at Farrandsville, the owners of Lonaconing aimed to replicate the Welsh manner of ironmaking, down to copying the style of Cyfarthfa's blast furnaces (fig. 48). Both companies recruited skilled European ironworkers and coal miners in hopes of speeding construction and production. They sought the glory of being the first to make iron with mineral fuel and to reap the commercial advantage of establishing dominance over the coal trade in their primary markets, Baltimore and Philadelphia.

Events unfolded rather differently at Lonaconing than at Farrandsville. The George's Creek Coal and Iron Company struggled but did not fail completely. By all accounts, Lonaconing Furnace produced coke iron that equaled British coke iron. The company's coal found ready markets from the beginning, and its mines remained in production until 1952.[109] Labor relations were fraught in both places, but at Lonaconing the mix of personalities, technical acumen, and skill gave labor-management relations rather different dynamics. The story of Lonaconing is captured in an unusually comprehensive and personal journal kept by company superintendents from 1837 to 1840. As a record of managers' views, the Lonaconing journal reinforces the impression that Americans trying to establish modern iron manufacturing in frontier conditions were feeling their way by trial and error in every aspect of the business.

At first Alexander and Tyson made swift progress. The Maryland state legislature granted them a corporate charter in March 1836 that allowed the George's Creek Company to mine coal and ore, " 'erect and carry on mills and manufactories of iron,' " purchase additional land, and build a railroad.[110] Three months after publishing their prospectus in October 1836, the partners had at least nine and perhaps as many as twenty-two investors, including prominent Baltimore businessmen Patrick Macaulay and Richard Wilson, president and secretary, respectively, of the American Life Insurance and Trust Company, and several directors of the Bank of the United States (later known as the National Bank). The stockholders also included Louis McLane, president of the Baltimore and Ohio Railroad Company and former secretary of the treasury under Andrew Jackson.[111] The new company had ample capital and strong political support.

By September 1837, Alexander had laid out the main street of Lonaconing.[112] Welsh immigrant John Steele was busy supervising the first crew of coal miners and burning test batches of coke. Steele's father was the longtime agent of the Plymouth Iron Company, where Edward Thomas worked before coming to

Figure 48. Welsh-style coke furnaces at Farrandsville (*top*) and Lonaconing (*bottom*). The two furnaces are very similar in size. Farrandsville photo by the author, 2004. Lonaconing photo by Philip C. See, ca. 1995.

Farrandsville. A manager at the neighboring Penydarren Iron Works in Merthyr Tydfil had recommended the younger Steele to Daniel Tyler, who passed the letter on to Alexander.[113] Welsh founder David Hopkins arrived in mid-October and immediately took charge of building the first blast furnace. Two more Welshmen, apparently friends of Hopkins's, arrived in November just before the first snowstorm of the season. John Thomas and John Phillips were to be keepers of the furnace, responsible for overseeing the filling and tapping of the stack in twelve-hour shifts. Work slowed over the winter but quickened again in spring. By the first of May, thirty-six men were employed in the mines, while a crew of Americans assembled the boilers that would power the furnace's steam engine and blowing engine, both of which had been manufactured at West Point Foundry. By the end of June, Hopkins's family reached Lonaconing. His brother John soon became supervisor of the mines, a position David Hopkins had briefly held after John Steele quit. David's son William Hopkins assisted his father in various positions. While the Hopkinses were settling in at George's Creek, the bark *Tiberias* sailed from Newport, South Wales, with seventy-three passengers bound for Lonaconing, including twenty-nine coal miners, two founders, one farmer, and forty-one women and children.[114] Ten weeks later, these Welsh families landed at Baltimore and began the final leg of their journey to Allegany County.

While the Lonaconing journal is not entirely clear on the ethnic division of labor, it appears that the company initially employed Germans in its ore mines and limestone quarries, Welshmen in the coal mines and in skilled positions at the furnace and blacksmith shop, and Americans as carpenters, masons, haulers, and general laborers. Advertisements in Baltimore newspapers recruited the latter workforce. The owners turned to recruiting agents for Irish laborers to mine ore the first winter but let them go at the first sign of discontent over wages in the spring of 1838. Most likely, recruiting agents in Britain had secured the labor of the *Tiberias* miners and ironworkers.

It was just as well that Irish workers were removed from the mix, for even before the *Tiberias* immigrants arrived, hard-drinking Welsh miners whom Hopkins had recruited in Pottsville had gotten into violent scrapes with American workers at Lonaconing. The Americans were certainly no strangers to the moonshine for sale at Buskirk's, a shanty tavern in the hills nearby that the company futilely tried to shut down. Welsh and American workers continued to fight through the summer. The incorrigible John Thomas and a number of other miners who had not been laid low by the summer heat and diarrhea were discharged for drunkenness. On September 21, Alexander tried to impose stricter control over the unruly workmen by issuing "Rules of the Lonaconing Residency." The fourteen rules were similar to many other antebellum factories' dicta. They set the hours of labor (sunrise to sunset, Monday through Saturday); ordered workers to obey their supervisors and to begin work promptly when called by the tolling of the company bell; and warned the men that they were employed at the sufferance of the works manager. These conditions were probably spelled out in the immigrant workers' employment contracts. The rules go on, however, to reflect the deterioration of workplace discipline in their detailed prohibitions against selling and drinking "distilled spirituous liquor" and "all brawling, quarreling, fighting, and gaming. The firing of guns, which has been more frequent of late than usual in the valley of the works, is also, as dangerous and unnecessary amusement, forbidden in future. The managers must aid in the enforcement of this rule by reporting to the superintendent all violations thereof which may come under their knowledge."[115] The last injunction is telling. David Hopkins, who by the end of September was the leader of a Welsh workforce numbering at least forty-five men, had failed to keep control of his kinsmen. His response to the rules was to open his own grog shop.

Through the fall and winter of 1838–39, drink and violence kept workers inflamed. Some of the fights sprang from ethnic animosity. In December, one of the *Tiberias* miners took umbrage at four Scottish miners' being accommodated in a house that had been occupied by a Welsh family. He threw out the Scots' household goods three times before being arrested and put under armed guard. In February, works superintendent Charles B. Shaw, an American engineer, noted that the Germans were "much annoyed by the Welsh miners." They complained that John Hopkins favored the Welsh crews by letting them use horses and manual laborers to clear waste rock at night so that all of their own labor went to mining, an important benefit for workers who were paid by the ton. Disagreements may have been aggravated by misunderstanding, since some of the German and Welsh workers spoke little or no English.[116] What Shaw called "riots" became a fairly regular feature of nightlife in Lonaconing. In February 1839, Alexander ordered the organization of armed night patrols of mixed ethnicity that continued through the winter. A month later he established what he called a "military company"—a garrison, in effect—of trusted workers to keep order in the settlement.

Ethnic conflict does not explain all the violence at Lonaconing. William Williams, a Welsh miner, lashed out at the Hopkins family during an evening of carousing at their home and nursed a grievance against them until he was forcibly driven off company lands. A few Welshmen and Americans beat their wives. (Alexander commented wryly in the journal, "No chivalry in Lonaconing."[117]) A few Welsh and German women exploded in anger as well or joined their menfolk in tavern brawls. One woman's affair with a Welsh miner resulted in her being transported by the bailiff to Frostburg, with her lover and his brother in hot pursuit. As in other isolated industrial places with few entertainments, plenty of cheap liquor, and a predominantly male population working long hours in all weather, domestic and communal violence was probably inevitable. Boredom and short pay contributed to the discontent. When heavy snow stopped work, men were not paid. When frigid temperatures froze the sand casting bed and engine parts at the blast furnace, no pig iron could be made. Idle hands became a recurring headache for the works managers.

Despite all the ructions, Lonaconing Furnace was completed in good time, particularly considering the weeks lost to bad weather. Eighteen months after David Hopkins arrived, the furnace was charged and put in blast. The first run of iron on May 17, 1839, inaugurated a campaign of pig iron production that continued into August. Although mechanical problems caused Hopkins, Alexander, and Shaw to decide to use a cold blast, the blast furnace made good foundry iron that boded well for the company's long-term prospects. The mines meanwhile continued to produce excellent coking coal.

Two factors were particularly important in the initial success of the George's Creek Company. One was the quality and accessibility of its coal. Coal outcroppings along the steep sides of the valley led to broad veins of bituminous coal up to fourteen feet thick. Lonaconing coal was well suited to coking and to smelting iron because it had little sulfur or phosphorus. The George's Creek coalfield had perhaps the best coking coal mined in the United States before the development of the Connellsville coke region some fifty miles to the northwest just over the Pennsylvania border.[118] George's Creek coal was far superior to the thinly bedded, shale-riddled coal at Farrandsville. The company was also fortunate in finding deposits of fairly good iron ore interbedded with the coal measures, as at mines in South Wales. Although Lonaconing ore proved less abundant and of somewhat poorer quality than the original surveys had suggested, there was plenty to feed the blast furnace.[119]

The second factor in the company's initial success was the ethnic solidarity of Lonaconing's miners and ironworkers and the leadership that David Hopkins assumed among the Welsh, whose skills were most important for the development of both the coal mines and the furnace. The Welsh at Lonaconing displayed the same intense ethnic-occupational identity and loyalty that typified Welsh ironworkers and coal miners throughout the antebellum period. The Hopkins family became the anchor of a fluctuating Welsh workforce. They attracted skilled workers to Lonaconing and helped retain them by defending their kinsmen against what they considered unfair treatment by American managers. They also disciplined and even fought fellow Welshmen who crossed the line of what they considered acceptable behavior. When Welsh workers committed "an outrage" or caused serious bodily harm, the Hopkinses did not oppose their dismissal from the works. David Hopkins remained at Lonaconing for at least three years, long enough to see the furnace into sustained operation. During that time he resisted Alexander's authority on a number of occasions and thumbed his nose more than once at the dyspeptic, condescending Shaw. Hopkins disputed technical specifications for the furnace design and flouted the residency rules against alcohol. But the company journal suggests that he and Alexander developed mutual respect. Alexander and Shaw called Hopkins to task several times for selling alcohol but never disciplined him. Alexander did not hesitate to leave furnace affairs in Hopkins's hands, though he later wrote that for hiring Welsh workers "we have had no great reason to congratulate ourselves."[120] Even Shaw, whose dislike for the Welsh grew with time, did not directly challenge Hopkins's technical knowledge or supervision at the work site. For all their troublesomeness, the skilled Welsh workers who came to Lonaconing included many able, reliable workers. Welsh colliers were the last miners fired by the company when Alexander suspended operations in February 1840.

Despite these advantages, the George's Creek Coal and Iron Company ceased operations after building only one of the four envisioned blast furnaces. Why? The chief culprit was the prolonged economic depression that deepened in 1839–40, just when Lonaconing Furnace had proved its potential. The collapse of credit halted many state and private infrastructure projects. Predictions that the Chesapeake and Ohio Canal would reach Frostburg in time to ship the company's first iron and coal to eastern markets proved to be wildly optimistic. The canal in fact did not reach Cumberland until 1850. The Baltimore and Ohio Railroad stalled six miles east of Cumberland from 1839 to 1842; it was not completed through to Wheeling until 1853.[121]

The company's financial problems were aggravated by the high cost of labor. When Shaw totted up labor costs a few days after the furnace was successfully put into blast, he was dismayed to discover that wages per ton at Lonaconing were nearly double those at Varteg Furnace in Glamorganshire, South Wales. Varteg's low labor costs were the envy of many antebellum ironmasters who read *Voyage métallurgique en Angleterre,* Dufrénoy's technical guide to British ironworks. However good Lonaconing iron might be, it could not compete with British iron that was produced and shipped at half what it cost to make iron in western Maryland.

The efficiency of placing several blast furnaces shoulder to shoulder had been demonstrated at Cyfarthfa by the end of the eighteenth century. That scale of construction and production eluded American iron manufacturers until the early 1850s. The first US company to have as many as four blast furnaces in a row was the Lackawanna Iron and Coal Company, which completed its fourth anthracite furnace on Roaring Creek in Scranton in 1853. By that time the industrial economy had revived from the depression of the late 1840s to foster a fresh boom in railroad construction, and the technical requirements and design for anthracite and coke furnaces had been sufficiently established to remove most of the risk associated with their construction.[122] All that came too late for Lonaconing's iron company. The furnace went out of blast for the last time in 1855 or 1856, after years of stop-and-start operation. Coal proved far more profitable than iron for the George's Creek Company. It remained a going concern as a coal company until 1910, and its successor continued to mine coal at Lonaconing until 1952.[123]

The problems at Lonaconing and Farrandsville highlight many of the obstacles that slowed modernization of the US iron industry in the antebellum period. Instead of conveniently layered resources, the Lycoming Coal and Iron Company's land held only coal in significant amounts, and the coal proved ill suited to making iron. The company resorted to quarrying limestone and mining ore at distant sites, which greatly increased transportation costs and undermined labor discipline. Lonaconing had better and more proximate resources, but the George's Creek Coal and Iron Company incurred steep costs transporting its pig iron to market over the Alleghenies. Both companies' distance from population centers, and the general scarcity of skilled ironworkers and coal miners in the United States, meant that the owners had to hire recruiting agents and cover the cost of overseas passage for dozens of workers and their families. The decision to hire immigrant labor was driven in part by the desire to be first to produce iron with mineral fuel. The fast-track approach backfired when construction delays left companies saddled with more workers than they needed or with workers who became fractious when stinted on their wages. The American companies' haste may have contributed to errors in judgment. Had the partners known more about the kind of manufacturing they were taking on, they might not have invested so heavily on the industrial frontier. George's Creek Company expenses reached $200,000 by the time the first and only furnace went into blast. Clearly, Alexander and Tyson's original estimate that it would cost $167,930 to build four furnaces and a rail mill was naive.

One must forgive the misjudgment of minerals by surveyors and investors in these early ventures, given the limits of scientific knowledge and methods for assaying coal and ore at the time. Historians of the iron industry have been right to point to resource problems as a major reason mineral-fuel smelting in western Pennsylvania and other parts of the trans-Appalachian west lagged behind the Mid-Atlantic by fifteen to twenty years. The impurities in bituminous coal in the Pittsburgh area, for example, prevented mineral-fuel blast furnace construction in that city until rail transportation linked Pittsburgh to the Connellsville coke region.[124] Local coal was adequate, however, to power the city's many rolling mills and for reheating charcoal pig and bar iron produced elsewhere. The biggest geographical advantage Pittsburgh possessed that both Farrandsville and Lonaconing lacked was its proximity and ease of access to a rapidly growing market region—the Midwest. Had all the other problems highlighted in this chapter been solved by financial wizardry, astute management, more favorable weather, and better luck, Farrandsville and Lonaconing might still have failed to make a profit because of distance and transport problems.

The labor problems that emerged in Lycoming and Allegany Counties highlight the great differences in the social circumstances of British and frontier American ironmaking communities. Identities based on class, occupation, language, and religious affiliation strengthened the cohesiveness of the industrial workforce in many parts of Britain. In the United States, so far as I have been able to determine, every antebellum iron company that sought to employ large-scale British technologies employed multiethnic workforces, which were also multiracial at southern ironworks. Immigrant workers hired for their valuable skills encountered varying degrees of ethnic prejudice from American works managers. Some found their technical knowledge dismissed out of ignorance or inexperience. Disagreements with managers and owners may have been particularly irksome to artisans who regarded themselves as authorities on modern ironmaking. When tempers flared on the frontier, there was not much one could do but fight or leave.

The labor conflicts at Farrandsville and Lonaconing also illustrate the effects of fragmenting the network of social relations that bound together workers, managers, and owners in Great Britain. Edward Thomas gladly accepted the invitation to emigrate to Pennsylvania because it offered a chance for him to escape oppression under "the Tories," meaning the powerful elites in South Wales who ruled industrial works and blocked workingmen's full participation in politics. When he arrived in central Pennsylvania, however, Thomas found himself almost entirely cut off from the supportive network of fellow artisans and workers among whom he had learned his craft. Even if he had been able to build and run a successful blast furnace at Farrandsville, he could not have reconstituted the world of work he had known. At Lonaconing, skilled immigrant ironworkers shared more connections, including those of kin. David Hopkins was the leader and spokesman for a group who knew how to work together and could help one another solve the social and industrial problems that they encountered in the woods of western Maryland. This comparison suggests that the more fully the networks among immigrant workers carried over to US ironworks, the more likely they were to apply their skills effectively.

It's a big business & ought not to be lightly touched by Amateurs like you & me.

JOHN MURRAY FORBES, December 21, 1859

4 The Elements of Success

The Lycoming Coal and Iron Company failed to produce marketable iron because its owners tried too literally to replicate the British model of ironmaking in an unsuitable location. The model fit less than perfectly at Lonaconing as well. Henry Alexander's dream of re-creating Merthyr Tydfil or Glasgow in western Maryland was more nearly realized at Mount Savage, an iron company chartered in 1837 about twelve miles northeast of Lonaconing at the head of George's Creek (fig. 49). The Mount Savage Iron Works was an audacious venture funded by British and American capitalists who built one of America's first fully integrated ironmaking operations. By 1845, a large rolling mill and two fifty-foot coal-burning furnaces (blown by an engine made at West Point Foundry) were producing about two hundred tons of pig and finished iron a week. The industrial community had three thousand residents, one in six of whom was employed at the ironworks. Mount Savage was one of the first two American mills to produce T rail, named for its distinctive shape. The T rail was more durable and stable than American strap rail, being made of solid rolled iron rather than wood with a layer of rolled iron on top. From 1845 to 1847, the Mount Savage rolling mill and

Figure 49. Mount Savage Iron Works, ca. 1850. Courtesy of the Mount Savage Historical Society. This rare view lays out the industrial core of one of the country's largest frontier iron towns. The large rolling mill, with many slim puddling furnace chimneys, stands on the valley floor, partly obscuring one or more blast furnaces. Ranks of workers' housing stand to either side, while the company store and managerial offices dominate the right foreground. The somewhat fanciful hills may show more forest than actually existed at this time, but they also convey a sense of hemmed-in isolation.

the Montour rolling mill in Danville, Pennsylvania, produced most of the T rail manufactured in the United States.[1]

This promising start soon faltered. One of the capitalists who invested in Mount Savage was Bostonian John Murray "J. M." Forbes, the exuberant and ordinarily very successful brother of John Bennet Forbes, William Lyman's partner in the Franklin Nail Works at Farrandsville. J. M. summed up his experience as a would-be iron entrepreneur in 1859 in a letter to their brother Paul Forbes. J. M. warned Paul not to risk his money in the iron industry. "The Iron trade requires a combination of skill & capital (& both are very rare) with general business ability to boot. The objections to the trade are in my mind so strong that apart from danger of want of skill & personal liability I would rather put money into 5% U.S. Stocks than into the best Iron Company in this country at ½ cost." J. M. had first invested "with other solid men" in the West Stockbridge Furnaces in western Massachusetts. That venture yielded no profit. Then he owned part of a rolling mill in Boston—"*the best mill in the world*"—and took an interest in several blast furnaces near the rich iron mines in Essex County, New York, along the western shore of Lake Champlain. "Result after laying out of money—endorsing paper—doing lots of work & having anxiety without end—I sold out at about half the original cost losing 5 to 8 years interest." He suffered similar losses from his investment in the Mount Savage Iron Works. When Congress lowered the tariff on British iron in 1846, Mount Savage could not compete with imported Welsh rails. The company halted construction of a third blast furnace and reduced wages. The workers went on strike. Unable to meet its obligations, the company was sold for $200,000, a fraction of the investors' stake. These bitter experiences, Forbes told his brother, had made him thrice cautious where iron was concerned.

> I would not therefore meddle with Iron even in compy. with people I know to be skilful & honest, & even if they would put in their share of capital. Far less would I go in with strangers who have not much money. If you put manufacturing out of the question & only look to buying Land for Pyrites I can only say—don't be deluded into the idea that some great chance is offered which must be snapped up without *examining the whole country* first for the best location. . . . Peter Leslie who married *Susan Lyman* & whose address is Philad. . . . has investigated the western ores & might for small pay give you valuable data—possibly his Book on American Iron may give you something "The Iron Masters Assistant" is I think its title. Finally if the advice of a burnt child wont keep you out of Iron Land, or Iron Making let me put you into *relation* with Joseph Tuckerman who has made a fortune out of Iron & would give you any hints that he could. Its a big business & ought not to be lightly touched by Amateurs like you & me.[2]

The failure at Farrandsville and the struggles at Lonaconing and Mount Savage constituted a catalog of what could go wrong for antebellum iron producers who tried to modernize the industry. What, then, was the formula for success? The parts of the British model related to refining iron and producing finished goods were not particularly difficult to replicate, as demonstrated by the concentration of rolling mills and foundries in antebellum Pittsburgh. To replicate British ironmaking in its entirety, however, Americans needed a fuel source for smelting that was sufficiently similar to British fuel in its physical and chemical properties, as well as knowledgeable artisans and managers who understood the relevant technologies, ample capital to build and staff large works, proximity to both good resources and markets for crude and finished iron, and a well-developed transportation infrastructure. The one United States region that possessed all these characteristics in the antebellum period was the Mid-Atlantic. The key resource that favored replication of British methods there was anthracite coal.

The remarkable growth of the anthracite coal and iron industries in nineteenth-century America has drawn the attention of many economic historians. Darwin H. Stapleton presented the duplication of Welsh anthracite smelting methods in Lehigh County, Pennsylvania, as a classic case of technology transfer. Alfred D. Chandler Jr. argued that anthracite enabled the leap from small-scale to large-scale iron production in the United States, and he noted the importance of anthracite-powered steam engines in driving the growth of American manufacturing in the late antebellum period. Peter Temin documented the productivity gains at anthracite blast furnaces, which could produce up to five times as much pig iron as the largest charcoal furnaces. He argued, along with others, that the greater efficiency of anthracite furnaces drove small producers using antiquated technologies out of business from Virginia to western Pennsylvania.[3] Albert Fishlow explained the reciprocal relation between railroad construction and the growth of the iron industry, particularly as demand for less expensive, domestically manufactured rails spurred investment in anthracite blast furnaces.[4] The increase in productivity with anthracite was tremendous: from their introduction in 1839 to 1860, anthracite furnaces vaulted to claim production of 57 percent of all US pig iron.[5]

Economic historians and historians of technology have generally portrayed events in the Mid-Atlantic states as the template for a modern American iron industry. Their emphasis on output and technology, however, has glossed over how anomalous the region was. Measuring other regions against the Mid-Atlantic has skewed historical understanding of what was a highly varied period of industrial development. Nor has the industry in the Mid-Atlantic region itself been fully understood. Although its geographical and social conditions were most similar to conditions in Great Britain, even here iron companies did not adopt the British model in its entirety. Manufacturers continued to use large amounts of charcoal iron, and the industry as a whole remained far more spatially disaggregated than in Britain.

The technology for smelting iron with anthracite was not perfected until 1837, fifty years after coke smelting was well established in Britain. It developed as a method peculiar to furnace companies in western South Wales, a part of the principality that possessed Britain's only sizable anthracite deposits. Ironically, anthracite iron in Wales never grew beyond the small anthracite belt in northern Carmarthenshire and Breconshire; by 1860 fewer than twenty Welsh anthracite furnaces were in blast. The anthracite iron industry was much larger in the United States. Anthracite pig iron production grew dramatically from 1840 to 1855. Output continued to increase until 1874 and remained significant to the end of the nineteenth century.[6] Anthracite smelting technology reached the Mid-Atlantic region when it had the country's most extensive system of canals. Some of the first American railroads were built to haul anthracite over the barrier of the

Appalachian Mountains to canals and improved river channels that connected the anthracite district to America's largest manufacturing centers. This sequence of development gave the Mid-Atlantic region a huge initial advantage in mineral-fuel iron production. The availability of low-cost freight shipment on canals, and the relative proximity of the anthracite deposits to Greater New York and Philadelphia, help explain why Mid-Atlantic iron production remained spatially disaggregated through the antebellum period.

Labor relations were less contentious for Mid-Atlantic iron companies than along the bituminous coal frontier to the west, partly because of managers' greater technical knowledge and shop floor experience. This accords with Philip Scranton's study of antebellum textile manufacturers in Philadelphia. Scranton found that owner-managers with mechanical backgrounds were better able to resolve labor disputes than were the merchant owners and more bureaucratic managers of large textile companies like those in Lowell, Massachusetts.[7] The rapid replication of anthracite technology in the Mid-Atlantic region, along with the growth of rolling mills feeding off anthracite pig iron, made the area a rookery for skilled iron artisans, to borrow a phrase from David R. Meyer. The region's distinctive industrial geography bears out Meyer's prediction that hubs of innovation would become more competitive and innovative over time, while locations with fewer, less diverse contacts would fall behind.[8] As we shall see, the degree to which networks of skill and technical expertise were extended across the Atlantic helps explain why the Mid-Atlantic region became a model for industrial takeoff in the antebellum period.

Lehigh Crane Iron Works

America's anthracite pig iron industry was first conceived as a source of demand for anthracite coal. One can trace the idea back to 1817, when Josiah White (1780–1850) and Erskine Hazard (1790–1865), owners of a rolling mill on the Schuylkill Falls in Philadelphia, leased six thousand acres of anthracite coalfields and timberland along the Lehigh River. They wanted to secure a reliable supply of anthracite for their mill, where it was used to reheat crude iron to roll into wire. They also saw an opportunity to sell coal in Philadelphia, where demand exceeded the supply shipped from Virginia or imported from England. Improving navigation along rugged stretches of the Lehigh River was essential for their plan to turn a profit.[9] In 1818 the Pennsylvania legislature granted White and Hazard's Lehigh Navigation Company "the opportunity of 'ruining themselves'" by improving the Lehigh River channel and constructing canals and locks down the Lehigh to its confluence with the Delaware River at the village of Easton. The stretch from Easton to the newly established settlement of Mauch Chunk opened in 1820, with improvements continuing over the next decade. In 1827 the company also built Pennsylvania's first railroad, a nine-mile gravity railway that carried coal down from Summit Hill to the river at Mauch Chunk.[10]

The coal measures that White and Hazard had leased from the Lehigh Coal Mine Company lay in the first, or southern, anthracite coalfield, along Panther Creek between Mauch Chunk and Tamaqua (fig. 50). In 1832, Prince Maximilian von Wied traveled up the Lehigh River from Easton to see Mauch Chunk, which was already known as the chief loading place of the strange new coal called anthracite or "stone coal." Maximilian described it in his journal as "very hard and shiny and . . . of excellent quality." The canal built to carry the coal to Philadelphia, he noted, "cost large sums of money, and people do not believe that it will pay for itself."[11] The plunging hillsides of the Lehigh Gorge, thickly wooded to the water's edge, made canal construction exceedingly difficult. The dramatic scenery struck a romantic chord in Maximilian and the young Swiss painter Karl Bodmer, who accompanied him as artist for the expedition (fig. 51). Bodmer's watercolor of Mauch Chunk views the landing from a slough upriver. Vegetation seems to boil from the steep, dark hills. The drowned trees in the foreground suggest the river's force during spring freshets, though they may in fact have been immersed after

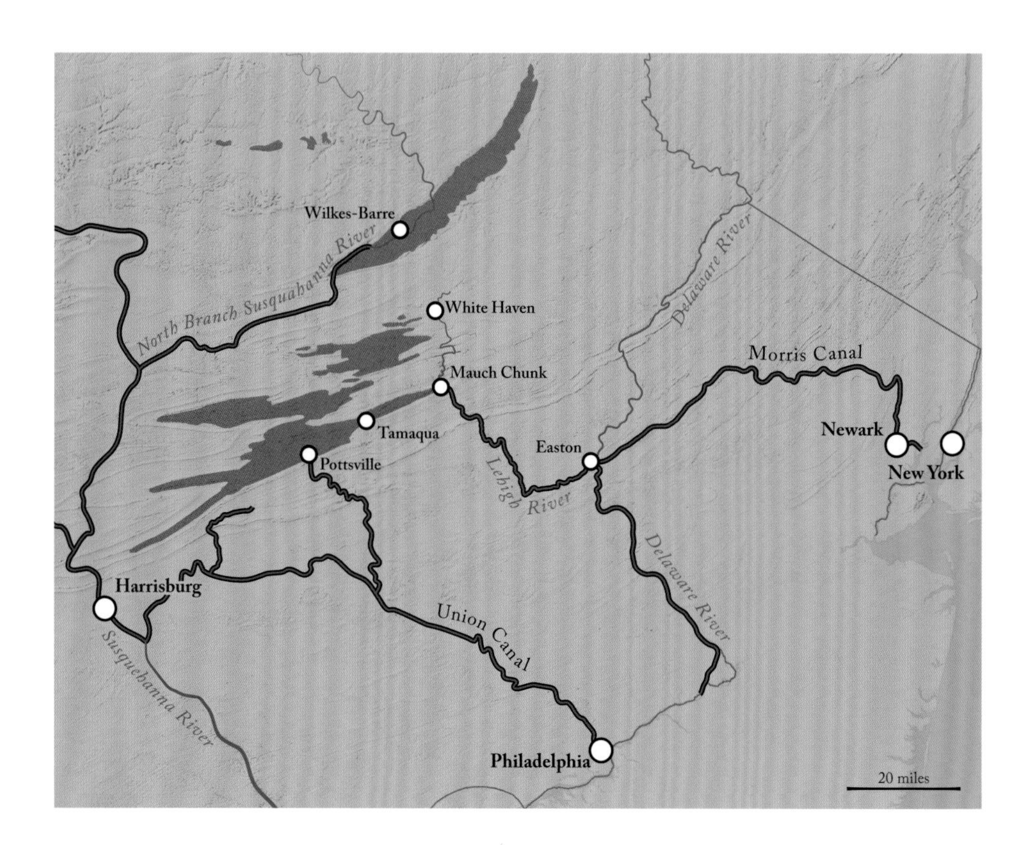

Figure 50. Mauch Chunk was established where the southern anthracite field (*in gray*) reached the Lehigh River. Channel improvements and canals connected the coal depot there to Philadelphia and New York.

the Lehigh Coal and Navigation Company built a dam downstream to improve water flow in its canal. Wisps of smoke or mist hanging above the far bank hint at scattered home fires or perhaps the oppressive humidity of Pennsylvania in August. In Bodmer's twilight scene, Mauch Chunk looks nearly deserted. During the day, however, Maximilian observed that the river was

> filled with small rafts and boats, all of them intended for the use of the colliery. On the bank one sees large supplies of beams, boards, etc., all of it material for the mines. Buildings of various kinds, such as warehouses, stores, stables, an inn, and the like, have been erected here. Barges are being built here to carry away the coal on the Mauch Chunk Canal. . . . In short, the production of coal brings to this hidden, lonely, wild nook of the wild valley an interesting, noteworthy activity that provides a most interesting drama for travelers.[12]

From the opening of the gravity railway in 1827, Mauch Chunk had become a tourist attraction for lovers of wild landscapes and the technological sublime. Travelers making the "Fashionable or Northern Tour" ventured up the Lehigh

Figure 51. View of the Lehigh River at Mauch Chunk. Watercolor on paper, 11⅞ × 17 in. Joslyn Art Museum, Omaha, Nebraska, gift of Enron Art Foundation. Swiss artist Karl Bodmer's *View of Mauch Chunk, Pennsylvania, with Railroad* evokes a wild, damaged landscape. The gravity railroad line that slopes down to the river (*right*) delivered anthracite from mines up in the hills.

Gorge to admire the magnificent forests and rapids, to see "a real bear's haunt," as Maximilian described it, and to coast down the nine-mile railway from the top of Summit Hill.[13]

To recoup their heavy investment in transportation improvements, White, Hazard, and the other investors in the Lehigh Navigation needed markets for anthracite. They were able to sell some coal to rolling mills in Greater Philadelphia, but the market for home heating coal developed slowly because of the difficulty of igniting anthracite and keeping it burning in grates designed for bituminous coal. The partners therefore tried to stimulate the development of anthracite blast furnaces along the Lehigh Navigation, reasoning that blast furnaces would burn vast quantities of coal. In 1835 and 1836, the company offered free coal and low shipping rates to anyone who built an anthracite furnace along the canal. (White had tried to smelt iron with anthracite at Mauch Chunk in 1826, but the experiment failed because of his "imperfect idea of the hot blast.") In 1838 they sweetened the offer with free waterpower and land for erecting industrial plant and workers' housing.[14] A company history cum prospectus published in 1840 makes plain the economic rationale for their generosity:

> No other business can probably be found which produces so much freight for a navigation as the manufacture of iron. To make a ton of pigs will require about two tons of coal, two to three tons of ore, and a half to one ton of limestone, or about five tons of freight for each ton of pigs. And the conversion of the pigs into bar iron will also create freights equal to three or four tons for each ton of iron. The *Lehigh Company* will not only derive an income from the tolls on the iron manufacture, but also from the profits on the sales of water power, coal, and iron ore, all of which they own in abundance.[15]

No one took up the company's offers because the riddle of how to smelt iron with anthracite on a commercial scale had yet to be solved in the United States. Anthracite was an enticing fuel for blast furnaces because it burned hotter than bituminous coal and contained less impurities. At the same time, the hardness of anthracite, which made it difficult to burn in home grates, made it even harder to ignite and keep burning at a steady heat inside the crucible of a blast furnace. Because anthracite is denser than charcoal, bituminous coal, and coke, its weight tended to cause the charge of burning material to collapse, smothering the fire inside the crucible and producing cinder that clogged the furnace, necessitating costly cleanout and repairs. A number of Americans tried to smelt Pennsylvania ores with anthracite in the 1820s and 1830s. Most of the trials, including those at a furnace in Mauch Chunk, failed to produce more than small samples of marketable iron. In 1839 William Lyman, former co-owner of the rolling mill at Farrandsville, won the $5,000 prize offered by Nicholas Biddle by running

Figure 52. *David Thomas*, ca. 1845, by J. Carlin. Oil on canvas. Courtesy of Larry Mouer. Copy photograph by Joseph E. B. Elliott, professor of art, Muhlenberg College. By the time of this portrait, Thomas was well known on both sides of the Atlantic as the man who brought anthracite iron to Pennsylvania.

his Pottsville Furnace on anthracite for the requisite three months, but the blast equipment broke down shortly after the trial. Lyman soon sold the business, and the furnace was out of production for several years.[16]

The two men who worked longest to crack the problem of anthracite iron were David Thomas (1794–1882) (fig. 52) and George Crane (ca. 1784–1846), superintendent and owner, respectively, of the Ynyscedwyn Iron Works in Breconshire, South Wales.[17] Ynyscedwyn was on the outskirts of Ystragynlais, a small town about twenty miles northeast of Swansea, in the heart of the South Wales anthracite coal measures. Thomas and Crane began experimenting with anthracite sometime in the 1820s in hopes of substituting local coal for the coke that had to be shipped from fifteen miles away. They finally produced good foundry iron primarily with anthracite in February 1837.[18] Like experimenters in the United States, Thomas and Crane had tried varying the proportions of anthracite and coke, without success. The key, they discovered, was to preheat the blast (the air forced into the furnace under pressure to aid combustion) using the hot-blast apparatus that a Scotsman named James Beaumont Neilson, manager of the Glasgow Gas Works, had patented in 1829. Neilson had demonstrated at the Clyde Iron Works that raising the furnace blast to three hundred degrees Fahrenheit greatly improved smelting efficiency, producing more iron with less coke. Ironmasters at other Scottish blast furnaces found additional savings by raising the blast to six hundred degrees Fahrenheit. This high temperature proved to be the technological tipping point for bringing a heavy load of anthracite to white-hot heat and keeping it hot enough to smelt a full charge of iron ore and limestone.[19]

It took several years from the time Neilson's patent and the Scottish furnace results were announced in the British technical press[20] for Crane and Thomas, as well as American ironmasters, to apply the hot blast to anthracite smelting. Particular technical knowledge may have been needed to construct Neilson's apparatus and fine-tune it for use with a fuel other than coke. Although Thomas had apprenticed in a machine shop and foundry, he hired "an expert mechanic who understood the construction of heating-ovens" to replicate Neilson's design. The costs may also have delayed them. The final, successful experiment required a large furnace, forty-five feet tall and eleven feet across at the boshes, as well as hot-blast apparatus, heating stoves, and a large blowing engine.[21]

Crane applied for a British patent in September 1836, several months before the trials at Ynyscedwyn produced good iron.[22] The following spring he advertised his success in hopes of capitalizing on his patent. In April 1837 Crane wrote to the manager at Dowlais:

> Have you heard that I have now successfully brought the Stone Coal question to a termination. The experiment has cost me the labour of some months, and has been attended with a serious expence [*sic*], but I am now making the Ton of Pigs with 31 to 33 Cwt. of Raw Stone Coal, the fuel 10 to 11 Cwt. for the Heat air Stoves to be added, and we hope to do better than this. The Iron *very much* stronger than *cold blast* Coke Iron. I am told that Mr. Crawshay and yourselves are forming New Works in the Forest of Dean, if you were to come down to see what I have now been doing for the last 8 to 9 weeks with Stone Coal only, . . . I think that you would recommend The Dowlais Company to apply for a Licence to me, and look out for some eligible spot in some part of the Stone Coal District.[23]

Dowlais did not pursue Crane's offer. With seemingly inexhaustible coking coal in its backyard, and given the difficulty of smelting with anthracite at that time, the company had no incentive to switch fuels. The works manager, however, passed on the news to Solomon White Roberts, a visiting American engineer who was inspecting rails at Dowlais for the Philadelphia and Reading Railroad. Roberts happened to be a nephew of Josiah White. When Roberts learned of the breakthrough at Ynyscedwyn he went out to see the process for himself. His very favorable report prompted White and several other partners in the Lehigh Coal and Navigation Company to propose building their own anthracite blast furnace. The company's managing board accepted the proposal and conferred on its affiliate, the Lehigh Crane Iron Company, all the benefits previously offered to others: free waterpower, coal, and land on which to erect the furnace.[24]

In November 1838, board member Erskine Hazard crossed the Atlantic to negotiate licensing with George Crane, who shortly before had purchased the US patent rights to smelting iron with anthracite from the estate of Frederick W. Geissenhainer, a Lutheran minister who ran successful trials at a furnace in Schuylkill County in 1836 but became ill and died before he was able to repair the furnace's faulty machinery. Crane agreed to the license, and he suggested that Hazard invite David Thomas to superintend construction of the Lehigh Crane Iron Company's new furnace. Crane saw an opportunity to profit from Thomas's expertise, for every American company that adopted their process would be obliged to pay patent royalties. "I should prove to be the greatest benefactor to Pennsylvania that ever lived," he later joked in a letter to Thomas. Crane nursed the same ambition that inflamed Edward Thomas's imagination when he went to Farrandsville—that he would grow rich from patent royalties as Americans built anthracite furnaces—though Crane had more solid grounds for hope. David Thomas accepted Hazard's generous terms and set sail for Pennsylvania with his wife and three children in April 1839.[25]

Thomas brought more to Lehigh County than his knowledge of how to build and run an anthracite blast furnace. The sites of production and the social networks through which he acquired his expertise are key to understanding his technological accomplishments in South Wales and Pennsylvania. David Thomas exemplifies the kind of skilled mechanic that Meyer calls a hub of technical knowledge and best practice. Thomas became a hub for the inculcation and dissemination of ironmaking expertise in the United States, after decades of acquiring skills and experience that made him ideally suited to the role. He developed extensive and diverse contacts over the course of his career in Britain partly because he spoke both Welsh and English, the two chief languages of the British iron industry. His Welsh-language education began at the strict, Bible-literate Welsh Independent (Congregational) chapel his family attended in Neath, Glamorganshire. At nine he was sent to the best school in Neath, where his English-language studies probably began.[26] In 1812 Thomas entered a five-year apprenticeship at the machine shops and foundry at Neath Abbey Ironworks. During his sojourn there, Neath

Abbey produced the largest steam engines yet manufactured, including high-pressure beam engines for Richard Trevithick and pumping engines for deep-pit mines in Cornwall, one of which Thomas installed. He gained knowledge of industrial engines of all kinds and worked with some of Britain's best machinists and engineers. He also worked at the company's two coke-fired blast furnaces. In 1817 Thomas was hired at Ynyscedwyn as general superintendent of the blast furnaces and iron mines. He stayed for twenty-two years.[27]

Like other men in his position, Thomas read the British technical press, which published news of technological innovations and other trends in the iron and coal industries. He would have known a fair number of other ironmasters and works managers in South Wales as well as hundreds of skilled and unskilled workmen from daily encounters at Ynyscedwyn over the years. When he traveled to Glasgow to examine Neilson's hot blast, Thomas widened his familiarity with leading iron companies in Britain, for he doubtless would have taken the time, as travelers did in his day, to call on the Clyde Iron Works, the Carron Foundry, and other ironworks in the Scottish Lowlands. Although David Thomas was not uniquely well connected, his breadth of training, experience, and exposure to different branches and regions of the British iron and coal industries knitted him into various networks through which technical information and best practice flowed in the early nineteenth century. He had intimate knowledge of industrial machinery and power systems and better knowledge than anyone of how coke iron smelting differed from making iron with anthracite. He had direct experience of three regional industrial cultures (South Wales, Cornwall, Lowland Scotland). In what Carol Siri Johnson calls the "prediscursive" world of ironmaking, where technical information was communicated by oral instruction and learned through physical experience, the knowledge Thomas possessed was extremely valuable.[28]

As a hub of technical knowledge and social networks that extended across the Atlantic, Thomas was a personal conduit for the transfer of technology. His experience and connections lent speed to the process of furnace construction. By replicating materials and equipment used at Ynyscedwyn, the new superintendent minimized the chance of mechanical failure, a problem that had plagued experimental American anthracite furnaces. Before he emigrated, Thomas oversaw the casting of hot-blast equipment for Lehigh Crane at Ynyscedwyn. He ordered firebrick for the furnace lining from a factory in Stourbridge, a town in the Staffordshire iron district. The furnace blowing engine from Matthew Boulton's famous Soho Works in Birmingham was the equal, if not the model, of the best engines made at the West Point Foundry in Cold Spring, New York.[29] Hazard's agreement to ship an engine from Birmingham to Pennsylvania indicates the company's willingness to defer to Thomas's judgment. Unfortunately, the cylinders for the Boulton engine were so large (five feet in diameter) that they did not fit down the hatchway of the ship commissioned to carry them. After several months' delay, Thomas and Hazard ordered a second set of cylinders from the Philadelphia firm of Merrick and Towne, which had to enlarge its boring mill to meet the design specifications.[30] Even with that delay, however, all the necessary equipment and materials reached the construction site on the Lehigh Canal by the spring of 1840. The first Lehigh Crane furnace, half again as tall as a typical charcoal furnace at forty-seven feet and more elaborately equipped, was put in blast little more than a year after Thomas's arrival. It ran well from the start.[31]

The presence of other skilled Welsh industrial workers in Biery's Bridge (later renamed Catasauqua) as early as 1840 suggests that Thomas's presence, if not his own efforts to recruit kinsmen, set chain migration in motion to the Lehigh Valley. Compared with the predominant population of German-speaking farmers, the Welsh were a small minority in Catasauqua. The 1840 manuscript census records just eight heads of household with typically Welsh names, including Thomas and the two keepers in charge of filling and tapping the furnace, William Phillips and Evan Jones. As was so often the case, the Welsh households were clustered in one part of the small town.[32] Ten years later, when the census noted place of birth, the Welsh were still less than 10 percent of Catasauqua's population, but they filled most skilled ironworking occupations. The 1850 census lists David Thomas as "iron master," his thirty-one-year-old son John as the works superintendent, and his son-in-law Joshua Hunt as company clerk. William Phillips had been promoted to agent in charge of pig iron sales. Other Welshmen worked as molders, patternmakers, machinists, and engineers. The iron inspector had also been born in Wales. Unspecified furnace workers and manual laborers were almost all German or Irish immigrants or native-born Pennsylvanians.[33]

Under Thomas's supervision the Lehigh Crane Iron Company became known as a leader in adopting innovative technologies. Each new furnace incorporated modifications to the original design. Thomas persuaded a reluctant Josiah White to shift from waterpower to steam engines, and he was one of the first Pennsylvania ironmasters to employ Christian E. Detmold's method of using hot gases expelled from the blast furnace to raise steam for power.[34] Lehigh Crane became a rookery for talented artisans, if in a minor way. In 1842 Giles Edwards, a young patternmaker from Merthyr Tydfil, emigrated to the coal town of Carbondale, Pennsylvania. By 1850 he lived down the block from the Thomas family in Catasauqua. A few years later he helped Thomas and his sons build the first furnace for the Thomas Iron Company in Hokendauqua, a few miles up the Lehigh River, before moving on to work as a supervisor for John Fritz at the new Cambria Iron Works in Johnstown, Pennsylvania. William R. Jones, born in Scranton, the son of Welsh immigrants to the coalfields, was apprenticed to Lehigh Crane in 1849 at age ten. He followed Giles Edwards to Cambria and then to Chattanooga, Tennessee, where Edwards supervised construction of the South's first coal-fired furnace and Jones was its master mechanic. Jones went on to become superintendent of Andrew Carnegie's J. Edgar Thomson Steel Works near Pittsburgh in the 1870s.[35]

One of the striking differences between Farrandsville and Lehigh Crane is the autonomy and respect extended to David Thomas virtually from the moment he was hired. Before leaving Wales, Thomas notified his new employers "that he did not feel satisfied with the mode of estimating his salary" in his contract with Erskine Hazard. He persuaded the company "to give him the United States mint value of the sovereign for a pound sterling," thus locking in a rate of exchange and eliminating uncertainty on a key term of employment that vexed immigrant workers at Farrandsville and elsewhere.[36] In Pennsylvania, Thomas supervised construction of the furnace and workers' housing and took charge of all aspects of furnace operations, from securing coal and ore supplies to making contracts for the sale of pig iron. Surviving company records are limited to minutes from meetings of the stockholders and board of directors. These brief, businesslike reports rarely reveal personal interactions, let alone the kind of friction evident in the letters between Daniel Tyler and Edward Thomas. Even so, the Lehigh Crane minutes suggest very amicable relations between David Thomas and his employers. The directors rarely disagreed with his recommendations, and only once did they curb his responsibility. The request in November 1840 that he obtain permission from the company president to contract for large sales of iron was soon quietly dropped.[37]

Thomas gave his employers good reason to trust his judgment. The first furnace he built was completed in just ten months. It went into blast on July 3, 1840, and ran continuously for seventeen weeks, longer than any previous anthracite furnace in the United States, producing an average of 41 tons of pig iron a week. Iron samples sent to manufacturers in Philadelphia, New York, and New Jersey immediately prompted substantial orders.[38] From November 1841 to November 1842, the furnace produced 2,332 tons of pig iron, far more than most charcoal furnaces made at that time. Over the next twelve months a second, larger furnace went into operation. Production increased year by year. In 1847 Lehigh Crane's four furnaces produced over 14,000 tons of pig iron, all of it finding ready buyers. With prices as high as $38 a ton, the company could well afford to pay Thomas his contractual bonus of $500 for each furnace brought online.[39]

Clearly, the furnace design that Thomas and Crane developed at Ynyscedwyn worked well with Pennsylvania anthracite and ore. The anthracite in eastern Pennsylvania was the chemical twin of Welsh anthracite. Thomas did not have to modify any of the machinery or manufacturing processes he had developed with Crane at Ynyscedwyn. Productivity increased during the 1840s as Thomas built larger furnaces and began to mix local hematite ore with higher-grade magnetite ore from New Jersey. This improvement led Lehigh Crane to purchase an ore mine in Sussex County, New Jersey.[40] Thomas also oversaw construction of a canal turning basin and a railroad trestle over the Lehigh River, which connected the manufacturing site to its raw materials supply and markets year-round. In 1845 Thomas purchased twenty shares of Lehigh Crane company stock, his first step across the boundary between management and ownership and a sign that the board of directors accepted him as an equal. Three years later, the board of directors noted with satisfaction, "We have now nearly every desired convenience at our works, and unless it shall hereafter be thought expedient to lay down rail road tracks to our prominent mines we know not what else can be necessary."[41]

Previously I noted Latour's observation that "every time you hear about a successful application of a science, look for the progressive extension of a network. Every time you hear about a failure of science, look for what part of which network has been punctured." He goes on to say, "When everything works according to plan it means that you do not move an inch out of well-kept and carefully sealed networks."[42] This insight illuminates several points of comparison between the failure at Farrandsville and success at Lehigh Crane. The networks of knowledge, business relationships, and skilled labor that supported ironmaking in South Wales were extended intact to Catasauqua, whereas in central Pennsylvania they became attenuated and frayed. David Thomas was a hub of expertise in the South Wales iron industry and quickly reestablished himself in that role at Lehigh Crane. Edward Thomas may have had a similar position at Plymouth Iron Works in Merthyr Tydfil, but in Farrandsville he was isolated and ineffective. With perhaps only two fellow Welsh artisans to help him build the furnace and an American manager who distrusted him and obstructed his efforts, Edward Thomas had little chance to prove his technical expertise.

The networks into which each company's owners were connected also differed significantly. Josiah White, like David Thomas, was a "practical" man who applied his mechanical skill to solving problems posed by his industrial undertakings. In this respect White was not unusual in greater Philadelphia, one of the nation's largest pools of mechanics and industrial inventors. A good number of industrialists there were former artisans, in contrast to the merchant financiers who dominated manufacturing in Greater Boston. White was a largely self-taught engineer with particular expertise in metalworking and water-powered millwork. He took pride in inventing devices related to his wire manufacturing business, the Lehigh Canal, and the Mauch Chunk railway.[43] By the time they formed the Lehigh Crane Iron Company, White and his partner Erskine Hazard had decades of experience with manufacturing, coal mining, canals, and railroads. They were well connected with merchants and manufacturers in Philadelphia's coal and iron trades. All these connections, as well as their deep familiarity with the Lehigh Valley and its natural resources, helped prepare them to venture into anthracite ironmaking. The Boston capitalists who funded Farrandsville Furnace were laced into other networks—first, those of the China trade, then textile manufacturing firms and financial institutions based in Boston. They were accustomed to operating as nonresident owners, investing at a social and physical distance. Although Thomas Handasyd Perkins and Patrick Tracy Jackson were very interested in industrial machinery and manufacturing processes, they had no direct mechanical experience as designers, builders, or operators of ironworks or coal mines. Their fateful error in hiring Daniel Tyler may or may not have been due to their inexperience in the iron industry. What seems certain is that they embraced him as a member of their most familiar social network, the educated elite of New England. White and Hazard's choice of a Welsh technical expert to build and manage their furnace reflects a different set of priorities rooted in different networks and experience.

The social circumstances at Lonaconing Furnace were more similar to those at Lehigh Crane. David Hopkins was respected as a figure of authority in the frontier community, and he apparently had fairly free rein in building the furnace and putting it into operation. The greatest initial difference at Lonaconing was the works' geographic situation in relation to resources and markets. Although coal, ore, and limestone were all available locally at Lonaconing, the furnace site was not connected by canal or railroad to its primary markets. It was 182 miles to Baltimore, approximately 80 over rough dirt roads that could handle only wagon traffic.[44] Lehigh Crane received anthracite from Mauch Chunk via canal and shipped its pig iron to seaboard cities by canal as well. Lehigh Crane's first known customers were Merrick and Towne, the Philadelphia machine shop that made the second set of cylinders for the furnace company's blowing engine, and two rolling mills, the Boonton Works near Dover, New Jersey (in easy reach along the Morris Canal), and the Ulster Iron Company in Saugerties, New York. All three companies were accessible from Catasauqua by canal and river transport.[45] Figure 53 shows the spatial arrangement of Catasauqua in about 1862, when the Lehigh Crane furnaces were connected by spur lines to the Lehigh Valley Railroad and the Catasauqua and Fogelsville Railroad. The map also shows the works' location on the Lehigh Canal, which had been completed all the way to the Lehigh Coal and Navigation Company's northernmost anthracite coalfields by the time David Thomas landed in Pennsylvania. The transportation system put in place by Lehigh Crane's parent company was an important factor in the iron company's success.

Canals and railroads aided the development of every branch of iron manufacturing in the antebellum period, but they were particularly beneficial for the anthracite iron trade.[46] Unlike the anthracite district in South Wales, Pennsylvania's anthracite coal districts had negligible deposits of iron ore. This meant that

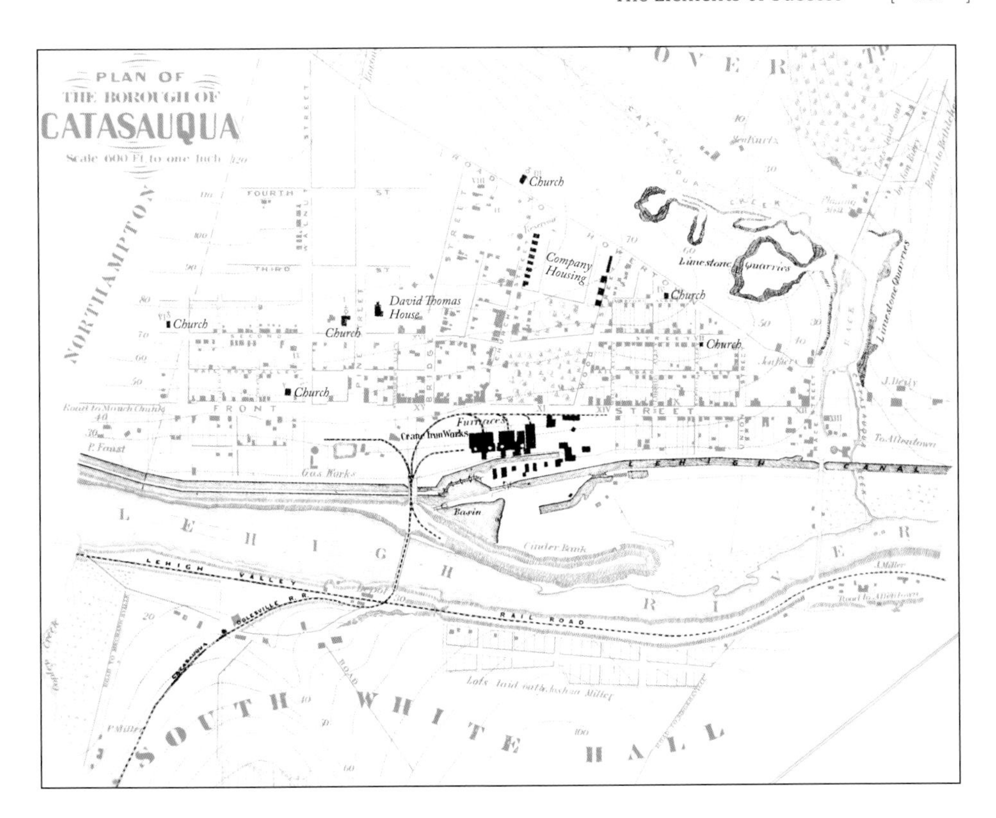

Figure 53. G. A. Aschbach's "Map of Catasauqua, Pennsylvania, 1862." Compare this map to figure 38. The original map has been modified to highlight the blast furnaces along the feeder canal, the Lehigh Crane Works' excellent rail connections, and the proximity of limestone quarries. It also picks out Catasauqua's six churches, which provided services in several languages for various denominations, including English and occasional Welsh services at the Presbyterian Church near David Thomas's house.

either ore had to be shipped to the coal, or coal to the ore. Because ore was the heaviest ingredient in the smelting formula, spatial logic argued for locating blast furnaces near ore deposits. Canals and railroads minimized the cost of transporting anthracite to Mid-Atlantic iron ore locations.

Cost calculations related to furnace location also had to consider distance to markets. As Lehigh Crane's early customers suggest, rolling mills proved to be the biggest consumers of anthracite iron before the Civil War. New anthracite furnaces were built along canals and railroads, first between the anthracite coal region and established iron manufacturing centers, then closer to rolling mills. By the early 1840s, when anthracite furnaces began to proliferate in the Mid-Atlantic iron district, the region had more rolling mills than any other part of the country. When Lesley completed his survey of the iron industry in 1858, the Mid-Atlantic had 31 percent of the country's rolling mills and 67 percent of its anthracite blast furnaces. The southeastern portion of Pennsylvania that included the anthracite district also held approximately 50 percent of the state's miles of railroad track.[47] These figures add a geographical dimension to Albert Fishlow's description of the reciprocal relation between railroads and iron production. Not only did growing demand for rails spur the construction of ever-larger anthracite furnaces, the construction of trunk and spur lines also promoted the diffusion of anthracite smelting technology throughout the Mid-Atlantic and the Hudson-Champlain iron regions.[48]

Railroads and steam power loosened the geographical constraints that had formerly confined manufacturing to sites of resource extraction and waterpower. For rolling mills, access to waterpower was essential until steam engines became commonplace, though many rolling mills continued to use waterpower to drive the rolls to save on fuel costs. As the rail network expanded in the 1850s, Great Lakes shipping developed and western coal deposits began to be exploited. It then became economical to build rolling mills near centers of population regardless of their proximity to raw materials. Hence the large rolling mills that Joseph Lesley observed under construction in Columbus, Ohio, Indianapolis, and Chicago, cities with neither significant waterpower potential nor local deposits of iron ore.

Before about 1855, however, the economic calculus of ironmaking pulled most American rolling mills to canals or railways near manufacturing cities. Unlike most in Britain, only a few US mills were part of an integrated ironworks. The first large-scale American ironworks that included coal and ore mines, blast furnace, and rolling mill in one place under single ownership was the Mount Savage Iron Works in Allegany County, Maryland, which began operation in 1839. The next ironworks with all stages of extraction and production in a single location was the Cambria Iron Works in Johnstown, Pennsylvania, whose first incarnation was built in 1853–54 (fig. 54).[49] Both Mount Savage and Cambria were near excellent coal and ore deposits but far from urban markets. They were built as rail mills intended to supply new railroads (the Baltimore and Ohio and the Pennsylvania Railroad, respectively). Furnace companies rarely seem to have considered building forward linkages to iron processing. The owners of the Lehigh Crane Iron Company briefly considered building or acquiring a rolling mill in 1844–46, when the Mid-Atlantic region was experiencing its first major railroad construction boom, but the board of directors let the idea drop.[50] It was much more common for rolling mills to capture crude iron supplies by acquiring an established smelting operation. Several Pittsburgh mills bought or built a furnace or forge in the Juniata district, though most antebellum rolling mills bought their crude iron from iron merchants.[51] Whichever approach they took, mill managers mixed and matched pig iron and blooms according to their quality. Customary as this practice was throughout the antebellum period, it entailed a number of uncertainties that could prove costly for manufacturers, as we shall see.

The other key difference between the contexts of production for British and American rolling mills was labor supply. While British mills had access to large pools of urban and rural labor, US mills had to assemble a workforce from more scattered, often distant places. At the largest rail mills in the prewar United States,

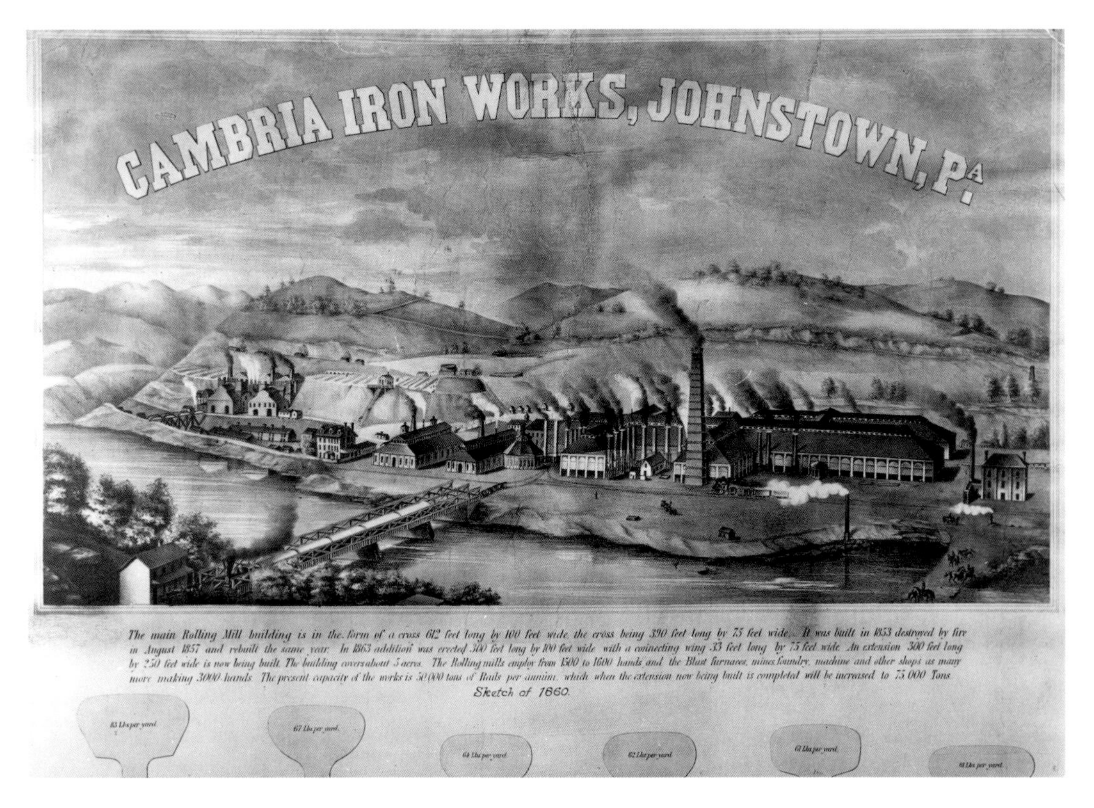

Figure 54. Cambria Iron Works, Johnstown, Pennsylvania, ca. 1860, Bethlehem Steel Corporation Collection, P20091218_001, Hagley Museum and Library. Cambria was one of the few fully integrated ironworks in the United States in the late antebellum period. This engraving shows the rolling mills with thin chimneys marking the puddling and heating furnaces in the right foreground. Smoke from two or three blast furnaces rises near the bend in the river (*left*).

up to five hundred or more employees were skilled workers—puddlers and their assistants, rollers and catchers, heaters, hammermen, machinists, engineers, blacksmiths, and patternmakers. At most northern mills and some in the South, most men in metalworking occupations were immigrants, chiefly from the British Isles, secondarily from Germany, France, and other continental countries. It was significantly easier to obtain large numbers of skilled workers, particularly highly sought British immigrants, near coastal ports of entry and manufacturing centers than farther inland.

These challenges could make the first few years at a rolling mill difficult. The exceptionally complete early records of the Trenton Iron Works offer an unusual window onto the travails of one group of entrepreneurs who tried to capitalize on the first railroad boom by replicating the British rolling mill model on American soil. Their success reveals the risks and benefits of large-scale enterprise in the Mid-Atlantic region as well as the steep learning curve of mastering a dispersed and complicated system of industrial production.

Trenton Iron Works

The Trenton Iron Works is known in the annals of American business history as a progressive company run by an unusually creative businessman, Abram S. Hewitt, who dodged the worst economic blows of the antebellum period by developing innovative products and experimenting with new technologies that paved the way for the leap to large-scale steel production after the Civil War.[52] Although the company began as a fairly typical Mid-Atlantic rolling mill, its advantageous location, ample capital, and exceptional owner-managers made the Trenton Iron Works' early years particularly successful.

The rolling mill was built on the shore of the Delaware River in 1845 by Peter Cooper (1791–1883), a New York City industrialist and inventor. Cooper had built a foundry in Manhattan in the early 1830s that he converted to a rolling mill to manufacture rails, merchant bar, and wire.[53] Further expansion was difficult at the site on Thirty-Third Street and Third Avenue. Cooper was interested in taking advantage of the anthracite coal and pig iron being shipped down the Lehigh Canal and the Delaware River, and he wanted to shift some of the responsibility of his iron business to his son, Edward, so he could concentrate more on his thriving industrial glue factory. He therefore decided to relocate the rolling mill to a far less developed site at Trenton, New Jersey. Although Trenton was a small manufacturing center in 1845, its proximity to the Delaware and Raritan Canal and the newly constructed Camden and Amboy Railroad promised heavy traffic in years to come. In addition to spending a significant portion of his personal wealth on the rolling mill's construction, Cooper bought nine-tenths of the waterpower provided by the Delaware and Raritan Canal Company to ensure that the rolling mill's power supply would not be usurped by other manufacturers.[54] As shown in an engraving of the Trenton Iron Works from sometime in the 1850s (fig. 55), the mill's location along the canal also put it next to the railroad, while the flat land between the river and canal left ample room for coal storage, turning basins, and expansion as the company grew.[55]

Cooper's site choice and his heavy personal investment in the mill and its power supply reflected his long experience in American industry. He first invested in ironmaking as one of several partners in a rolling mill outside Baltimore that produced strap rails for the Baltimore and Ohio Railroad. In 1829–30 he designed and built an experimental steam locomotive called *Tom Thumb*. Not long after establishing his foundry, Cooper chaired the committee that oversaw construction of the Croton Aqueduct and the system of cast-iron pipes that delivered fresh water to New York City beginning in 1842.[56] Each of these projects taught Cooper about the uses and qualities of iron as well as the inadequacies and opportunities for development in American iron manufacturing.

Figure 55. Trenton Iron Works. Image courtesy of the Cooper Union Library. Judging by the locomotive in the foreground and the extent of the ironworks, this image probably dates from the 1850s. The view is looking southeast. Note the turning basin to the far right, beside the rolling mill along the Delaware River.

Although Peter Cooper was the first president and chief investor in the Trenton Iron Works, he soon handed much of its management to his son, Edward Cooper (1824–1905), and Abram S. Hewitt (1822–1903). The two young men were in their early twenties when the Trenton mill was built. They had become good friends as students at Columbia University. They became even closer friends after surviving the shipwreck of a cargo vessel on which they were returning from a postgraduation Grand Tour of Britain and the Continent.[57] Edward, who had something of his father's gift for mechanical engineering, worked alongside the mill's first superintendent to learn puddling and rolling, production management, and how to deal with the problems of waterpower. Hewitt had briefly taught mathematics at Columbia University, but he quickly passed the role of company clerk to his younger brother Charles. Before the first summer was out, Abram was traversing the East Coast to cultivate business as the company's agent. Although he did not marry Edward's sister Amelia for some years, Abram was already considered a member of the Cooper family.[58]

Peter Cooper's capital, his experience with the iron industry, and his reputation as a prominent businessman and industrialist gave the new firm a strong start. Construction proceeded swiftly under the supervision of John S. Gustin, who had worked at Cooper's New York City wire mill. The main buildings of the Trenton mill were completed in about six months; waterpower was applied to mill machinery not long after.[59] The company's ambitious plan to build a very large rail mill attracted the attention of potential suppliers and customers. Cooper signed the mill's first rail contract, with the Camden and Amboy Railroad, in January 1846. Workers rolled the first rails in June, little more than one year after breaking ground for the mill—fast indeed by contemporary standards.[60]

Like other rail mills, the Trenton Iron Works was a puddling mill; that is, it converted pig iron to wrought iron by puddling (and the associated processes of rolling, heating, and shingling) to remove impurities and to blend various grades of iron to achieve the qualities required for any particular product. Wrought iron had to be soft enough to absorb the blows of a helve hammer or even more forceful steam-powered Nasmyth hammer without shattering, and strong enough for the iron to hold its shape. Malleability, or ductility (the ability to absorb energy while being deformed), was crucial for rolling wire and rails because of the great pressure exerted on hot iron as it was pulled and squeezed between the rolls.[61] If the iron was brittle, rails could break when holes were punched for the bolts that secured the joints between rails.[62] It took the right combination of crude iron with certain qualities and the skills of experienced puddlers, rollers, heaters, and hammermen to produce rails that railroad companies would accept. The stakes were high for rail mills because one railroad company's order for miles of iron rails could amount to more than half a mill's business for a given year. The Baltimore and Ohio Railroad, for example, awarded contracts totaling 2,400 tons to the Mount Savage and Avalon rolling mills for one section of its road in 1846. At $80 a ton, the value of the rails alone was $192,000. One estimate of total demand for rails in the East that season was 30,000 tons, about twice the total capacity of American rail mills at that time.[63]

Softness and malleability were the hallmarks of charcoal iron. This was why the Trenton Iron Works, like other rail mills, purchased charcoal pig iron and blooms as well as anthracite and coke pig iron. Figure 56 maps the location of blast furnaces and forges that sold pig iron and blooms to Trenton from 1845 to 1847. Although most of the crude iron came from smelting operations within one hundred miles of Trenton, the company also purchased iron directly and through commission merchants from suppliers much farther away. These included charcoal blast furnaces whose names were synonymous with quality in the antebellum industry, such as Cornwall Furnace, as well as a number of anthracite furnaces. Trenton's biggest single supplier for much of the 1840s and 1850s was the Lehigh Crane Iron Company. The mill also took small deliveries of imported Sheffield steel and high-grade Swedish and Russian iron for particular uses, such as making puddlers' tools.[64]

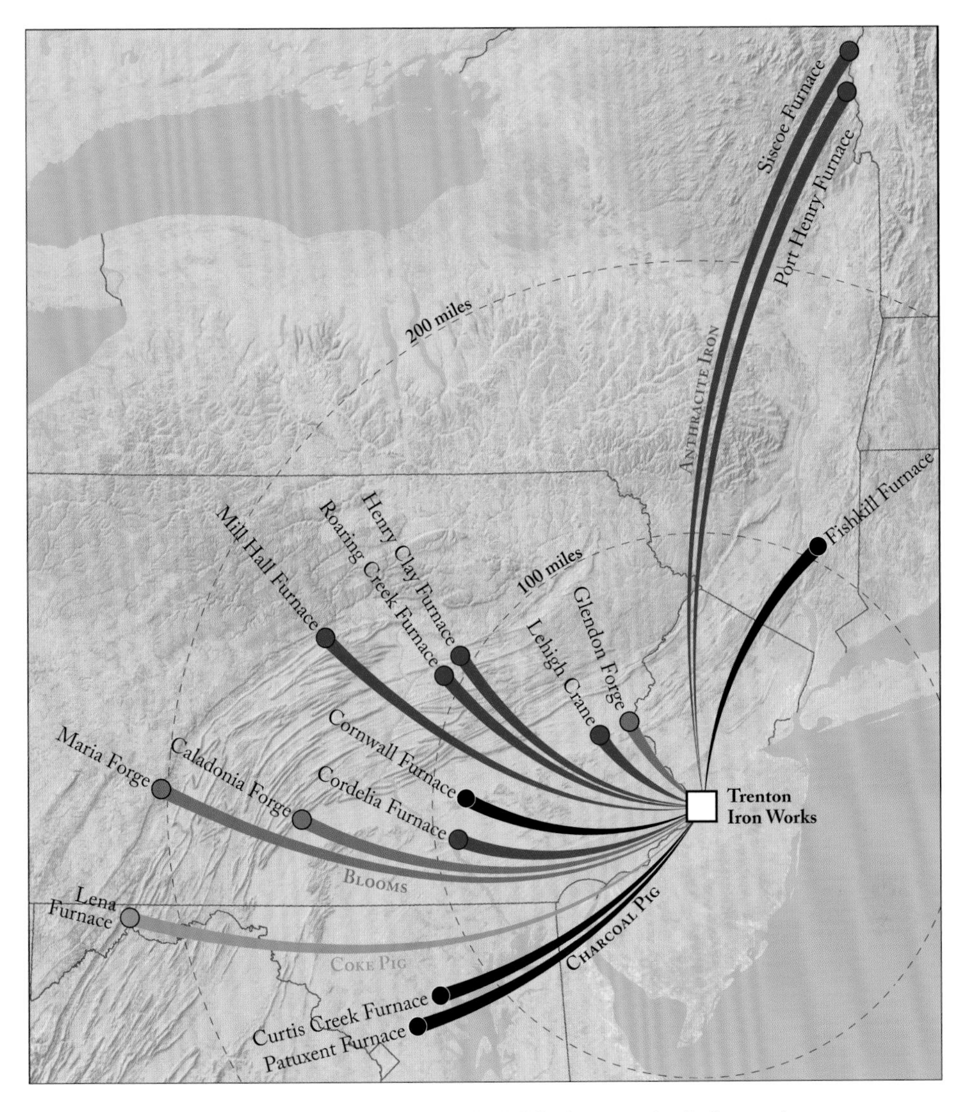

Figure 56. Iron suppliers to Trenton Iron Works, 1845–47. Before building its own anthracite furnace, the Trenton works purchased charcoal, anthracite, and coke pig iron and blooms from a wide region. The farthest suppliers, Siscoe and Port Henry Furnaces in Essex County, New York, were about 250 miles from Trenton, roughly the distance from Merthyr Tydfil to Glasgow as the crow flies.

There were some advantages to having many suppliers. One was having numerous types of iron in stock to blend as needed. Multiple suppliers gave mill managers greater assurance of an adequate supply in case some contractors failed to deliver. It could be difficult to keep track of each kind of iron, however, as Charles Hewitt discovered when a load of costly Norwegian iron was misplaced. "It is impossible for us in the office to do more than inform those in the mill of distinct lots," Charles explained to the ever-watchful Peter Cooper, "and it remains for them to keep them separate till the time of shipping." Loss of the Norwegian iron was, Charles noted, "a very unpleasant mistake," for it could take weeks or months to secure another lot of the same quality from the importing merchant, and by then Trenton's customer might have shifted its order to another mill.[65]

This incident points to one of the chronic problems of the antebellum iron industry: the lack of standardization in both supply and demand. The quality of crude iron that mills received was crucial to their providing finished goods that met customers' specifications. Imported Norwegian bar iron was exceptionally pure and reliable, as were Swedish bar iron and pig iron from a few American producers. As a rule, however, rolling mill workers and managers had to try various combinations of pig iron, blooms, bar, and sometimes ore to see how well the resulting mélange would perform in the end product. Then they had to hope that the lots in stock were adequate to fill a given order. This was almost impossible for the biggest rail contracts, such as the Baltimore and Ohio's tender for enough rails to lay thirty miles of track, which Benjamin Latrobe estimated would require 2,405 tons of rolled iron. Product specifications also varied constantly. In the early years at the Trenton Iron Works, every order for rails specified a slightly different design, which often required the mill to purchase a new set of rollers to produce a particular company's rails. After twenty years in the business, Charles Hewitt reckoned that the Trenton mill had accumulated 250 large roll trains for making rails and other products.[66]

One way iron companies could try to limit the unpredictability and unwanted variation in their iron supply was to expand operations to include smelting and mining. Backward integration may have been part of Peter Cooper's plan from the start, for his correspondence hints at activity related to blast furnace construction less than a year after mill construction began. The company could have built its anthracite furnace in Trenton, but it would have incurred heavy tolls shipping both ore and coal to the site. Instead, it purchased land in the fall of 1847 in Phillipsburg, New Jersey, fifty miles north of Trenton. The first of three furnaces at the new works was fifty feet tall and twenty feet wide at the boshes, the largest yet built in the United States. It burned anthracite to smelt local hematite ore and a richer magnetic ore from the Andover Mine in north-central New Jersey, which the company had also bought in 1847. The furnace was equipped with a "blast of great power," according to Peter Lesley, and it produced prodigious amounts of iron—7,980 tons in 1855. Only the Lehigh Crane and Thomas Iron Company furnaces produced more.[67]

The location of Cooper Furnace split the difference between resource distance and proximity to market, in this case the internal market of the Trenton rolling mill. Phillipsburg was roughly equidistant from the coal depot at Mauch Chunk and the Andover Mine, which lay six miles north of the Morris Canal. Beds of hematite ore that the company also owned were much closer to the furnace. From Phillipsburg, the company could ship its pig iron to Trenton by canal. By 1855, railroads connected Phillipsburg to Trenton and linked the Andover Mine to the Morris Canal.[68] The rolling mill was also at the nexus of transportation lines to New York City and Philadelphia, antebellum America's largest manufacturing centers and leading ports for immigrant ships.

Hewitt's biographer, Allan Nevins, credits him with finding the Andover Mine. According to Nevins, Hewitt's indefatigable drive kept him scouring the New Jersey woods until he rediscovered the mine, which had provided much of the iron ore used to make American cannons during the Revolutionary War.[69] The mine opening was probably already familiar to G. G. Palmer, a geologist who accompanied Hewitt on his forays during the summer of 1847 and who became the mine's first superintendent.[70] That Hewitt went into the woods with Palmer, however, shows his eagerness to tackle whatever the business required. Abram, Charles, and Edward Cooper seem to have relished learning every aspect of the iron industry from the inside, on the job, though their education at Columbia University had also given them useful skills. In addition to mathematics, Abram learned enough German to negotiate the furnace land purchase with a Pennsylvania Dutch farmer. "If I did not speak German," he wrote, "I might give it up in dispair [*sic*], as that is the only way to get at the old Dutchman's heart, which is in his pantaloon's pocket." Charles used his French to translate letters from an engineer whose water turbine design interested Peter Cooper.[71] Even in the antebellum era, the iron industry was international. More important, the Hewitts and their friend Edward Cooper had a practical turn of mind that made engrossing work of the details of ironmaking technology, marketing, and even double-entry bookkeeping.

They gained confidence as their hard work and managerial decisions guided the rolling mill through a number of early crises. As the second in command on the shop floor, Edward learned the importance of iron quality to rolling rails and the need to lay in ample stocks of iron for large orders. "We are using up the old stock rolled for the small rail," he wrote to his father in November 1846.

> This iron is now nearly exhausted. . . . I think we can finish two hundred tons of good rails of this pattern this week. But to make good rails *we must have some good charcoal iron*. The Christian [a brand of charcoal iron] that is here is nearly exhausted and we will not have enough good iron to work up the anthracite which we have on hand. It is absolutely necessary that you purchase all the good iron that you can get if we mean to make rails. . . . The matter of one or two dollars in price is no object. We must have iron of this quality and a good proportion of it.[72]

The same problem threatened the mill's wire production. Charles wrote that a purchasing trip to Philadelphia revealed "a great scarcity of good wire blooms, with scarcely any probability of the market being better stocked as the season is so far advanced that the canal boats are making their last trips."[73] He was referring to the shipping season. Winter's cold brought the close of navigation on many northern canals and rivers. In some years Trenton's southerly, low-lying location minimized problems with ice. In October 1845 Edward predicted that if the ice were "properly managed," it would not hinder the mill's waterpower. The first incident to prove him wrong came on New Year's Eve. As Charles reported,

> I arrived at the Mill today just in time to see the vast body of ice in the river move. It was a wonderful spectacle. The water rose six feet in a very few minutes, since which time it has been rising or falling alternately as the ice would cease its motion or rush on. The mill has been stopped since 12 O'clock of the New Year's eve, and will not probably run again this week. . . . [The] small water-wheel was broken, and the mill forced to stop.[74]

That winter, ice in the river and the power canal halted production repeatedly until the middle of February. Then came the potentially more damaging spells of high water during the spring freshets. On March 16, Charles again described the scene:

> The greatest freshet since 1841 is now subsiding. When I arrived at the mill last night I found the water at the highest point to which it has been washing against the timbers that project over the river at the rear of the mill. The lower end of the addition has settled somewhat from the partial falling away of a small portion of the slope wall. I sent for Cutter who got one of his men and they put up some props which I think have rendered it secure. I have been on duty all night part of the time at the mill and the remainder on the W. [water] Power just below the stop-gates. A break had taken place there and the water from the river was pouring into the canal. When I left there this morning at 4½ O'clock the water was running in with diminished force and I think (judging from the height of the water at the mill) that the river and the W. Power at the break must be now at about the same level, so that no further danger is to be apprehended.[75]

The tone of Charles's letter is typical. He, Edward, and Abram took charge. They portrayed themselves as courageous problem solvers—even heroes—when trouble threatened the mill.

Periodically, shortages of skilled workers posed one of the most difficult problems for the Trenton Iron Works, as for other rolling mills. When demand for rails and other rolled iron escalated in the mid-1840s, demand for puddlers and rollers rose commensurately, and with it their wages. No rolling mill could afford to lose many of these essential workers. Experienced men minimized the chance of losing time, money, and a firm's reputation to poor workmanship. After the prolonged spring freshets of 1847, Charles noted that some of the mill's idled artisans had left to seek work elsewhere. "The men [still at the mill] principally being new hands at the business, many having left from the long stoppage, the rails were not as handsome yesterday and last night as we could wish."[76] Puddling was the key bottleneck in rolling mill production, a fact of which puddlers were keenly aware. Their pivotal position in the flow of work through a rolling mill gave them greater bargaining power than men in any other metalworking occupation in the iron industry.

In June 1846, the puddlers and rollers at Trenton went on strike. The action may not have lasted long; a single, brief letter in the company correspondence notes that Peter Cooper came down from New York while Charles traveled to Manayunk, an iron and textile mill town outside Philadelphia, to find replacements.[77] By early 1847, the general scarcity of skilled rolling mill workers inspired serious discussion among mill owners about fixing wage rates to prevent puddlers from going on strike or migrating to mills that offered higher pay. Some owners proposed pooling funds to pay for a recruiting mission to Britain to alleviate the shortage.[78] Trenton's managers cautiously sided with them. Edward gave notice to the mill's puddlers that their pay would be reduced from $5 to $4 per ton of 2,240 pounds on March 1. "At the reduced rates," he argued, "puddlers will make at least $2.25 [per day] when they work or certainly $12.00 per week and I see no reason why a class of men such as they should make more than this when mechanics generally of much more intelligence and who have spent a much longer time in learning their trade cannot command an amount considerably less than this."[79] As the March 1 deadline approached, Charles observed to Peter Cooper that "the Puddlers here that is to say the oldest, seem to have the confidence that if they cannot do well at $4, after a fair trial, you will increase the pay." On March 1, he reported that the mill was operating normally. "The Puddlers have gone cheerfully to work, Edward having raised the price to $4½ the same as at Phoenixville. It would have been impossible to run at a lower rate."[80]

The willingness to meet workers halfway reflected the tone of labor-management relations at Trenton throughout its early years. Superintendent John Gustin and Charles Hewitt, in his capacity as company cashier, moderated customary practice and paid workers part of their week's wage when stoppages in the winter shut down the mill. Once the freshets of 1847 subsided and the mill was again able to roll rails, Charles noted that "Mr. Gustin is among the Puddlers with his coat off." If the company had rules of behavior, they do not seem to have been strictly enforced. "One of the Puddlers here who came from Haverstraw states that Evan Simmons was considered to be Peck's best Puddler," Charles told Peter Cooper. "He is not a tee-totaller but never gets intoxicated. There will be work for him at Puddling in about two weeks." When they could, Trenton's managers avoided hiring newly landed immigrants; they preferred employing local residents, "as their attachments bind them more strongly here, making the liability to lose them much less than if they had come from abroad." But the need for dozens of puddlers and equal numbers of men in other skilled positions necessitated hiring Welsh, English, and Irish workers.[81]

At the Lehigh Crane Iron Company, David Thomas was a hub of best practice and a technical innovator. Peter Cooper played a similar role at the Trenton Iron Works. His influence accounted for much of the company's initial success. It was his younger colleagues' ability to respond creatively to the problems that iron manufacturing constantly presented, however, that made the firm successful over decades of operation. The alacrity with which the Hewitt brothers and Edward Cooper learned their business and responded to problems makes a striking contrast to the rigidity and stubbornness of Farrandsville's Daniel Tyler. Like John Fritz, superintendent of the Cambria Iron Company, they learned as much or more on the shop floor as from formal education. In the early 1840s, when Fritz worked as chief mechanic at a Norristown, Pennsylvania, rolling mill,

> the mill men, such as puddlers, heaters, and rollers, were generally English and Welsh. . . . In the evenings between heats, while they were smoking their pipes, cutties as they generally called them, I would sit down on a charge of pig iron and listen to them describing their mills in England and Wales, and their method of working. In all of this I was greatly interested, and at the same time I gained their confidence, which is so essential in the management of workingmen. In all my experience I have ever sought to secure and retain the good will of the workmen. With confidence fully established between the workmen and their employer, strikes rarely occur.[82]

The most important factor in the lives of most rolling mill workers was the volatility of national and international markets for iron. Of the 220 rolling mills in Lesley's survey, only 19 (less than 10 percent) were abandoned between 1785 and 1858. This is a much smaller proportion than the more than 40 percent of blast furnaces Lesley listed as being abandoned by 1858. Of the mills that shut down, 9 closed either during the depression years of 1848–49, when inexpensive British rails inundated the American market, or in the depression of 1855–58. Several mills went back into production once prices lifted, such as the two Juniata rolling mills in Pittsburgh.[83] Rolling mills weathered economic slumps better than blast furnaces for several reasons. Their start-up costs were high enough that the price of entry into that branch of the industry tended to limit investors to individuals and corporations with enough capital to survive slow periods. Mills could reduce production without shutting down entirely, whereas blast furnaces had to remain in or out of production; furnace output could not be fine-tuned in response to market demand. The same constraint meant that a furnace company could significantly increase production only by building a new furnace stack, whereas rolling mills could grow incrementally by adding a few more puddling and heating furnaces and another roll train or two. Mill managers could also change their products to meet shifting demand by changing the mix of crude iron and substituting one set of rolls for another—to produce wire instead of rails, for example. A blast furnace could produce different grades of iron, but its basic product (pig iron) remained the same, and the quality of that product was chiefly determined by the raw materials the furnace was designed to smelt. While rolling mills' relative flexibility helped them survive sharp economic downturns, it could come at the cost of sharply cut wages, as in Pittsburgh in 1850. If that was not enough, mills fired large numbers of workers.

The managers of the Trenton Iron Works minimized these problems by responding quickly to changes in the iron market. The Trenton mill was originally built to manufacture "heavy" rails for long-distance trunk lines, using the rapidly growing demand for such rails in the mid-1840s as a springboard to large-scale production. The mill produced T rails in June 1846, just eight months after the country's first ones were manufactured at the Montour Rolling Mill in Danville, Pennsylvania.[84] When economic downswings halted railroad construction projects, the Trenton mill found new customers for its iron wire. One of the most important was John Augustus Roebling, who with Peter Cooper's help established a factory near Trenton that would later make the cables for Roebling's suspension bridges over the Niagara gorge, across the Ohio River at Cincinnati, and most famously the Brooklyn Bridge across the East River in New York City.[85]

The Trenton Iron Works further diversified its products by making railroad "chairs," the iron pieces that fastened railroad ties to the base of T rails.[86] The market for rails revived in 1850–53, but by 1854 the country's rail production again exceeded demand, and prices fell. This time lowered tariffs opened the door to British imports, notably Scottish coke pig iron and Welsh rails and bar iron. The Trenton managers saw an undeveloped market in structural iron beams, a new building method that Peter Cooper wanted to use to frame the large structure he had in mind for Cooper Union. Instead of using cast-iron beams, Cooper had the foreman at the Trenton mill experiment with adapting rolling technology to produce hollow, flanged beams that could be bolted together as lightweight, relatively inexpensive columns to support fireproof masonry construction. In 1853 the mill rolled beam sections twenty feet long and seven inches deep. Larger beams soon followed. The success of this venture attracted national interest (fig. 57). Over the next few years, Trenton beams were used in a number of monumental government buildings, such as the General Post Office (also known as the Tariff Commission Building) in Washington, DC, and the US Custom House in New Orleans.[87]

Figure 57. Design for structural iron for a market in New Orleans, 1866. Courtesy of the Library of Congress. Bennett and Lurges Foundry to Cooper, Hewitt and Co., September 22, 1866, Cooper, Hewitt Papers. The rolled structural iron beams pioneered by the Trenton Iron Works contributed to a new era of prefabricated metal framing for large buildings. This design sketch for roof supports for a marketplace suggests the loft and visual lightness of iron construction in the late nineteenth century.

In the historiography of the iron and steel industries, the Trenton Iron Company and Lehigh Crane Iron Company are among the handful of firms singled out as harbingers of large-scale modern industry. Allan Nevins, whose biography of Abram S. Hewitt is also an excellent history of the Trenton Iron Company, argued that Hewitt's vision of integrated iron production inevitably swept aside the "tiny ironworks everywhere, but particularly in Pennsylvania, with poor equipment, and an uneconomic force of men." Some other historians of the antebellum iron industry have shared Nevins's view. "A handful of talented manufacturers perceived the competitive advantages of integration and were flexible enough to adopt the new structure before the Civil War," wrote business historian Harold C. Livesay. "Successful rail manufacture required integrated mass production, and few antebellum furnace or rolling mill proprietors had resources or imagination enough to adapt to the new conditions."[88]

I have argued in this chapter that more was required than capital and imagination. The geographic circumstances of production, both physical and social, strongly conditioned the choices available to American industrialists and the outcomes of those choices. The trajectory of every US iron region's development was influenced by its resource endowments and topography, potential for transportation improvements, proximity to centers of population, and access to skilled labor. New technologies were most quickly and effectively put into operation by firms whose workforce included skilled, experienced men in most if not all key positions. Chain migration of skilled Welsh ironworkers to Lehigh Crane followed links already established between industrial South Wales and industrial towns across Pennsylvania. Trenton was within hailing distance of two of the country's greatest immigrant ports, New York and Philadelphia, while both cities' machine shops and foundries provided pools of native and immigrant skilled metalworkers the ironworks could draw on. The geographic advantages of this chapter's two case studies were actualized through exceptionally intact or rapidly constituted networks of workers whose skills and knowledge of technical processes were supported by appreciative owner-managers. The latter's extensive industrial experience, or their willingness to make a study of the processes they supervised, promoted the successful implementation of smelting and rolling technologies.

As the antebellum period progressed, the industry expanded in the wake of westward settlement but also became increasingly concentrated in a few locations where geographical and social conditions favored large-scale iron production. Pittsburgh was one of those places. By 1860 its rolling mills included several of the largest works in the country, such as Jones and Laughlin's American Iron Works, and the average of 185 men employed in western Pennsylvania rolling mills was more than double the average of 86 east of the Alleghenies.[89] Conditions in the Mid-Atlantic region were more favorable to the replication of the entire British model, though more spatially dispersed than in Britain because of the geography of resources.

On the eve of the Civil War, the volume and diversity of iron manufacturing in the Mid-Atlantic region and in the Northeast more generally stood in stark contrast to the state of the industry in the South. Until 1858, all blast furnaces in the South made charcoal iron, though some used hot-blast technology and matched the productivity of northern charcoal furnaces. Because of the much greater output of anthracite iron furnaces and the concentration of rolling mills in the Mid-Atlantic region, however, by 1858 the North produced nearly 85 percent of the nation's crude iron and 87 percent of its rolled iron. A few of the South's established rolling mills were comparable in size to those in the Northeast, but most southern rolling mills were smaller. Only the Gate City Rolling Mill in Atlanta, built in 1858 to roll rails, rivaled the capacity of the rail mills in Trenton, Johnstown, Danville, Scranton, and Troy and a new rail mill in Indianapolis. Antebellum southern railroads bought most of their rails from northern mills, from northern merchants, or from Great Britain.[90] Decades of heavy investment in plantation agriculture had given the South one of the world's wealthiest economies and a culturally sophisticated planter class, but the lack of investment in heavy industry meant that few members of the southern elite were familiar with modern methods of iron manufacturing.[91] Economic specialization in the late antebellum period gave the industrial North a huge initial advantage over the agricultural South when the nation went to war in 1861.

The iron was wanted more than anything else but men.

JOHN B. JONES, August 11, 1863

Iron for the Civil War 5

The American Civil War has been called the first industrial war because of the scale of the conflict, the importance of railroads, and the arms race that produced ever-larger cannons, the Gatling gun, and ironclad battleships that became emblematic of a new age of naval warfare. As a rapidly growing industrial region, the North entered the war with an enormous advantage in every branch of manufacturing, including production of heavy ordnance and small arms (fig. 58). Most American-made pistols and rifles were produced in the Northeast at the Colt, Sharps, Whitney, and Remington factories and the federal armory in Springfield, Massachusetts. Three of the country's four largest cannon foundries were in Union territory: Cyrus Alger's works in South Boston, West Point Foundry on the lower Hudson River, and Fort Pitt Foundry in Pittsburgh. The Confederacy's ordnance manufacturing was concentrated in Richmond, whose small-arms factories and cannon foundries included the Virginia Armory, Tredegar Foundry, and nearby Bellona Iron Works. A factory in Palmetto, South Carolina, produced good rifles. The Virginia state assembly approved the seizure of the federal armory at Harpers Ferry and the removal of its machinery to the Virginia Armory

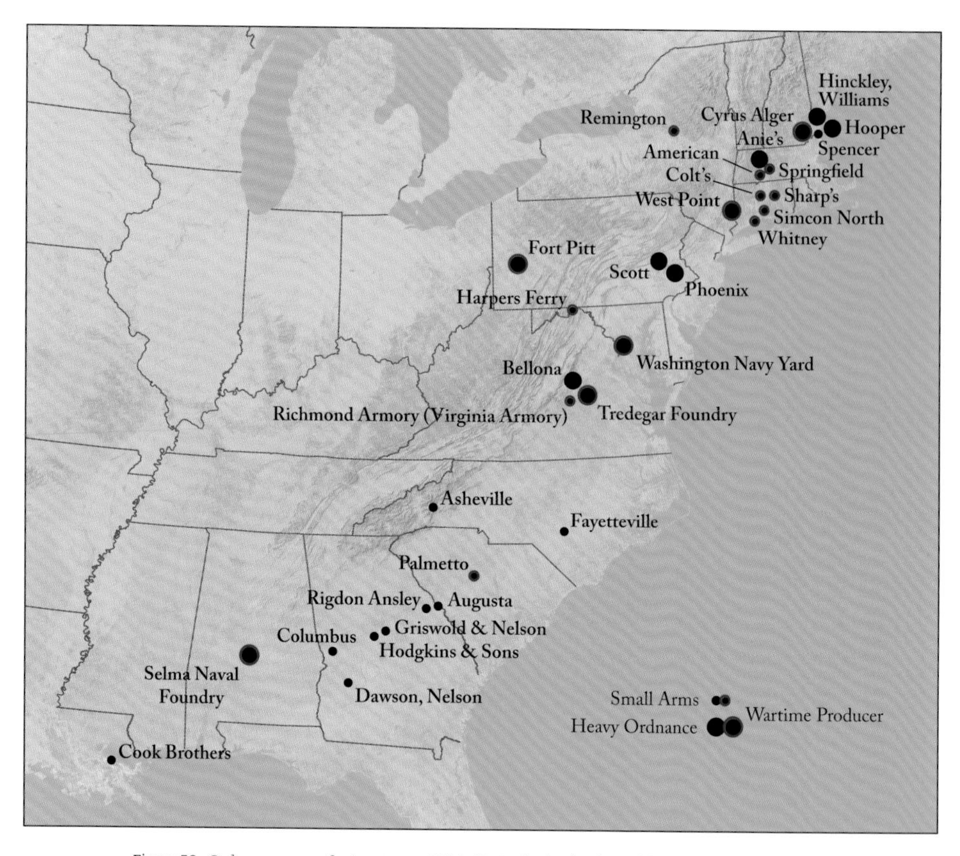

Figure 58. Ordnance manufacturers, ca. 1861–65. At the beginning of the war, the Confederacy possessed only a few heavy ordnance manufacturing facilities. The Union held the lion's share of cannon producers and small arms manufacturers, including the Alger, Fort Pitt, and West Point Foundries, the country's leading private pistol and rifle factories, and the federal small arms factory at Springfield Armory.

just days after the Confederate attack on Fort Sumter. Other than these facilities, when the war began the South had no established, large-scale weapons manufacturers and precious little of the equipment needed to make arms.[1]

According to historian James McPherson, the Confederacy's efforts to meet wartime demand for ordnance were remarkably successful. "Although often less well armed than their enemies," he writes, "Confederate soldiers did not suffer from ordnance shortages after 1862." McPherson credits Confederate chief of ordnance Josiah Gorgas, "a genius at organization and improvisation," for scrapping together necessary supplies from unlikely sources and creating a domestic arms industry almost from scratch. Gorgas declared in 1864, "Where three years ago we were not making a gun, a pistol nor a sabre, no shot nor shell (except at the Tredegar Works)—[not] a pound of powder—we now make all these in quantities to meet the demands of our large armies."[2]

While Gorgas sounds confident in this declaration, his correspondence on behalf of the Ordnance Bureau was not so sure. He noted, for example, that small arms were always in short supply; by September 1864, most were being imported from Europe.[3] The arsenals, nitre mills, and laboratories that the Bureau established mainly produced ammunition and powder, not weapons. The Confederacy gained at most three new rolling mills and three new heavy foundries during the war. That so few facilities were built is not surprising, given that it took up to a year to erect a blast furnace and longer to build and equip a rolling mill, even when capital, machinery, and labor were readily available. From producers' perspective, the Confederacy struggled constantly to meet wartime demand for heavy guns, iron plate, and the crude iron needed for other war materiel.

Industrial warfare in fact strained the capacity of the iron industry in the North and the South. Demand for iron increased rapidly as the war engrossed more men, animals, and supplies. Quartermasters needed more horseshoes, wagon rims, and camp equipment. Union and Confederate ordnance bureaus urged foundries to increase the output of heavy weapons while they also imposed stricter standards of inspection to improve safety. Manufacturers competed for iron supplies, particularly for high-quality charcoal pig iron, blooms, and wrought iron. Small-arms manufacturers and foundries needed the best grades of iron to meet government requirements for rifle barrels and cannons, while wrought charcoal iron made the toughest armor plates for Union "monitors" and the Confederacy's iron-sided ships.[4] Railroads demanded more wrought iron as both governments and private companies hastened to extend military supply routes and scrambled to repair damaged track and equipment.[5] These were the direct demands for iron related to the machinery of war. Indirect demand also increased as textile factories and other industries providing war materiel needed more steam engines, machine tools, and iron implements to fill government procurement contracts.

Wartime demand drove up the price of iron to double, triple, even quadruple what it had been during the prewar depression. Iron companies reaped the benefits of inflation in windfall profits. Infusions of foreign capital helped build railroads and revived some dormant ironworks. At the same time, inflation put government contractors in a bind when payment on contracts failed to cover their operating expenses. The Confederate government tried to control costs and regulate iron supplies by essentially nationalizing most of the Southern iron industry. Federal contracts and direct investment funded the expansion and construction of Northern ironworks, and the government claimed all output from many works.[6] The growth of iron manufacturing created acute scarcities of raw materials, refined iron, and skilled labor. By autumn 1863, even the Cambria Iron Works in Johnstown, Pennsylvania, was sending labor recruiters to Castle Garden immigrant station in New York to engage miners fresh off the boat from the

British Isles, with plans to recruit directly in Wales and Scotland if landed immigrant labor proved insufficient. As Confederate war clerk John B. Jones noted in his diary on August 11, 1863, "iron was wanted more than anything else but men."[7]

The scarcity of iron artisans was particularly vexing for the Confederate industry. The old problem that white skilled workers would not "stick" now had more serious implications, for discontented workers might defect to the enemy. Southern ironworks managers pleaded with conscription officers to leave vital workers where they would be most valuable to the cause, often to no avail. In the North, first patriotism and then the draft swept ironworkers into the army. Those who remained at work were well aware of the heightened value of their skills. The Civil War saw skilled workers reach the apogee of their power in labor relations. At the same time, the extraordinary pressures of wartime production and the opportunity perhaps to make a fortune spurred some owner-managers to experiment more aggressively with laborsaving technologies such as the Bessemer process. Federal ordnance inspectors' drive to improve safety and impose government standards also threatened the shop floor authority of skilled workers.

These contradictory forces played out differently in each of the country's main iron-producing regions. I begin with the general problems Gorgas faced as he struggled to arm the Confederacy and secure industrial labor. Those issues are illustrated in detail through the example of the Shelby Iron Works in central Alabama, which grew from a small charcoal furnace to one of the Confederacy's largest smelting and rolling operations. I then turn to the tensions between escalating demand, production standards, and labor-management relations at Northern ironworks.

Iron for the South

In 1860 the South had fewer and generally smaller ironworks than the North (table 5). Producers in Maryland, Virginia, Kentucky, Tennessee, and Missouri, along with companies along the north bank of the Ohio River, met much of the region's need for agricultural implements and machinery, stoves, and other domestic wares. Iron manufacturers in New England also supplied these items to southern merchants, notably the metal parts for cotton gins. Nail mills in Wheeling supplied the building needs of the upper South, the Mississippi Valley, and newly settled areas in the westward-moving Cotton Belt. Most of the iron required for southern agriculture was tellingly called "small iron," which included hand tools, nails and spikes, and the small lots of bar iron and blooms that southern bloomeries and forges sold to blacksmiths for local use. War required iron smelting and manufacturing on a very different scale. Each Rodman or Dahlgren cannon took two to five tons of iron. A single plate of ship's armor could weigh

Table 5. Southern rolling mills and blast furnaces, circa 1858

	Number active	Annual production (number of works reported)		Mean capacity of works reported (tons)
Blast furnaces				
Alabama	3	1,326	(3 furnaces)	442
Georgia	7	2,508	(6)	418
Kentucky	30	38,298	(26)	1,473
Maryland	21	30,602	(18)	1,700
Missouri[a]	4	3,213	(2)	1,607
North Carolina	4	675	(3)	225
South Carolina	3	1,156	(3)	385
Tennessee	35	40,225	(35)	1,149
Virginia	36	17,960	(28)	641
Southern states	143	135,963	(124)	1,096
Northern states	306	485,196	(258)	1,881
Rolling mills				
Georgia	2	12,900	(2 mills)	6,450
Kentucky	8	19,989	(6)	3,332
Maryland	11	11,738	(6)	1,956
Missouri	5	7,053	(4)	1,763
North Carolina	1	215	(1)	215
South Carolina	2	810	(2)	405
Tennessee	3	2,780	(3)	927
Virginia	11	21,191	(10)	2,119
Southern states	43	76,676	(34)	2,255
Northern states	159	366,119	(131)	2,795

Source: J. Peter Lesley, *The Iron Manufacturer's Guide to the Furnaces, Forges and Rolling Mills of the United States* (New York: John Wiley, 1859). Figures for each state and region exclude ironworks listed by Lesley as having been abandoned by 1858.

[a] Missouri's blast furnaces were in territory under Confederate control at the beginning of the Civil War (see fig. 63).

from five to twenty-five tons. Before 1861, not even Tredegar's large rolling mills were adequate to produce plate for ironclads. The Confederacy's only rail mills were Tredegar, Etowah, and Atlanta's Gate City Rolling Mill, which began production in 1858.[8]

As table 5 shows, the South had roughly one-quarter the productive capacity of northern ironworks on the eve of the war.[9] All the large coal-burning blast furnaces were in New England or the Mid-Atlantic region, as were the great majority of rolling mills. The only coal-fired furnaces in the South—Bluff Furnace in Chattanooga, Tennessee, and three furnaces in Virginia—were all small-scale producers.[10] All other southern furnaces were single-stack charcoal operations. War soon severed southern access to the extensive iron region along the north bank of the Ohio River and the rolling mills and foundries in Philadelphia, Baltimore, and St. Louis. The sudden deprivation of supply compelled Gorgas to find new sources of iron and to site any new iron manufacturing facilities in areas that would be relatively safe from Union attack. Existing blast furnaces in the Valley of Virginia—crucial suppliers for the Confederacy's largest ironworks in Richmond—became essential strategic assets, one of several reasons the Confederate Army concentrated its forces in that industrial-agricultural corridor.

Efforts to increase the Confederate industry's productive capacity were constantly undermined by the abiding structural problem of labor scarcity. The geography of industrial capacity and ironmaking technology before the war had important consequences for regional labor markets during the war.[11] Particular production processes and machines required particular skills and work relations. Although men anywhere could learn to puddle, roll, and heat iron, training them took time. Regions without large rolling mills lacked the artisans needed to operate the mills effectively. The smaller scale and generally less mechanized technologies at antebellum southern ironworks had produced a regional workforce of isolated artisans, mostly slaves, at scattered furnaces, forges, and bloomeries. They included highly skilled hammermen, forge carpenters, refiners, molders, and founders,[12] but very few puddlers, rollers, heaters, or furnace men who understood coal-fired smelting. While northern ironworks routinely sought out and hired European labor at all levels of skill, few southern firms did so before the war. The Etowah Iron Works in northwest Georgia was a rare exception. When they added a rolling mill to the forge and blast furnace complex in 1848–49, Etowah's managers hired a mix of skilled workers from England, Wales, Scotland, Germany, Massachusetts, Pennsylvania, and southern states.[13] The workforce in Stewart County, Tennessee, was more typical. In 1850, employees at a number of the county's ironworks included only two northerners and one European, a Scottish mechanic. All other white workers were natives of Virginia or Tennessee. The overwhelming majority of ironworkers in Stewart County were slaves who were owned by iron companies or iron company managers or whose labor

Figure 59. *Rees Davies, Mechanic, Hirwaun*, ca. 1835, by W. P. Chapman. Private collection. Itinerant artist W. P. Chapman painted a series of portraits of ironworkers in Merthyr Tydfil. The man depicted here may be the Rhys Davies who emigrated from Wales to France in the mid-1820s and then to Virginia in 1835.

was rented from local farmers.[14] The lack of immigrants at most rural southern ironworks contrasted sharply with the cosmopolitan laboring population in the South's largest cities before the war.[15]

The tension between managers' ideological commitment to industrial slavery and their need to secure skilled workers was greatest at the South's most advanced ironworks. The Tredegar Iron Works initially hired many skilled iron artisans, including Welsh immigrants such as Rhys Davies (fig. 59), who supervised construction of Tredegar's new rolling mill in 1838. Davies was hired to guarantee that the works were built and run to the standard of ironworks in Tredegar, Wales, an iron town near Merthyr Tydfil.[16] At the same time, the Virginia company's managers were determined to lessen their dependence on white workers and minimize the potential disruptiveness of independent-minded immigrant labor. Between 1842 and 1847, Tredegar's owner-manager Joseph Reid Anderson attempted to replace white puddlers and rollers with their slave apprentices. His efforts met stubborn resistance that culminated in a strike led by Welsh and American workers. By 1850 Anderson was obliged once again to hire skilled northerners and European immigrants so that his firm could fill orders for rails, locomotives, and heavy ordnance.[17]

Throughout this difficult period, Anderson continued to advocate industrial slavery. In 1850 he claimed to have trained thirty-five slaves to be puddlers, heaters, and rollers. In 1852 he magnanimously offered to help train slaves for the rolling mill at Etowah. In 1859, Anderson urged the owner of the Shelby Iron Works in Alabama to employ slaves rather than white men. "I have used both white and slave labor many years in a mill," he wrote. "A city you know is bad for slaves but in the Country, I would use only negroes in a rolling mill besides the manager. The great advantage is that you can rely on slave labor, whilst you will find it hard to get good white workmen to come to you and then they will quit when you want them most."[18]

Historian Kathleen Bruce, taking Anderson at his word, credited him with "revolutionizing" labor in the Virginia iron industry by employing slaves in top skilled positions. A more recent study by Gregg D. Kimball argues a very different interpretation—that the strike at Tredegar "had an immediate and devastating effect on rolling-mill production," and that Anderson essentially gave up his vision for industrial slavery out of the practical necessity for highly skilled labor at the ironworks. In 1850 he considered selling all his slave artisans because of the difficulty of having them work beside whites on the shop floor. Kimball notes that Tredegar continued to rely on immigrant and northern workers even for apprentices. "The white sons of Virginia," he writes, "seem to have had little taste for the hard work of forging iron."[19] Anderson's rhetoric also may have inhibited the diffusion of ironmaking technologies in the South more generally by supporting cultural barriers against the migration of skilled workers and encouraging other ironmasters to use slaves in skilled positions. His vision of a tractable, rooted, artisanal workforce was increasingly at odds with the realities of heavy industry.[20] Tredegar's need for experienced artisans compelled Anderson to employ more Europeans, not fewer, as the works expanded. During the Civil War, European immigrants accounted for more of the skilled workforce at Tredegar than at some Northern ironworks. In 1864, 47 percent of rolling mill artisans, 43 percent of machinists, 33 percent of those in engineering and design occupations, and 29 percent of skilled foundry and forge workers at Tredegar were foreign born.[21] The high proportion of immigrants may have been partly due to the influence of A. G. Osterbind, the German superintendent of Tredegar's rolling mill, and Englishman Peter S. Derbyshire, who supervised ordnance production at the Tredegar foundry.

The trade-off that Anderson never acknowledged, though his own hiring practices suggest he understood it perfectly, was that the geographic immobility that made slaves a reliable workforce also limited their skills and technical knowledge. British metalworkers in the nineteenth century, like engineers and other industrial artisans, traveled extensively both to keep themselves employed and to move up the occupational ladder.[22] On the American iron frontier, roving white artisans learned to adapt techniques and machinery to suit local conditions, becoming expert problem solvers and innovators in the process. Although skilled slaves became masters of their trades, they had fewer opportunities to develop the range of skills that became the hallmark of "tramping" white artisans. Slaves' knowledge was further limited by the predominance of older technologies at southern ironworks. When the Confederacy suddenly needed more men to cast cannons and roll iron plates for warships, slaves could not quickly fill the breach, and experienced white workers were hard to come by. These limitations on domestic labor supply set the stage for a crisis in Confederate iron production during the Civil War.

The Confederate Labor Crisis

Confederate president Jefferson Davis foresaw the problem of skilled labor shortages in the spring of 1862, when he told the Confederate Congress that "the want of mechanics" to manufacture small arms "does not permit us to hope for such extensive results as would satisfy existing necessities."[23] Secretary of the navy Stephen Mallory warned the Congress on November 30, 1863, that ironworks at Richmond, Charlotte, Atlanta, and Selma would not be able to supply all the large guns needed unless "the proper amount of skilled labor can be concentrated." In April 1864 he wrote that "the want of skilled labor is severely felt" in ordnance production. In November 1864, Mallory called the lack of skilled labor a "serious evil."[24] The general scarcity of labor hampered the construction of the new munitions laboratory in Macon, Georgia. Once the facility was built, the lack of skilled labor severely limited production.[25] Gorgas told secretary of war James A. Seddon in October 1864, "The limited number of mechanics left to the Confederacy, must be retained; or the best interests of the Government will be hazarded. Already large amounts of machinery are lying idle, in all parts of the country, for want of workmen to operate them; while three years since, the want felt was of machinery."[26]

Iron manufacturers and Ordnance Bureau officials tried to retain skilled workers by keeping wages high. At the Shelby Iron Works in early 1862, white artisans were paid from $2 to $5 a day, common laborers from 75¢ to $1. By the end of the war, wage rates at Shelby had quadrupled and more, with the furnace founder receiving $17 a day, the mine supervisor $14, and other furnace and rolling mill workers $5 to $12 a day. Puddlers, rollers, roughers, and heaters, who were paid by the tonnage they produced, received weekly pay ranging from $84 to $122 ($14 to $20 a day).[27] According to Mallory's estimates, labor claimed up to 36 percent of naval ordnance production costs in 1863–64. Tredegar's wages advanced more slowly, which cost the company some workers until Gorgas imposed controls in an effort to equalize wage rates at government ordnance installations.[28]

High wages, however, were no defense against military conscription or workers' efforts to evade military duty. By the best estimates, which admittedly are anecdotal and incomplete, Confederate heavy ordnance producers lost from one-quarter to one-half of their workforce to conscription and desertion by the end of 1864.[29] Some of the most eloquent protests against the conscription of ironworkers were written by John M. Brooke, commander of the Confederate Navy's Office of Ordnance and Hydrography in Richmond and a close adviser to Mallory. Brooke reported time and again that the lack of properly trained workmen at ordnance factories resulted in delays and second-rate products. Confederate warships continued to use old-fashioned smoothbore cannons that did not hurl projectiles with enough force to penetrate the iron cladding on Union ships. Their own cladding, Brooke reported, gave way to the Union's more powerful rifled cannons. Contrary to James McPherson's suggestion that being "less well armed" did not significantly hamper Confederate forces, Brooke argued that inferior shells put the Confederate Navy at a serious disadvantage. The Confederates knew how to make the new shells, but their ordnance factories lacked the artisans required to produce them.[30]

Brooke explained in 1864 that the problem was not an absolute lack of skilled ironworkers, but the difficulty of protecting them from being conscripted into military service, a problem that grew more acute as the war dragged on.

> There are in the Southern States more than a sufficient number of mechanics to work these establishments to their full capacity and to supply all the heavy ordnance required to arm the iron clads and other vessels completed . . . and to furnish guns for the defence of our ports against which the iron clads of the enemy cannot stand. But these men have been swept into the Army en masse and their services can only be obtained by special and individual detail, months are generally occupied in the process and so rarely are applications granted that the services of not more than one in ten are secured. Constant effort is being made to supply the deficiency of labor, but with slight results. . . . a considerable number of boys are employed, who are gradually acquiring skill but their services will be more valuable hereafter than they are at present.

Brooke estimated that "two hundred machinists, blacksmiths, pattern makers, and moulders would be sufficient to accomplish all that is desired." He knew one hundred of them by name. "And it may be safely assumed," he concluded, "that the service, which these 200 men can render in the field[,] is incomparably less than that which they could render" using their skills to make heavy weapons.[31] Brooke was particularly frustrated that efforts to build and run a new, well-equipped cannon foundry and rolling mill at Selma were repeatedly stymied by the removal of skilled hands.[32] Confederate naval historian Raimondo Luraghi contends that the lack of munitions artisans posed a more serious threat to the Southern cause than did the shortage of sailors.[33]

While the navy was generally the least well-equipped branch of the Confederate military, the labor problems that plagued naval ordnance facilities cropped up at most of the South's ordnance factories. Throughout the war, Joseph Reid Anderson peppered military officials with pleas and demands that individual artisans be detailed to the Tredegar Iron Works so that the company could fulfill government contracts for field cannons and other ordnance. Few of his requests were granted.[34] Anderson's most successful move was creating the home guard unit known as the "Tredegar Battalion" in June 1861. The battalion sequestered up to three hundred workers from battlefield duty by confining them to Tredegar under the command of (now) General Anderson. Other ordnance manufacturers attempted to copy the Tredegar model, though Anderson's forceful personality and his local influence in Richmond made the home guard strategy particularly effective for his company.[35] Elsewhere, military recruiting officers and government officials routinely violated Confederate laws that exempted skilled ordnance workers from field duty. Even Jefferson Davis was unable to enforce the exemptions meant to safeguard arms production.[36] In 1864 Gorgas proposed legislation that would attach all ordnance workers to plant-based military units. This, too, was only partly successful. As we shall see, ironworks' ability to retain crucial labor ultimately depended on the determination of works managers and a works' exposure to the pressures of the war.[37]

Rumors circulated that General Ulysses S. Grant was offering Southern workers a bounty of $600 to switch their allegiance to the Union, where ironworkers reportedly were truly exempt from military service. (In fact, they were not.)[38] Workers who were conscripted and taken to Confederate training camps would be more likely to defect, reasoned Shelby Iron Company partner James W. Lapsley. He argued that not only should skilled workers be put to their best use, they should be given preferential treatment to guarantee their loyalty to Southern industry.[39] Such arguments did not sit well with conscription officers. Confederate records hint that industrial workers may have been targeted for conscription by officers who resented the special treatment they received. Confederate naval officers, for example, complained bitterly that ordnance workers' pay kept pace with wartime inflation while officers' fixed wages reduced them and their families to "the point of destitution, or of charitable dependence."[40] It is also likely that Confederate officers and their commanders did not understand the potential consequences of removing skilled ordnance workers from their posts. Had Southern gentlemen been more familiar with heavy industry, Lapsley, Anderson, and other industrialists might not have had to expend so much effort to secure the labor they required.

The Shelby Iron Company

The managerial approach taken by Lapsley and his partners at the Shelby Iron Company mirrored Anderson's practice at Tredegar without the veneer of slave ideology. Shelby's owners made little money during the war, but the division of labor they established and generally maintained proved well suited to running an industrial operation in wartime. They hired an experienced Welsh immigrant to supervise daily operations and recruit a skilled workforce to operate the rolling mill. Slaves provided all manual labor and occupied lesser skilled positions. As the war progressed, the owners increasingly had to serve as intermediaries between workers and Confederate officials to protect the workforce.

Shelby was among the most self-sufficient of Confederate iron companies. Its six thousand acres included deposits of limestone and excellent iron ore. The company's woodlands and bituminous coal deposits offered unlimited supplies of fuel for smelting and steam power. Springs in the area provided a year-round water supply. The furnace stood six miles from a railroad junction at Columbiana, which connected Shelby to manufacturers and ordnance depots in Selma, Montgomery, Atlanta, Mobile, and New Orleans.[41] Horace Ware, co-owner of the furnace, had been making pig iron at Shelby since the late 1840s. He sold most of it to Daniel Pratt's cotton gin factories in Autauga County, Alabama.[42] When Benjamin Smith Lyman visited the state in 1857, he reported that Shelby was Alabama's only hot-blast charcoal furnace. "The works are well situated," Benjamin wrote to his uncle, Peter Lesley, "and the ore is rich and an immense quantity of it is very convenient to the furnace. The owner, Horace Ware, has not enough capital to carry on the business as he wishes to, and he desires to sell out either all or a part of his property, and he promised to give me a thousand dollars if I will find him a purchaser. He asks for the whole $80,000 cash."[43]

Not long after Lyman's visit, a fire destroyed the wooden superstructure and sheds around the furnace. Between 1858 and 1861, Ware rebuilt and enlarged the works, adding a midsize rolling mill. He hired British workers from the Cumberland Iron Works in Stewart County, Tennessee, and elsewhere.[44] Ware began to manufacture "cotton ties," rolled iron wire that planters used to cinch bags of cotton when rope became scarce during the Union blockade. Shelby was the only integrated ironworks in Alabama, with the only rolling mill. When the company's board of directors approved a three-year contract to produce iron for the Confederate Ordnance Bureau in spring 1862, the advantages of Shelby's geographic location were clear. It possessed the best resources for making iron of any works in the Deep South, had among the best transportation connections by virtue of its proximity to existing rail lines and the Coosa River, and was closer to the western theater of the war than any large works deep in Confederate territory. The existing facilities could easily fulfill the company's contractual obligation to deliver one thousand tons of cold-blast, foundry-grade charcoal pig iron for making cannons. Producing eleven thousand tons of wrought iron in the form of bars, bolts, rods, and "plates for covering ships," however, was a very tall order. To meet this production quota, Shelby's owners planned to double or triple the rolling mill's capacity by adding up to eight new puddling furnaces, five heating furnaces, and a rotary squeezer. Given that the company's one charcoal furnace produced about one thousand tons a year before the war, its commitment to produce so much rolled iron meant finding new supplies of crude iron, whether by building another blast furnace on site or by shipping in pig iron and blooms from elsewhere.[45]

Company records and other manuscript evidence provide somewhat contradictory impressions of how fit the ironworks—or the iron company—was to become a major Confederate contractor. Robert Hall, a brick mason who superintended part of the expansion in 1862, described the state of Shelby's machinery at the beginning of the renovation as "all old, and in a shocking condition." The dilapidated blast furnace was badly cracked, and workers' houses were "rotting." Others involved in the expansion were less critical: the furnace stack, they recalled later, simply needed repairs; the company soon built many new houses to hold the growing workforce; and the machinery, though secondhand, worked well enough.[46] The company's attorney claimed in 1862 that the investors were "gentlemen, merchants, planters, and professional men" who "were about to embark in a business to which they were strangers." His description was a somewhat disingenuous effort to persuade Ordnance Bureau officials to more favorable terms. In fact, Shelby's board of directors included two of the most experienced ironmasters in the South. Horace Ware had run the Shelby Furnace for fourteen years, and Moses Stroup, an "expert furnaceman," had been a leading figure in the southern iron industry for forty years or more, having built and managed a rolling mill and several blast furnaces in various southern states.[47]

In one crucial respect, however, the company was genuinely ill-prepared. Like virtually all Southern ironworks that entered into Confederate contracts, Shelby was undercapitalized. Lyman's comment in 1857 that Ware "has not enough capital to carry on the business as he wishes to" was indicative of the general problem of underinvestment in heavy industry in the South. Most wealthy southerners were planters who sank their surplus capital into agricultural improvements and slaves. According to the 1860 census, only 6 percent of wealthy slaveowners in the South (those who owned twenty slaves or more) invested in manufacturing.[48] Horace Ware had only one partner until war suddenly made investing in iron manufacturing appealing. Even then, when seven new partners joined Ware in

1862, their initial investment totaled only $50,000, less than the capital stock of much smaller charcoal iron furnaces.[49] The Confederate government advanced $150,000 in 1862 and another $200,000 in February or March 1863, yet Shelby was forever short of funds. Although Lapsley warned his partners to expect wartime inflation, no one could have anticipated how swiftly Confederate currency would depreciate or how difficult it would become to supply the company's voracious need for raw materials, provisions of all kinds, and labor.[50]

The company had a sizable workforce in place when it entered into its government contract. Ware's former partner Samuel Clabaugh recalled that in spring 1862, the company had "from one hundred and fifty to two hundred operatives, mostly negroes." The company's first step toward increasing its workforce was to hire Giles Edwards in March or April 1862 (fig. 60). Edwards was brought in to recruit skilled workers and supervise the rolling mill expansion. He was born in Merthyr Tydfil, Wales, in 1824, the son of a collier. He grew up in the shadow of the Cyfarthfa and Dowlais ironworks and may have served an apprenticeship at one of them. When Edwards emigrated to America with his sister and their

Figure 60. Giles Edwards, ca. 1870. Photograph courtesy of the Alabama Historic Ironworks Commission.

widowed father in 1842, he was already an accomplished patternmaker and soon got work in Carbondale, Pennsylvania.[51] He later moved to Scranton and then to Catasauqua, where he lived near ironmaster David Thomas and worked as a patternmaker in the Lehigh Crane Iron Works. A few years later he assisted in the construction of the Thomas Iron Works in Hokendauqua.[52]

By 1855, Edwards had moved again to become a blast furnace supervisor at the Cambria Iron Works in Johnstown, Pennsylvania. It was there that he met John Fritz, a leading ironmaster who was then developing his first major innovation, the three-high rail mill. Fritz apparently befriended Giles Edwards and helped him through a couple of rough patches. In 1858 Fritz recommended him to supervise construction of the Bethlehem Iron Company's rolling mill at Bethlehem, Pennsylvania. When the severe depression in the iron industry threw Edwards out of work, Fritz provided recommendations that landed his Welsh friend another job, this time to superintend the modernization of Bluff Furnace in Chattanooga, Tennessee, a business in which Fritz was personally interested. Over the summer and fall of 1859, Edwards rebuilt Bluff Furnace to smelt ore with coke, making it the first furnace in the southern Appalachian iron region to burn mineral fuel.[53]

Thus Edwards brought to Shelby a wide range of technical expertise, practical experience, and a network of contacts that extended from Pennsylvania to Tennessee and beyond (fig. 61). As soon as he was hired, Edwards returned to Chattanooga to recruit workers from Bluff Furnace. He contacted Welsh, English, and Irish artisans in Atlanta, Selma, and Montgomery. Nine men soon joined him, including Evan Thomas (furnace builder), D. James (roller), H. T. Beggs (molder), Florance Donovan (finisher), and W. G. Moyle (engineer).[54] Puddler and roller David J. Davies, also recruited in 1862, may have been the Welshman of the same name who participated in the Tredegar strike of 1847. Several English-born artisans from the Etowah Iron Works came to Shelby later in the war.[55]

Giles Edwards's career embodied the British tradition of the tramping artisan recast in a new context. While working for various American employers, he gained technical skills and broad knowledge about the operation and design of furnaces and rolling mills. He became familiar with the technologies involved in hot- and cold-blast smelting with anthracite, coke, and charcoal. Each new job required him to solve localized problems, such as finding the right mix of materials and techniques to produce good iron from local ore and coal. As a former patternmaker, Edwards would have been able to draft plans for machinery, a very useful skill at an isolated industrial location such as Shelby. During his time in Tennessee, he encountered the managerial issues that arose in mixed-race workforces in the South, though he was also typically "clannish": he hired friends and acquaintances and showed a preference for British artisans. As a native Welsh speaker, he would have had a special bond with fellow Welsh immigrants.[56]

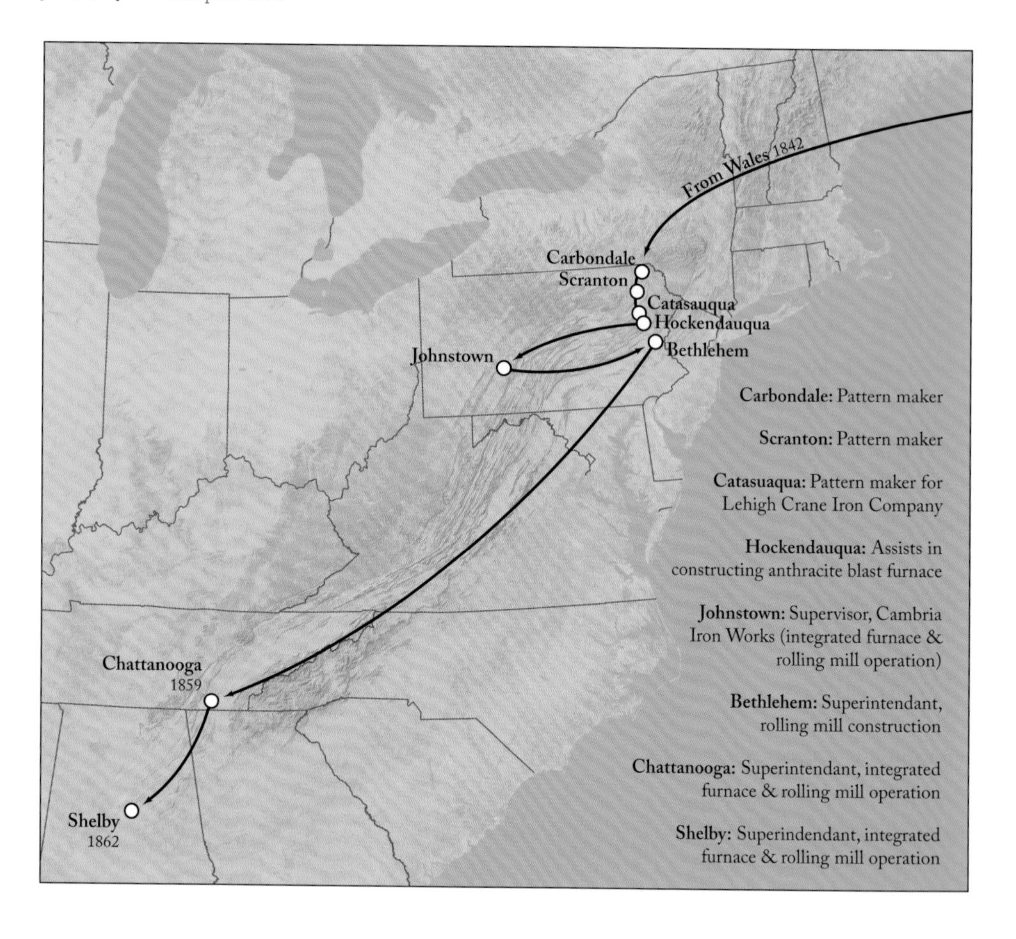

Figure 61. Career migration of Giles Edwards. At each location, Edwards acquired new skills and experience as a manager of ironworks. By the time he reached Shelby, Alabama, few men in the industry had such broad knowledge of the technology and labor processes of smelting and refining iron.

Edwards's technical expertise, personal contacts, and understanding of the work cultures of British, northern, and southern ironworkers would prove invaluable in Alabama. His importance to the company was quickly acknowledged. In its September 1862 meeting, the board of directors voted to permit Edwards to become a shareholder, his purchase of $12,500 worth of Shelby stock "to be paid for by him out of his wages, and out of his share of the profits of the concern." In other words, without costing Edwards a penny out of his own pocket.[57]

In addition to rebuilding the furnace stack and enlarging the rolling mill, the company erected houses for white artisans and superintendents and enlarged its slave accommodations. In a deposition taken after the war, Robert Hall described the Shelby iron community as a classic iron plantation, if unusually large.[58] The largest house, built of brick, one and a half stories high, with six rooms, was probably home to Shelby president and general manager Andrew T. Jones and his family. One-story frame dwellings of two to four rooms each accommodated skilled white workers, including engineers, carpenters, molders, a furnace mason, patternmaker, and bookkeeper. Blacksmiths and carpenters, as well as white laborers, probably lived in the "mechanics sleeping rooms" in the boardinghouse. "Eight residences in Welch Town[,] first class, one story frame buildings with necessary outbuildings for each," were probably set aside for the elite workers who manned the rolling mill.

Slave laborers lived in rougher cabins and "slave quarters," most likely simple plank buildings lined with bunks. The largest cluster of slave cabins, "sufficient to accommodate 150 Negroes" as Hall recalled, covered more than five acres "enclosed with a tight plank fence eight or ten feet high." Another 125 or more slaves were housed near the coaling grounds in the woods where charcoal was produced, along with small houses for the chief colliers. Hall also listed "commodious stables for two hundred head of stock" and 225 acres of pasture, plus a grain mill, wells and springhouses, smokehouses, a farm with sugar mills, a shoe shop, company store, brickworks, and storage sheds, not to mention the blast furnace and rolling mill at the heart of the sprawling industrial community. At the height of its population and productivity, Shelby was home to at least 60 white male workers and perhaps as many as 350 slaves, including some women and children.[59]

The workforce was strictly segregated by race and skill. White workers smelted, refined, and finished iron. Most slaves did heavy manual labor in support of producing iron or feeding the community. Under the careful eyes of white overseers, slave work crews chopped wood, made charcoal, hauled coal and provisions, built roads, and charged the furnace. There were perhaps a dozen slave craftsmen, including carpenters, shoemakers, tanners, and brick masons. The few female slaves at Shelby presumably cooked and laundered for the men and tended agricultural plots (company managers tried to limit the proportion of women to no more than one in ten of the resident black population).[60] The number of slaves employed at Shelby fluctuated over the course of the war, depending on the availability and cost of slaves for hire, but the total probably never dropped below 200 men, women, and children, of whom no more than 10 to 20 percent were owned by the company.[61]

A few hired slaves described as "good furnace hands" may have been experienced forge carpenters and blacksmiths. A slaveowner claimed that "one of [his] boys is an *engineer*—Ja[me]s Hunt was a rock blaster [and you could] put him [to work] at same."[62] As is typical for slave labor records, Shelby company records are frustratingly mute on the actual work done by James Hunt or any other slave. Surviving rosters merely log the number of days worked by male slaves. The most detailed records note that slaves earned extra money by cutting wood for charcoal, hewing ore, hauling provisions, and serving as night watchmen. Two work entries note that slaves produced a small number of spikes or nails, probably as blacksmiths. There is no other evidence that slaves worked iron at Shelby during the war.[63]

Only once did the company attempt to breach the divide between slave and free labor. In the autumn of 1864, Shelby's officers drew up a plan to build two more puddling furnaces and a railroad spur to expedite coal shipments to the rolling mill.[64] To enlarge the workforce commensurately, Andrew T. Jones intended to hire slaves to fill at least a few positions in the rolling mill. Jones wrote to Major Thomas Peters, commander of the Confederate Quartermasters Department in Selma, that he needed "a practical rolling mill manager, a real active fellow who can manage negroes [for whom] we would pay a big price." A month later, Peters wrote to Jones, "Col. Hunt informs me that you had, sometime since, two negro puddlers whom you could not use on account of opposition from your white puddlers. If you have these men still I think I can exchange them . . . [for] two white puddlers if you so desire."[65]

I do not know how hard Jones tried to convince white mill workers to accept black artisans. In his letters he comes across as a practical man, not one to make a stand for industrial slavery on the kind of ideological grounds espoused by Joseph Reid Anderson before the war. Jones may have employed slave puddlers because of the extreme scarcity of white workers. In July 1862 he had complained to Confederate chief of ordnance Josiah Gorgas about "the great scarcity of efficient labor."[66] Under pressure to produce weapons-grade iron in great quantities, he could not risk shop floor discontent or the loss of a single skilled worker. If anything, Jones did all he could to placate white workers, from providing good housing and competitive wages to paying for magazine subscriptions and allowing artisans to borrow the labor of slaves for their own use.[67] Most important for the works' continued productivity, the company struggled throughout the war to maintain an adequate workforce, to keep workers fed, and to prevent their conscription into the Confederate Army.

The Pinch of War

Shelby hired most of its slave labor from plantations in northern Alabama, Mississippi, and western Tennessee. Early in the war, cotton planters whose market access was threatened by the Union blockade welcomed the opportunity to rent their surplus labor to the ironworks on monthly or annual contracts. In January 1862 planters told company agent John M. Tillman that they were "anxious to move their negroes south" to keep them safe from the Union forces that were massing in western Tennessee as Grant prepared to attack Vicksburg. Tillman hired sixty-eight slaves for the year at $150 each, plus food, clothing, and medical care. He expected another seventy-five slaves from the Kirkman plantation in Grenada, Mississippi.[68] Six months later, Tillman found the mood changed. "I inquired at every place I visited in North Ala.," he told Jones in July 1862, ". . . but I found the people afraid to hire their negroes fearing they would run away or

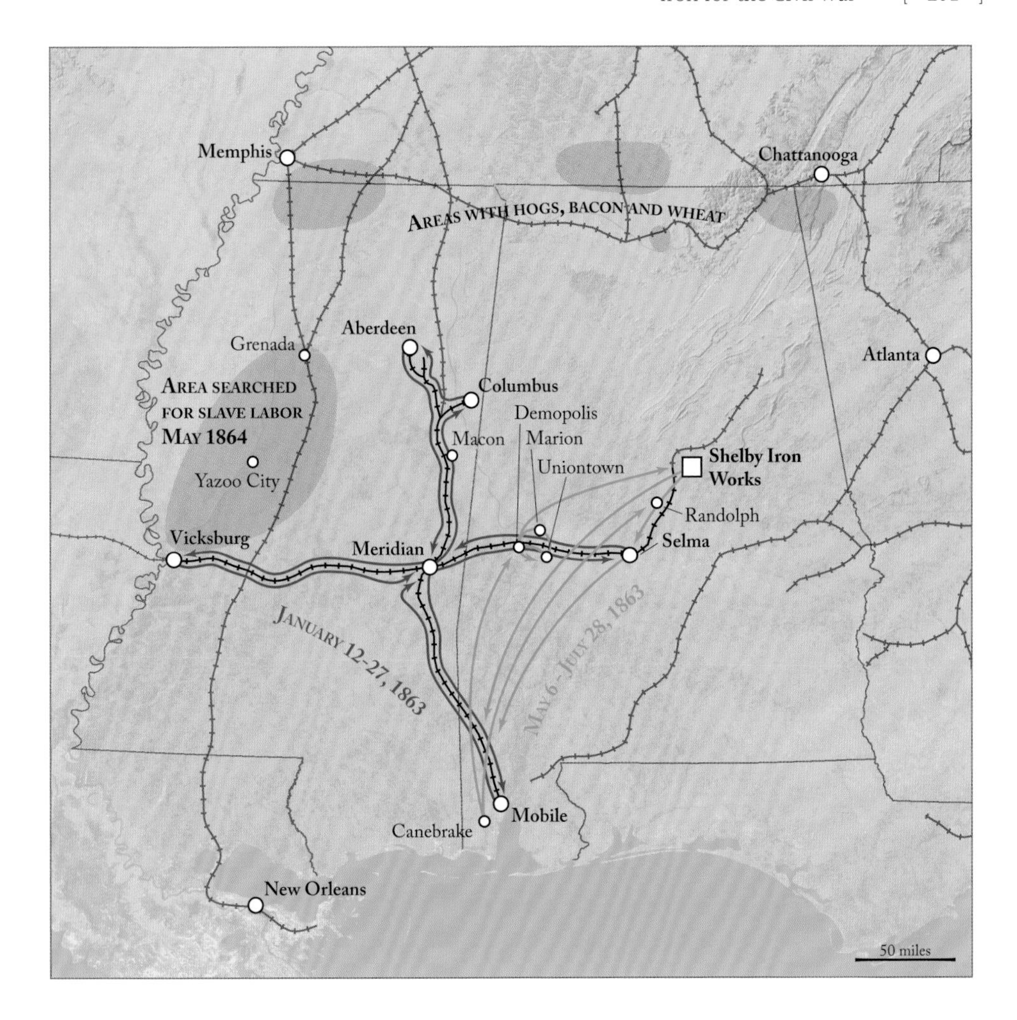

Figure 62. The search for slaves and provisions, 1862–64. Shelby agent John M. Tillman traveled more extensively each year trying to procure labor, food, and fodder to keep the Shelby Iron Works in operation.

report to the Yankeys [*sic*] who would confiscate their property." Until the Union Army was driven north of the Tennessee River, he thought, much of the region's supply of slave labor would be inaccessible.[69]

In December 1862, Tillman again set out to secure slave workers for the coming year (fig. 62). The rates for ordinary hands were still $150 or less at Selma, a major slave market, but the rate for carpenters had reached $350 and a slave patternmaker commanded $450 for a year's labor. Tillman was more worried, however, about the scarcity and cost of provisions along the Alabama-Tennessee border. Around Huntsville, Alabama, "government agents had impressed all the large lots of Bacon killing hogs. . . . I had to go to secluded neighborhoods . . . remote from any thoroughfares of travel that had escaped [their] attention." The pork he could purchase cost 15¢ a pound.[70] Slaves and provisions were even scarcer in early January 1863, as Grant intensified his campaign to control the lower Mississippi River. Tillman found "no one willing to hire" out their slaves in

Yazoo City, Mississippi, some forty miles northeast of Vicksburg. His journey by rail and buggy through the heart of cotton country in Mississippi and central Alabama yielded only twenty-nine slaves.[71] Fearing that slaves would "try to escape to the Yankeys" if they were removed from their home plantation, planters were now more inclined to rent out their slaves to Confederate military authorities than to Shelby.[72]

As Grant's Vicksburg campaign dragged on into the spring, Tillman had to look farther afield for provisions. In April 1863 he raised the idea of traveling to Florida for bacon, the main source of protein for all workers at Shelby. By the end of the year, food supplies were so depleted that it took Tillman a month to scrounge up 250 hogs fat enough to drive overland from Memphis to Shelby. Wagons and teams were scarce, and once again planters were unwilling to hire out their slaves. "You will find all the negroes that the owners wished to move [are] already in Ala., Ga., and S.C.," Tillman advised Jones. Another problem had reared its head by this time: the accelerating depreciation of Confederate currency. Pork on the hoof had risen to 25¢ a pound. Butchered meat sold for $2 a pound in Atlanta, corn for $2 to $3 a bushel. "I feel alarmed on the meat question," Tillman told Jones in October 1863, "after witnessing the destitution where I have traveled in North Ga. and Ala."[73] Tillman suffered a personal blow that month as well. He took a rare leave from Shelby to rush to a battlefield hospital at Chickamauga in hopes of saving the life of his wounded brother-in-law, who had nursed Tillman through typhoid fever years before. He arrived too late; all he could do was give the man a proper burial.[74]

Obtaining slave labor and provisions became increasingly difficult as federal forces penetrated farther into Southern territory. In early February, General William Tecumseh Sherman mounted a slow, destructive march across central Mississippi from Vicksburg to Meridian, a major railroad junction that connected central Alabama to plantations in the Tombigbee River Valley. Union troops tore up railroad track for miles around Meridian, and they devoured supplies. At the same time, Sherman fought to protect the rail lines south and east of Memphis that his supply train relied on.[75] With train service severed at Meridian and federals in control of most of the region's rail network, Shelby's access to labor and supplies was severely compromised. Previously, Tillman had used the Mobile and Ohio Railroad to transport grain and workers from Mississippi and the Memphis and Charleston Railroad to ship supplies from the Tennessee River Valley. Now he had to find alternative ways of getting provisions to the ironworks while dodging federal troops. The 250 hogs he found near Memphis in December 1863 had to be driven on the hoof roughly 150 miles, including a dangerous crossing of the Tallahatchie River.[76]

The following spring, no amount of asking turned up planters willing to hire out their slaves. Planters felt their slaves were safer in Mississippi than in northern

Alabama, Tillman explained, "as all the Yankee troops have been withdrawn from Miss. and sent to Georgia to reinforce Sherman," who was working his way from Chattanooga toward Atlanta. Slaves were safest if kept at home, out of sight of Confederate agents who were impressing Negroes for military work. Tillman had heard that "some of the planters made their negroes hide in the swamps to evade them." To make matters worse, food stores were so decimated by Union depredation and destruction that bacon was selling for up to $2.50 a pound. The situation was becoming impossible.[77]

In June 1864 Tillman told Jones that officers in the Confederate Ordnance Bureau had promised to use their influence to help Shelby because "they are anxious for your works to get all the laborers you want as the government is very much in want of iron." With Sherman's forces closing on Atlanta, the Bureau had lost the Etowah Iron Works and feared losing the Gate City Rolling Mill as well. The new Atlanta mill was now the Confederacy's only manufacturer of railroad iron except for Tredegar, which was also under threat as the Army of the Potomac neared Petersburg and the eastern approaches to Richmond. "If Johnston drives Sherman back" across the Tennessee River, Tillman wrote optimistically, "all the negroes from northern Georgia will be to hire as the plantations and crops are destroyed and it will be too late to plant a new crop." His hopes proved fruitless, as the superior number of Union troops eventually overwhelmed Confederate defenses. Sherman took control of Atlanta on September 2, 1864. Ten weeks later, the Union general began his march to the sea.[78]

Tillman's correspondence chronicles his ongoing struggle to provide the manual labor and the calories required to keep Shelby Iron Works in operation. He had much closer contact with the enemy than did those at the works site, which remained under Confederate control until the very end of the war. Yet superintendent Giles Edwards and general manager Andrew T. Jones faced another kind of depredation close at hand: Confederate conscription officers' efforts to remove men from the works for military duty. Their nemesis in this struggle was Lieutenant P. L. Griffiths, an officer at the local enrolling office in Talladega, Alabama. In June 1863, Griffiths brought formal charges against the Shelby Iron Company. He accused Jones of failing to report the names of those detailed to the works and charged the company with "harboring or keeping in their employ a deserter." The offense grew worse when he sent six men to arrest the deserter. "Giles Edwards Superintendent of said Company refused to give up any man that was there employed," Griffiths wrote, and "he, Edwards, said that they had orders from Major Hunt not to pay any regard to my orders whatever and if any man or men came there after any of the men employed by said Company not to give them up, but to resist, if necessary, with arms and that they would not give up any man." Last, Griffiths declared, Jones had "transferred three Men from his works . . . without my knowledge or consent."[79]

Jones's reply to the charges reveals the frustration that industrial managers experienced dealing with the bureaucratic red tape of Confederate military law. His letter also suggests that his managerial approach was similar to the respectful, practical attitude that prevailed at the Trenton Iron Works and Lehigh Crane. After explaining patiently that his inquiries had led him to believe that a list of already detailed men was not required, Jones "emphatically denied" harboring a deserter. The man in question, "a helper at the puddling furnace," showed "no evidence . . . that he is a deserter." He had a certificate proving that he had been detailed to work at Shelby. To the last charge—that he had transferred men without permission—Jones replied that he had indeed agreed to two men's requests to be transferred to other government ironworks "for the reason that they were not working satisfactorily for this company. It is impracticable to get proper services out of men, when [they are] forced to remain against their will; and although the company really needed their services, yet for the reasons stated, I gave my consent."[80] Only "efficient," willing labor was productive at an ironworks.

From manufacturers' perspective it was profoundly counterproductive and wasteful for the military to remove skilled men from their best use in order to add their insignificant numbers to battle. In February 1864, James W. Lapsley, one of Shelby's leading partners, summarized this argument in a letter to Confederate secretary of war James A. Seddon. The interference of conscription officers at Shelby and other ironworks must be stopped, Lapsley argued, for it threatened to drive away workers whose labor was critical to the Confederate war effort.

> These men do not feel identified in any great degree with the South, and are not imbued with sentiments and feelings calculated to impress them so strongly in favor of our cause, as to induce them to make any great sacrifices of interest or feeling in its behalf. They are generally without families. . . . So far, those of them who have remained, have been induced to do so by the very high wages paid them. . . . If these men were left to draw their own conclusions from the facts, uninfluenced by other causes, it is but reasonable to conclude that more or less of them would be induced to leave this country.
>
> These men . . . are apt to regard their position as insecure, and to conclude that if they can be thus summarily taken from their work to a conscript camp, they may some day be suddenly summoned into the field. . . . The effect of such proceedings is unfortunate, and any thing but assuring to these workmen. They are as a class, very clannish, and what they regard as harsh treatment of any of their number, is resented by all. True policy would I think dictate the most liberal and assuring course toward men so indispensible [*sic*] to our cause.[81]

Despite a general order issued in the fall of 1863 that specifically prohibited conscription of "employees at any furnace or other establishment under the supervision or control" of the Nitre and Mining Bureau (which oversaw production under government contracts at Shelby),[82] enrolling officers continued to pursue iron workers—and no wonder, for by 1864 Confederate forces had dropped to barely one-third as many men as the North still had in uniform. That Shelby secured details for between forty and seventy white men, far exceeding its allotted number of about thirty skilled workers and supervisors, attests to company managers' determination if not their diplomatic skills.[83]

Giles Edwards's refusal to give up any workers suggests a strength of character that the ironworkers under his supervision must have appreciated. At the same time, Edwards believed in disciplining the workforce. In November 1862, when an Irishman working as a puddler's helper quit Shelby without permission, Edwards asked Jones if he could "get him arrested as a deserter" and have him brought back to the ironworks. "If an example of this kind can be made," Edwards argued, "it will have a very beneficial effect in the future."[84]

Edwards's firmness contributed to what appear to have been stable work relations at Shelby throughout the war. While the absence of evidence never proves anything, it is interesting that the large body of surviving company records contains very few signs of friction between the company's managers and its workers, free or slave. At some point in 1863 or 1864, skilled rolling mill workers "stopped work on account of the low price paid for our labour and the high prices of provisions and all the nesesaries [*sic*] of life," but the courteous letter to Edwards in which the men announced their conditions for returning to work suggests no acrimony between them and the superintendent.[85] The company received many letters from iron artisans, miners, engineers, carpenters, and brick masons seeking employment—and military detail—with Shelby. The ironworks was a very appealing employer, since it offered a relatively safe haven to those granted a military detail to work there. The number of letters addressed directly to Edwards also suggests that he was familiar to the small network of industrial artisans in the South, many of whose members hailed from the British Isles or Pennsylvania.[86]

Among the numerous letters from slaveowners in the Shelby correspondence, only three note that slaves complained of mistreatment. One master asked that his slave Charles be transferred from filling the furnace—a job that made him ill—to keeping, or watching, the furnace. Another wrote more urgently asking Jones to look into an alleged incident in which a Shelby overseer had driven a sick slave to his death. In the third instance, planter Samuel Kirkman, the company's most regular supplier of slave labor, noted that "the negroes appear well satisfied, except they complain that some of your white operatives impose upon them. I will be glad," Kirkman concluded, "when you are able to keep them seperate [*sic*] from the white laborers."[87] The racial division of labor, at Shelby as at Tredegar, minimized conflict.

The company's white rolling mill workers went on strike in part because they were not being paid regularly. Having no cash in hand, and no idea when they might be paid, was as great a concern to Shelby workers as it had been for iron artisans and coal miners at Farrandsville during the cash-poor years of the late 1830s.[88] Shelby's failure to pay wages points to the company's chronic shortage of funds. Like all Confederate iron producers, Shelby's managers found themselves caught in a constricting financial noose as the prices at which they had contracted to sell iron covered less and less of their operating costs. Andrew T. Jones estimated that his company's costs had increased from about $18,000 to $42,000 a month during the fifteen months since Shelby entered into its contract with the Confederate government. "This increase is attributed to the larger increase in everything . . . required in the business, including labor," he explained to the chief of the Nitre and Mining Bureau.

> To give some instances by way of illustration[:] The price of bacon has increased from 35 to 40 cents per lb. to $1.50 . . . corn from $1.00 per bushel to $2.50. Oznaburgs [heavy cotton cloth used for slave clothing] from 35 cents to $1.50 per yard. Negro shoes from $4.50 to $5.00 increased to $15.00 to $16.00 per pair; coal from $6.00 per ton to $20 per ton; the cost of all which has been considerably enhansed [*sic*] in addition, by a larger increase in the prices of freights by railroad and the increased expenses in hunting up supplies. . . . Negro labor has advanced about 40 per cent; and mechanical labor (white) such as puddlers, heaters, rollers, engineers &c, has increased about 75 per cent, the prices paid being fully up to those of Richmond.[89]

Faced with soaring expenses, understaffed, and unable to operate at all when coal shortages shut down the rolling mill, the company fell behind in delivering promised shipments of iron to the Nitre and Mining Bureau. The Bureau responded by deducting 10 percent from the amount due—about $50,000—in 1863 and 1864.[90]

Shelby's managers tried to reduce costs by becoming as self-sufficient as possible while maximizing iron production to generate income. In July 1863, slave crews began leveling the roadbed for a six-mile rail spur from the ironworks to the railroad depot at Columbiana. Many hopes were pinned on this rail line. It would eliminate the problem of transporting heavy supplies in, and iron out, on muddy roads. It would enable Jones and Edwards to "enlarge the business" by tapping Shelby's own coal deposits instead of competing with other companies for coal from more distant mines. The greatest hope was that Shelby's bituminous coal could be used in a new coal-fired blast furnace that would increase the ironworks' overall capacity. As a side benefit, surplus pig iron could be exchanged for bacon, flour, fodder, and other provisions.[91]

The railroad spur was completed, but not in time to save the iron company. In early 1865, Shelby's tattered supply network was ripped apart by Yankee raids in central Alabama. On March 3, Jones told Shelby treasurer Thomas Peters that the company was $100,000 in arrears. On March 20 the company ran out of blasting powder for mining. Eight days later, the mill was "standing idle for want of coal." The same day, Peters telegraphed Jones urging him "to pack up valuables and be ready to run. . . . The enemy are making to Tuscaloosa or Elyton," which could only mean that "their object [was] the RR[d] & Iron works."[92] On March 31, the works were largely destroyed by Union cavalry under Major General James H. Wilson, who went on to capture Selma on April 2.[93]

Although Shelby remained in operation longer than most Confederate iron manufacturers, its experiences were in most respects quite typical of the difficulties, constraints, and conditions of production at Southern ironworks during the war. The Confederacy's initial disadvantages for iron production became more acute every year, as Union territorial gains cut off access to established iron companies in border states and threatened works deeper in Confederate territory (fig. 63). Scarce labor became scarcer over time as planters sequestered slave workers and white workers were lost to military conscription and desertion. One of the few silver linings for the Confederate iron industry was that the region's blast

Figure 63. Confederate ironworks and Union territorial gains, 1861–65.

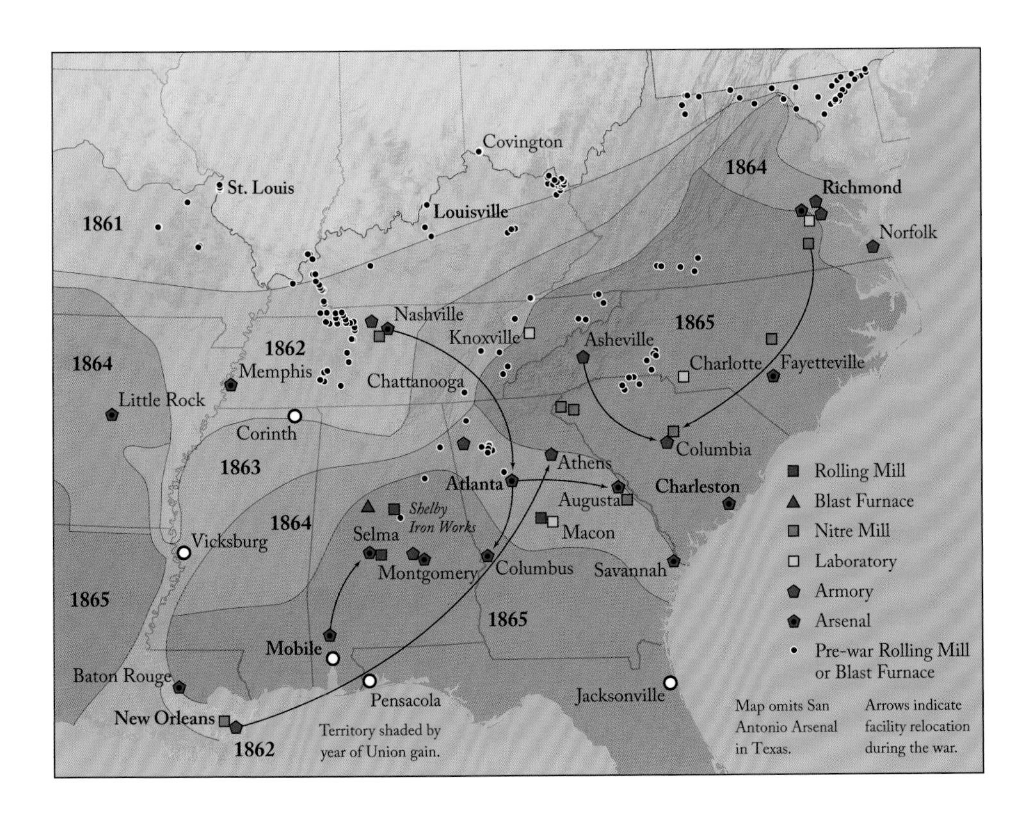

furnaces and bloomeries generally produced charcoal iron of adequate quality for ordnance production, though never in sufficient quantities to satisfy the military's voracious appetite.

Evidence from Shelby and the Ordnance Bureau casts doubt on Gorgas's boast that the Confederacy met "the demands of its large armies," but one must also ask how the Confederacy managed to engage in industrial warfare for four years given its small and constantly declining industrial capacity. It may be that the complaints of Commander Brooke and James Lapsley were akin to farmers' complaints about the weather—always gloomy because the conditions of production were so unpredictable. Shelby records do not state how many white workers were lost to conscription or desertion, or how many Negro workers the company was actually able to hire after 1863. Somehow the company managed to produce iron as late as March 1865. It supplied iron rods and plate for Confederate battleships on the Mississippi River and the floating battery that defended Mobile. Many tons of Shelby pig iron were shipped to the Confederate arsenal at Selma. The company made its own rails and may have provided track for railroads elsewhere in the South.[94] The company records do not say how much iron Shelby produced, exactly where it went, or how important it was to the Confederate war effort. More research needs to be done to determine how much iron Confederate manufacturers produced, and for what purposes, before we can fully understand the relative weakness and resourcefulness of the Southern industry during its most challenging years.[95] There is no question but that the Confederacy was perpetually losing ground to the North where iron was concerned.

The Northern Iron Industry

By the end of 1862, the Union had gained control of about half of the Confederacy's initial ironmaking capacity. The North could muster more raw and finished iron than the South at every stretch along the sectional border. It had many more managers and workers with long experience in the iron industry. There were far more proven deposits of workable iron ore and coal in the North, as well as more extensive transportation systems. While the Southern industry scrambled to grow, then fell into a spiral of worsening crises of supply and production, the war revived iron production across the North and provided the means for many firms to expand operations. Increased demand for iron, coupled with labor shortages, heightened owners' and managers' interest in laborsaving technologies. The same conditions gave ironworkers a new sense of political identity, which found expression in the growth of the country's first ironworkers' labor unions.

For wartime manufacturing, particularly heavy ordnance production, ironworks in the North were able to draw on considerably more managerial experience than those in the South. West Point Foundry in Cold Spring, New York, fifty miles up the Hudson River from the port of New York, had been making cannons for the federal government since about 1820. By 1848, the foundry employed from four to six hundred men. Many were European immigrants, like the first artisans recruited from northern Ireland and England to man the works in 1818. The 1850 federal census of Putnam County lists a strikingly international workforce in the skilled metal trades, with engineers, finishers, molders and patternmakers, turners and wire drawers from England, Ireland, Scotland, Germany, Switzerland, and France.[96] The workers made cannons of various kinds as well as steam engines and boilers, pipes, iron wheels, and machinery. Gouverneur Kemble and his family were involved in running the business from its beginning in 1818 through the Civil War, a continuity of resident owner-mangers that contributed to the company's long-term success.[97]

The foundry's most influential manager was Robert Parker Parrott, a onetime physics teacher and an 1824 graduate of the US Military Academy at West Point, New York. Parrott served as a captain in the Ordnance Department of the US Army. After one year as ordnance inspector at the foundry, Parrott resigned his commission to become the works superintendent. Technically knowledgeable and inventive, Parrott instituted many improvements. Most important for the Union, he developed a method for reinforcing the breech, or firing, end of heavy cannons to prevent them from exploding (fig. 64). Parrott guns became the largest cannons in the Union's arsenal, able to hurl projectiles weighing up to three hundred pounds. West Point Foundry operated almost exclusively as a cannon foundry during the war, turning out up to one hundred guns a month, for a total of more than three thousand cannons and 1,600,000 rounds of shot and shell. At the height of its wartime production, when the foundry employed up to one thousand men, it was probably the largest heavy ordnance manufacturing facility in the United States.[98]

Cyrus Alger, founder of the South Boston Iron Works, had even more years of involvement in ordnance manufacturing than Parrott and Kemble. He first became engaged in the foundry business in 1809. During the War of 1812, he worked with Thomas H. Perkins, perhaps in relation to the Vergennes Iron Works in Vermont, casting cannonballs. In 1817, Alger incorporated the South Boston Iron Company, a foundry built on mudflats in Dorchester, which gradually became a center of machine shops and foundries. An early experimenter with metallurgy—what Peter Lesley would have called a good practical man—Alger developed methods for strengthening cast iron that he put to use making mortar guns for the government. He also invented the process for hardening the outer surface of the rolls used in roll trains (called chilled rolls), which prolonged the machinery's life and made it possible to apply more pressure. Alger's son, Francis, took over the foundry when his father died in 1856. Under his management, the South Boston works nearly doubled in size by adding a large ordnance foundry

Figure 64. Proving a Parrott gun (three hundred–pounder) on the grounds of the West Point Foundry. Courtesy of the Putnam History Museum. A note on the back of this photograph dates it to May 1, 1864.

that went into operation in March 1863. It chiefly made Rodman cannons, including large coastal defense guns. An industrial history published in 1864 notes that Francis Alger continued his father's policy of "retaining men through a series of years, often keeping many on half-pay when their services were not needed," which "secured for this Company an unexcelled force of employees who fully understand the requirement of every department of labor in which they are employed." The wartime workforce of four hundred men worked in day and night shifts, as did workers at West Point Foundry.[99]

The North's enormous advantage in rolling mill capacity carried with it a depth of managerial experience that was rare in the South. Half of the rolling mills clustered along the lower Schuylkill River and Brandywine Creek in southeast Pennsylvania had been built by 1840. Some, including the family-owned Lukens Iron Company, had been rolling iron plate for boilers and steam engines since the 1810s. In Pittsburgh, rolling mills and foundries had been producing plate and parts for steamboats since 1815, and many mill owners in the "Iron City" had first worked as skilled artisans or trained as apprentices under skilled workers before assuming managerial responsibilities. Some mills in Cincinnati,

Louisville, and St. Louis similarly had been in operation for many years before the war. While a handful of managers at Southern rolling mills had comparable experience, notably Tredegar's Joseph Reid Anderson and Daniel Hillman at the Cumberland Rolling Mill in western Tennessee, they were exceptions.[100]

The North also had more ironworks that could be brought back into production after having been shut down during the prewar depression. The revived fortunes of Katahdin Furnace illustrate how the tide of war lifted all boats in the iron industry. Situated deep in the woods of northern Maine, Katahdin sold 5.5 tons of charcoal pig iron in 1859, then 1,793 tons in 1863. Even low-grade Katahdin pig iron that had sat unsold since 1853 in warehouses in Philadelphia and Charlestown, Massachusetts, found buyers during the war.[101] The Maramec Iron Company, a midsize, integrated blast furnace and refining operation on Missouri's agricultural frontier, made better iron than Katahdin, but high transportation costs before the war ate up most of the company's profits. At the peak of wartime prices in 1864–65, Maramec pig iron sold for $60 or more a ton, blooms for up to $100 a ton. These prices not only more than covered the cost of production and shipment, they netted the works owner, William James, as much as $40,000 in 1865.[102]

Prices doubled and tripled from 1861 to 1864 for all grades of crude and finished iron in key Northern markets (table 6). Rising prices and the lure of lucrative government contracts provided powerful incentives to construct new ironworks. Between 1861 and 1865, entrepreneurs built seventeen anthracite blast furnaces and twelve rail mills (fig. 65). Because of the time it took to build a new coal-fired furnace or to equip a rolling mill, few of these large works contributed much to the Union effort. Among the investments that had little impact

Table 6. Iron prices during the Civil War (dollars per gross ton)

Year	No. 1 foundry pig at Philadelphia	Best refined bar iron at stores in Philadelphia	Iron rails at mills in Pennsylvania
1860	22.75	58.75	48.00
1861	20.25	60.83	42.38
1862	23.88	70.42	41.75
1863	35.25	91.04	76.88
1864	59.25	146.46	126.00
1865	46.12	106.38	98.63

Source: Peter Temin, *Iron and Steel in Nineteenth-Century America: An Economic Inquiry* (Cambridge, MA: MIT Press, 1964), table C.15, p. 283.

on the war was a government-owned rolling mill designed by John Fritz in Chattanooga, Tennessee. It was built to reroll worn and damaged rails. The mill went into operation in April 1865, too late to support Sherman's supply train during the Atlanta campaign. Only the Union Ironworks in Buffalo, built in 1862, another mill of the same name built in Chicago in 1863, and the Bethlehem Rolling Mill in Bethlehem, Pennsylvania, also built in 1863, produced substantial amounts of rail before the war was over.[103]

For the most part, the federal War Department and private contractors had to turn to existing producers for the iron they required. This created new opportunities for firms that were willing to diversify their products, seek new sources of supply, form new partnerships, and expand their facilities, all of which meant gambling on the potential return of government contracts. Iron companies' involvement with the manufacture of ironclad warships illustrates all these dimensions of wartime production.

The shipbuilders who succeeded best in building iron ships, according to historian William H. Roberts, were those "at the center of a cluster of well-developed metal production and engineering industries."[104] The first Union ironclads built to inventor John Ericcson's specifications were constructed in just

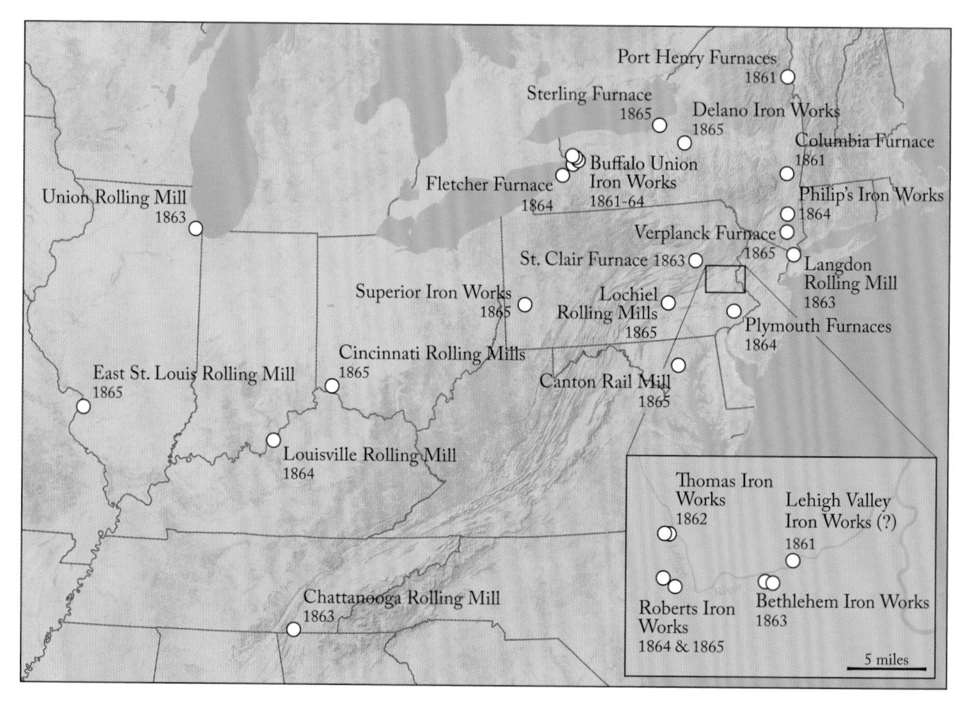

Figure 65. Anthracite furnaces and rolling mills built during the Civil War. The dates mark when new ironworks went into operation. As the map shows, the South gained not a single anthracite furnace and only one new rolling mill during the war. Southern manufacturers did, however, modify existing facilities such as Leeds Foundry in New Orleans to roll iron for ironclads.

such locations. The *Monitor*, which famously battled the Confederate *Virginia* to a draw at Hampton Roads on March 9, 1862 (fig. 66), was built at the Continental shipyard in Greenpoint, Brooklyn, amid the thickest nest of machine shops and foundries in the nation. Ironworks as far away as Ohio's Hanging Rock iron district later claimed to have provided "the" iron for the US Navy's most famous ironclad. Her cladding probably came primarily from works in southern New England and the Hudson Valley (fig. 67). One documented supplier was the Albany Iron Works in Troy, whose owners partnered with Ericcson and lobbied heavily in support of the congressional bill that authorized funds to build the first twenty ironclads.[105]

While a number of ironclads were built at Atlantic ports such as Jersey City, Boston, and New York, by the fall of 1862 the navy was also contracting with firms west of the Alleghenies to "spread the wealth of government contracting," to ease demands on East Coast shipyards, and most critically to speed delivery of ironclads to aid General Ulysses S. Grant's campaign against Confederate strongholds on the Mississippi River. As in the East, western ironclad construction

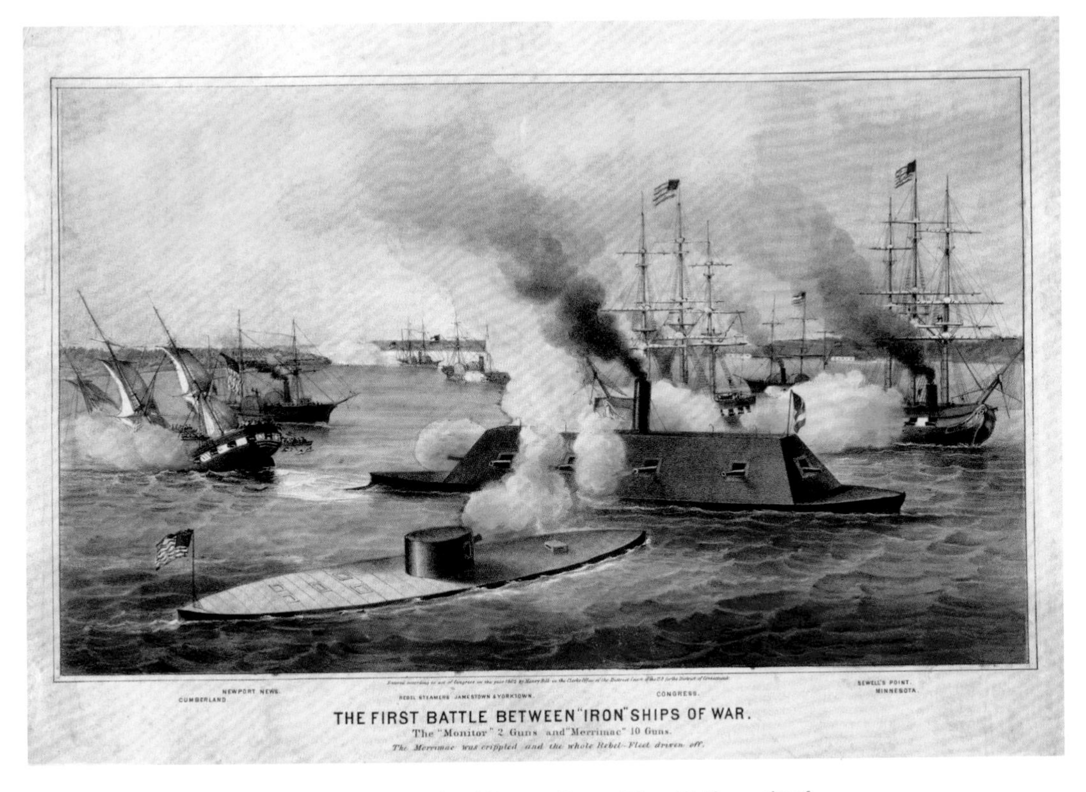

Figure 66. The *Monitor* and the *Virginia* face off. The Mariners' Museum, Newport News, VA. The novelty of the two ironclads' battle, and the military implications of this new technology, gripped Americans' interest. The battle was depicted in many engravings.

Figure 67. Iron plating on the deck of the original *Monitor*. Photographed between 1861 and 1865. Courtesy of the Library of Congress. The plates' dimensions reflect the limitations of American rolling mills at the beginning of the Civil War.

was clustered at sites of existing rolling mills and machine shops. For example, John Snowdon and Albert G. Mason, the machine shop owner and boat builder who built the *Manayunk*, "shut down their Brownsville [Pennsylvania] shops and brought their workmen to [a] new 'gunboat yard' in Pittsburgh." They built the vessel just three hundred yards from their main iron supplier, Lyon, Shorb, and Company's Sligo Rolling Mill on the south bank of the Monongahela River. The Sligo mill rolled plate up to three inches thick from cold-blast charcoal iron produced at the company's own furnaces in the Juniata district. In Cincinnati, Covington, and Newport, shipyards had been building steamboats for more than thirty years, but it was the presence of four rolling mills and hundreds of metalworkers, as well as the leadership of iron entrepreneurs Miles Greenwood and Alexander Swift, that made the region a center for ironclad riverboat construction.[106]

Western companies that took the plunge into building ironclads faced more difficulties than those on the Atlantic seaboard. Large as Cincinnati's labor pool was, it was dwarfed by the number of skilled ironworkers, shipwrights, and machinists available in Greater New York and Boston. Western firms had higher start-up costs and operating expenses. River towns had to build large drydocks from scratch. The navy's design specifications changed with irritating frequency and were sometimes counterproductive. For example, the navy asked builders to use more powerful engines and thicker plate, which made vessels much heavier, at the same time that builders were trying to make iron ships navigable in the shallow, snag-filled Mississippi. Iron cost more in the West as well. In March 1863, bar iron sold for $100 a ton in New York, $135 in Cincinnati. By December the price had risen $10 a ton in New York but $40 in the western city. Wage increases put further strain on producers. The final twist, which affected all ironclad manufacturers, was that long-term government contracts were extremely difficult to renegotiate to cover rising costs. For many companies, it was a devil's bargain to take on "the many annoyances attending a contract with the Government."[107]

Similar problems afflicted every branch of the supply chain for military iron in the North. The constraints on industrial capacity meant that iron supplies increased much more slowly than prices (table 7). As in the South, the rapid rise in iron prices constricted cash flow through the manufacturing system. Contract prices did not cover actual costs, which made it more difficult for companies to purchase supplies, which delayed production, which slowed payment on delivered goods, and so on. Ordnance manufacturers were most concerned about the limited supply of the best grades of charcoal iron, which was preferred for rifle and pistol barrels, heavy cannons, and ships' cladding.[108] Of the approximately 100,000 tons of charcoal iron produced annually in Union-controlled territory during the war, only a small fraction was considered military grade.[109] Companies that produced such iron demanded premium prices, and some sold virtually all their product to ordnance manufacturers. For example, Greenwood Furnace, owned by Robert Parrott's brother Peter, was the favorite supplier of West Point Foundry and Alger's cannon works in Boston. Along with Fort Pitt Foundry, the Union's three major heavy cannon manufacturers required about 8,500 tons of

Table 7. Iron production during the Civil War (thousand gross tons/percentage increase or decrease over previous year)

Year	Charcoal pig	Bituminous coal and coke pig	Anthracite and coke pig	Iron rails
1860	249 (–2%)	109 (+43%)	464 (+10%)	183 (+5%)
1861	174 (–30)	113 (+4)	365 (–21)	169 (–8)
1862	167 (–4)	117 (+4)	420 (+15)	191 (+13)
1863	189 (+13)	141 (+21)	516 (+23)	246 (+29)
1864	216 (+14)	188 (+33)	611 (+18)	299 (+22)
1865	234 (+8)	169 (–10)	428 (–30)	318 (+6)

Source: Peter Temin, *Iron and Steel in Nineteenth-Century America: An Economic Inquiry* (Cambridge, MA: MIT Press, 1964), table C.2, p. 266, and table C.6, p. 274.

top-grade charcoal iron a year. Much more iron was used for cannonballs, shot and shells, caissons, and the other accoutrements of heavy ordnance, though iron of lesser quality was adequate for those products.[110]

Iron quality mattered most in rifle and pistol barrels. Blast furnaces in the Salisbury iron district in western Connecticut and Massachusetts had been the primary suppliers of barrel iron for the federal armory at Springfield, Massachusetts, and sometime suppliers to Harpers Ferry since the beginning of the nineteenth century. Springfield Armory became the Union's sole domestic supplier of handguns and rifles after Harpers Ferry was sacked by the Confederates. Unlike European small arms manufacturers, who began to use steel barrels in the 1850s, Springfield continued to make gun barrels exclusively from charcoal iron plate throughout the war. It used half or more of all the iron the Salisbury district could produce.[111]

Some rolling mills also specialized in supplying military iron. The Lukens Iron Company in Coatesville, Pennsylvania, one of the country's oldest manufacturers of iron plate, was an early provider of cladding for Union monitors. By 1863 a small number of firms in St. Louis, Greater Cincinnati, Portsmouth, Ohio, and Pittsburgh were specializing in producing heavy plate for ironclad construction west of the Alleghenies.[112] Ironclads took more iron than any other branch of federal arms manufacturing. A single light-draft (*Casco* class) ironclad used 462 tons of iron for the hull, armor, and turret, plus 110 tons or more for boilers, bolts and bracing, pipes, and machinery. The Union's twenty *Casco* class ironclads required approximately 11,440 tons of iron. The entire Union ironclad program required something on the order of 50,000 to 60,000 tons.[113] Iron for armor plate had to be malleable enough to roll well but also strong enough to absorb the tremendous force of naval cannons and mortar shells. "Red short" iron, which tended to be brittle, could shatter under fire. Metal made from a mixture of pigs, blooms, and scrap iron produced cladding of uncertain quality. This problem frayed nerves at the conservative Lukens Company when government demand for armor plate compelled it to expand its supply network. In 1862 the company's Quaker partners reluctantly began to purchase blooms through brokers in New York and New Jersey, breaking their custom of contracting directly with forge masters. They also looked for more distant sources of supply, including a forge several hundred miles north in upstate New York, probably in Essex County. At Pittsburgh, military demand stimulated mills to seek ores from distant suppliers at Pilot Knob, Missouri, and the Upper Great Lakes iron region. As it happened, these regions possessed some of the country's richest and largest iron deposits—deposits whose potential the war brought more fully to light.[114]

Western iron deposits had been discovered many years before the Civil War, and some companies—such as the Trenton Iron Works—established far-reaching networks of supply in the 1840s. The war did not so much change the fundamental geography of iron production as accelerate the expansion of networks and awareness. Heightened demand, in the form of pressure from government supply agents and the lure of opportunity, also pushed ironmakers to consider changing other aspects of their operations. To innovators like Abram Hewitt and Horatio Ames, the war was an inspiration for new techniques suited to military markets. Hewitt saw the Union's need for small arms as a splendid opportunity for reviving business at the Trenton rolling mill. In spring 1862 Hewitt visited Staffordshire, England, where the craft of rolling gun barrel iron had been perfected, and where the firm of Marshall and Mills produced much of the iron plate used to make rifle and pistol barrels at the Springfield Armory. After months of trial and error, Hewitt secured a contract from secretary of war Edwin M. Stanton for Trenton Iron Works to provide two thousand tons of barrel plate.[115] Horatio Ames hoped to capitalize on the war by producing cannons at his Connecticut forge and foundry. Rather than casting cannons as established heavy ordnance manufacturers did, Ames developed a new design that welded together concentric rings of wrought iron. The guns withstood extreme proof and earned praise for their exceptional strength, but the company won few contracts, and only a handful of Ames cannons ever reached the battlefield. Ordnance inspectors were skeptical of the unusual design. John A. Dahlgren insisted that one set of Ames's one hundred pounders be tested for thirteen months. At the end of the trials, the war was over.[116] As a competing producer of heavy ordanance, Dahlgren has personal reasons to block production of Ames's cannon. But he was also obsessed with safety standards, which he saw being compromised under the pressure to produce. The need for safe, reliable ordnance spurred innovations of another kind.

Struggles for Control of Production

The US Navy's attention to ordnance production had been focused on quality control since 1844, when a disastrous explosion aboard the navy's first screw-propeller steamship, the USS *Princeton*, called into question the standards of manufacture for all heavy ordnance. The *Princeton*'s two enormous wrought-iron cannons, called the Peacemaker and the Oregon, were the largest yet made in the United States. On February 28, 1844, Captain Robert F. Stockton, general supervisor of the *Princeton* project, hosted the ship's inaugural launch at the naval shipyard in Washington, DC. Historian Robert J. Schneller Jr. relates what happened:

> Over five hundred people including President John Tyler, cabinet members, foreign ministers, congressmen, senators, army and navy officers, and their wives thronged on board the *Princeton* for a dinner party and a demonstration of the ship. Stockton fired the Peacemaker twice to show off its power. Someone asked

> him to fire the gun again. He did. Without warning the Peacemaker exploded, killing, among others, the secretary of the navy, the chief of the Bureau of Construction, Equipment, and Repairs, and [Abel Parker] Upshur, who was then secretary of state. . . . The next day the House Naval Affairs Committee crushed Stockton's proposal to build more ships like the *Princeton*. No more guns like the Peacemaker were made. The explosion of the Peacemaker was a disaster unparalleled in the history of American naval ordnance.[117]

The *Princeton* disaster galvanized the navy and the army to take a much more scientific, centralized approach to ordnance manufacturing. In the late 1840s, cooperative experiments involving military ordnance experts and private gun founders determined that the strength of iron varied directly with its density and tenacity. They also found that the method used to cast cannons affected those crucial qualities, which led to the invention of new casting methods and cannon designs, including the Parrott gun, the Rodman, and John A. Dahlgren's bottle-shaped design for heavy naval guns. Both branches of the military also established more rigorous standards of inspection and cadres of trained inspectors to ensure that standards were met before heavy guns were risked in battle.[118]

As the new system was put into place in the 1850s, foundry owner-managers and skilled workers chafed at the many ways it challenged their control over production. Ordnance inspectors arrived at Fort Pitt, Tredegar, West Point, and the Alger works with precision calipers and certain ideas that did not sit well. Navy inspectors, for example, pushed foundries to use only cold-blast pig iron rather than hot-blast iron. Dahlgren, the navy's most determined and scientifically astute inspector, insisted that Joseph Reid Anderson cease using "soft" metal for cannons at Tredegar in favor of harder iron that would "carry the strength of our Heavy Ordnance to the highest degree of excellence."[119] Military requirements that sample guns withstand "extreme proof" of five hundred firings or more also delayed delivery—and payment—for large orders of cannons. Charles Knap, the owner-manager of Fort Pitt Foundry, complained to Commodore Charles Morris, chief of the Bureau of Ordnance, that the navy should pay for the fifty guns the foundry had produced despite samples' having failed under extreme proof because the navy's contract had specified higher density and tenacity than Knap thought best.[120]

Inspector Andrew A. Harwood summed up the conflict between artisanal and scientific manufacturing as a struggle for consistency and reliability. The best way to achieve uniform results, he believed, was to follow standards "systematically and with all practicable exactness. No stronger proof can be offered of the insecurity of dependence upon the mere experience of workmen than the well established fact that guns of comparatively weak construction have sustained the severest trials while others of better model have burst prematurely. The interests of the founders warrants [*sic*] the supposition that in every case their experience was employed and the result shows that experience to have been frequently at fault."[121] William Rogers Taylor wrote an unusually sympathetic assessment of the value of artisans' experience after a visit to the Alger Foundry in 1859. He told Captain Duncan Nathaniel Ingraham, chief of the Bureau of Ordnance and Hydrography, that final decisions about the production process should remain with the skilled workers in charge of casting cannons precisely because their experience enabled them to assess the quality of iron, which varied unpredictably even from the most reputable furnaces.

> The blast furnace is constantly subject to influences beyond the control of the smelters and the character of the iron [that the furnace produces] varies from time to time, and not unfrequently from day to day. . . . The founders here inform me that they would be glad to change the proportions of Nos. 1, 2 & 3 Greenwood [pig iron] . . . as their judgment might lead them to think best. But they are restrained by the terms of their contract, which require them to follow the trial gun No. 1070 "as closely as possible in the proportions and quality of the metal" [quality that the contract set by brand and grade of iron, not by on-site testing of the metal]. Any deviation from the proportions of the different classes might, therefore, be regarded as a breach of contract.[122]

Founders' understanding of the metallurgy of heavy casting was imperfect throughout the Civil War and beyond. Gun crews on both sides were killed and maimed by their own cannons. A Senate committee reported at the end of the war that more men were lost by navy cannons' exploding and killing their own crews "than the loss sustained on the entire fleet from the fire of the enemy."[123] The need to improve safety drove efforts toward technological change. As in small arms manufacturing, some government standards looked to older, tried-and-true methods, such as the requirement that foundries use cold-blast charcoal iron. The mentality embodied in the military system, however, was decidedly new, based as it was on a faith in mathematical precision and instrumental measurement instead of craft tradition. Almost incidentally, the new scientific model of inspection, testing, and experimenting with new methods of casting challenged the conventions of artisanal production at large foundries, though both sides immediately understood the shift in power relations that such changes carried with them. At rolling mills, the engineer-managers who most vigorously pursued technological change during the Civil War also saw scientific methods as offering solutions to particular problems, most notably those related to manufacturing railroad rails. The biggest issue at rolling mills, however, was not the quality of rails but how to increase the volume of production beyond what existing technologies allowed.

The most serious shortages of iron for the Union, according to historian Thomas Weber, were generated by demand for railroad track, locomotives, and other kinds of "railroad iron," including wagon wheels, couplings, spikes, and the "chairs" that spikes were driven through to seat track firmly on the wooden ties. Track mileage in the North increased by over 19 percent from 1860 to 1865, from 21,276 to 25,372 miles. Most railroad iron went not to new lines, however, but to maintaining and repairing existing track. Wrought-iron rails that had been laid in the 1840s and 1850s tended to give way under long trains carrying heavy loads of troops and military freight—much heavier traffic than the rails had been designed to carry. Higher train speeds and heavier locomotives "play[ed] havoc with iron rails . . . of inferior manufacture," Weber writes. By 1863, "Scarcity of labor and interruption of work in rolling mills meant even less replacement of worn rail than was expected, with the consequent rough and increasingly dangerous track. Heavy traffic sometimes caused the iron to peel off in flakes, and occasionally rail ends would curl up and plow into cars."[124]

The best solution was to use steel rails, or to cap iron rails with a topcoat of steel, as European railroads had begun to do. Steel, being harder, resisted wear much better than iron. The first Bessemer steel rails imported from Great Britain arrived at Philadelphia in 1862. John Edgar Thomson, president of the Pennsylvania Railroad, ordered cast steel rails for the most heavily used tracks at the railroad's termini and stations in mid-1863. The company also ordered steel tires for locomotives and steel axles for passenger cars.[125] Very little steel was being made in the United States, however; it remained a specialized commodity that was used in limited amounts during the war.

The idea of steel, however, with its promise of being a form of iron that could withstand massive weight, pressure, torque, and tension without breaking or losing its shape, had a powerful impact on the thinking of some of the most influential men in the iron industry, among them Abram Hewitt and engineers Alexander L. Holley and John Fritz. Hewitt avidly followed the growth in Bessemer steel production at the Krupp works in Prussia and elsewhere as early as 1860. In early 1862, while visiting Europe to learn the techniques involved in ironclad ship construction, Holley "visited Sheffield, saw Henry Bessemer's steel making process in action, and was deeply impressed by its powers." He negotiated licensing arrangements for use of the patented process at the Albany Iron Works in Troy, where Fritz went to see the process. In 1868 the two men began transforming the Bethlehem Iron Works into one of the largest steel rail mills in the world.[126]

Although rail shortages during the Civil War put spurs to engineers' and entrepreneurs' pursuit of Bessemer steel, they had been trying for years to find a technological solution to the bottleneck in production represented by puddlers and rollers. The process for refining large amounts of iron that Henry Cort had developed in southern England and that reached its apogee in South Wales in the early nineteenth century consisted of the twin technologies of puddling and rolling. Puddling was the arduous manual process that rid molten iron of most of its impurities by stirring it to expose all the metal to oxygen. Rollers and their crews performed the only slightly less physically demanding work of manhandling pasty, white-hot masses of puddled iron through rapidly revolving rollers, which forced out molten slag that had formed during puddling. Puddlers and rollers could work only so many pounds (at most, so many hundredweight) of iron at a time. The limits of human strength and endurance set limits on the amount of finished iron a rolling mill could produce.[127]

A company could increase output arithmetically by adding puddling furnaces, roll trains, and skilled workers to operate them. The largest American rolling mill in 1858 had thirty single puddling furnaces and twelve heating furnaces, with total capacity of 17,808 tons. In 1865, the largest mill had seventeen double furnaces, thirty-one singles, and fourteen heating furnaces, with a total capacity of 34,000 tons.[128] Significant increases could also come from the intensification of labor. Historian Chris Evans estimates that the Dowlais rolling mill in Merthyr Tydfil increased the output of puddling teams by about 50 percent by modifying the puddling process. Lining the furnace with oxide-rich slag hastened decarburization, producing a "chemical reaction in the furnace bowl [that] was spectacularly rapid and violent (hence 'pig boiling' as it was known)." This allowed a puddler to produce more iron during his shift—which increased the amount of iron he loaded into the puddling furnace, manipulated with his rabble, and lifted out of the furnace for each "heat." The work of boilers, as they were called, was even more physically punishing than that of puddlers.[129]

Enlarging a mill and using the boiling method could increase output up to 150 percent. Making the leap to exponential growth—growth sufficient to gird a nation the size of the United States with railroads, for example—required a technology that replaced manual labor with machines that refined much greater volumes of iron while maintaining the desired qualities of whatever kind of metal the producer needed to make. The Bessemer process was patented by English engineer Henry Bessemer in 1855 and put into operation at his steelworks in Sheffield in 1858. It was the first technology to produce large quantities of "mild steel," which was tougher than wrought iron while still malleable enough to make many of the products traditionally made of wrought iron. It is not surprising that Bessemer began his quest for metal with these qualities while attempting to develop burstproof cannons during the Crimean War.[130] Transplanting the Bessemer converter to the United States was easy. The difficulty came from the chemical differences between British and American iron ores. The Bessemer process required ore that contained very low levels of phosphorus and sulfur. Low-phosphorus ores were rare in the eastern United States, and deposits were small.

One of the few exceptions was the rich lode at the Cornwall Mine in Lebanon, Pennsylvania.[131]

The obstacles to Bessemer steel production in the United States bought time for rolling mill artisans precisely when demand for iron peaked in the North. Labor historian Grace Palladino notes that the war "provided the economic leverage that allowed Northern workers to assert their rights," including ironworkers. "Demand for skilled workers so exceeded supply" in Pittsburgh, historian Francis G. Couvares writes, "that, even with the importation of metal workers from England, the Vulcans needed only to strike or threaten to strike to win their point quickly." Striking puddlers at one plant "won the exceptional rate of $6 per ton after only a few days out."[132] The war also instilled a new feeling of common cause among ironworkers and politicized their sense of identity. Labor historian David Montgomery notes that rolling mill artisans exemplified the new class consciousness that emerged among American industrial workers during the Civil War. On one hand the war kindled "intense nationalism" among artisans in places such as Philadelphia, Johnstown, Danville, Troy, and Pittsburgh, where skilled ironworkers volunteered for the army in disproportionately high numbers in 1861 and 1862. The Apollo Rolling Mill, near Pittsburgh, had to shut down for a spell when most of the workforce joined the Union Army.[133] Enrollments initially weakened the embryonic union movement in urban iron communities, beginning with the Iron Molders Union founded in Philadelphia in 1858 and the puddlers' and boilers' union, the United Sons of Vulcan, founded in Pittsburgh that same year. By 1863, however, "rising rents and higher food prices triggered a wave of trade organizing and strikes" among industrial workers. Where employers tried to enforce their control through antistrike legislation and military suppression of strikes, ironworkers responded with a new kind of militancy, particularly immigrant artisans in large iron towns.[134]

A few strikes by ironworkers were very successful. Pittsburgh puddlers went on strike in 1863 and won a rate of $6 a ton. After spring 1862, the Troy branch of the Iron Molders International Union grew to four hundred members. In March 1864 their strike produced a 15 percent increase in wages. In May and June puddlers in Troy led a series of strikes that "crippled iron production" at large mills. By the winter of 1864, Troy had six active branches of metalworkers' unions, including the Iron Rollers Association and the International Union of Tin, Copper, and Sheet Iron Workers. The limit to workers' power, however, was also reflected in their wages. Puddlers in Troy earned annual wages of from $600 to almost $800 in 1860. In 1865, they earned $1,000 to more than $1,200.[135] While these were hefty increases, they lagged well behind the rate of inflation. In some places, employers dealt with strikes more harshly than in Troy, where ironworkers' and garment workers' unions were particularly strong. In March 1864, for example, laborers at the West Point Foundry struck for a daily wage of $1.50, an increase of 20 to 50 percent over their stagnant level of pay. According to Iron Molders International Union president William H. Sylvis, four of the foundry strikers were arrested, held without trial for seven weeks, and banned from the works. Sylvis reported that "two companies of United States soldiers were ordered to the place, and martial law proclaimed, and the men forced to resume work at the old prices."[136]

Such forceful opposition to workers' protests was common in the anthracite coalfields during the war. As Palladino explains, "Because anthracite played a vital role in the Union's war effort, fueling not only the steamships that blockaded the South, but iron manufactories in the North, the Civil War offered Pennsylvania miners their first real chance to organize successfully. War-related demand for coal rendered 'combinations' and strikes especially effective. At the same time, however, military demand enabled coal operators to link work-related struggles to disloyalty." Owners branded strikers as a threat to the security of the Union, to bolster political support and justify calling in police and military force to compel miners to return to work.[137]

I have found no evidence that strikes by skilled ironworkers turned violent during the war. That the union included boilers indicates that some rolling mills had intensified labor in ways that workers in Wales resented, yet I find no record of workers at American mills protesting the change.[138] Several factors may have contributed to the relatively good relations between Northern iron workers and their employers. Many mill owners protected skilled workers through exemption from military service. Those who enrolled in volunteer units often went to war under the command of mill managers. Their common experience in the field forged close bonds that remained strong when workers returned home. Foreign-born workers, who still accounted for a large share of ironmaking artisans during the war, were exempt from the federal draft if they had not yet voted or declared their intention to become American citizens. Although the Enrollment Act of March 3, 1863, eliminated occupational exemptions in the North, War Department policy typically granted individual requests to excuse "skilled gun-makers" at private firms and exempted "all artificers and workmen employed in any public arsenal or armory."[139] In short, aside from the still distant threat of the Bessemer converter, the worst that most Northern ironworkers had to fear during the war was late payment of their wages and a declining standard of living.

The Civil War had profoundly different impacts on the iron industries of the South and the North. Although the South possessed a few modern, large-scale rolling mills and foundries, the general underdevelopment of manufacturing in the region seriously disadvantaged the Confederacy, and it lost ironmaking capacity from the moment the war began. The Confederate government and private investors poured millions of dollars into expanding iron production, particularly for ordnance and railroads, yet few works were newly established or significantly enlarged. The chronic difficulty of securing skilled workers became an insuperable crisis at many Confederate ironworks. The ironworks in the Valley of Virginia, Lynchburg, and Richmond were fiercely protected by Confederate forces, yet even in Virginia, shortages of crude iron and artisans became acute. As the Union Army claimed Southern territory, its major lines of advance tended to follow industrial corridors. Union troops destroyed or disabled Confederate ironworks wherever they went. The Tredegar Iron Works was spared in recognition of its value to the federal government, but few others in Virginia, Georgia, and Alabama remained in operable condition by April 1865.

Quite the opposite happened in the North, where military demand spurred private investors to enlarge existing works, build new ones, and revive many that had declined or gone out of operation during the prewar slump. Labor shortages were a problem in the North, particularly in the first two years of the war when iron artisans enlisted to support the Union cause. Northern firms also competed for the kinds of iron most needed for military ordnance and railroads. Shortages of men and iron, however, were never as severe in the North as in the South, and they benefited workers more. Increased demand for wrought and cast iron gave free white rolling mill and foundry workers greater leverage in their relations with employers. A heightened sense of occupational identity helped bridge the ethnic gap between English, Welsh, Irish, and American-born iron artisans in Pennsylvania and New York, where workers successfully, if episodically, organized branches of the Iron Molders Union and the Sons of Vulcan and won some concessions from owners. Only one ironworks in the North was damaged during the war.[140] Northern works and their laboring communities never faced the absolute scarcities that sent Shelby agent John M. Tillman up and down the lower Mississippi Valley searching for slave labor, coal, hay, and bacon.

Other aspects of the war's effect on labor relations do not so neatly follow the North-South divide. Shared experiences in battle created bonds of brotherhood between Johnstown rolling mill workers and their supervisors. Similar loyalties developed between Giles Edwards and Shelby's white artisans under the pressure of many difficulties and perhaps their common sense of ethnic-occupational identity. In both sections, the imposition of quasi-military discipline sometimes inflamed worker-management conflicts, as at West Point Foundry and the anthracite coalfields of eastern Pennsylvania, while at other times it provided a greater sense of security for workers, as at Tredegar.

The evidence I have found supports Walter Licht's conclusion that the Civil War "spawned few technological or organizational advances."[141] It was all that Confederate iron companies could do to produce iron of adequate quality to make reasonably good battleship cladding or to cast cannons that would not explode after firing at the enemy a thousand times. In the North, a few exceptionally innovative ironmasters conceived of new products or recognized opportunities to hasten the adoption of mass-production technologies such as the Bessemer process. Northerners had more capital at their disposal to pursue innovations and the luxury of safe passage across the Atlantic to study British technologies in person. Even they, however, hit the same invisible drags on technological change that had plagued the American iron industry throughout the antebellum period, including the chemical differences between US and British iron ore and coal deposits. One is also struck by how swift and short the Civil War was compared with the pace of change in the iron industry. Even the young industrial behemoth of the Union managed to bring only three new rail mills into operation before the war was finished. Solving the multitude of problems posed by wartime production was more a matter of improvisation and making do than technological innovation.

The North emerged from the war with a significantly larger iron industry. It was more diverse in technology and scale than in the 1850s, with the addition of several integrated ironworks that approached the size of Dowlais and Cyfarthfa even as small charcoal furnaces and bloomeries caught a second wind. The geographical extent and distribution of the industry had also changed. The antebellum centers of gravity in iron production in the Mid-Atlantic region and the Pittsburgh area became the twin hubs of America's late nineteenth-century industrial boom, while new works in Chicago and other western cities seeded heavy industry in the agricultural heartland. The war had shaped new landscapes of production where one could glimpse the future of ironmaking, labor, and technology in the American iron industry.

> I have no doubt that if our mill were moved to Chicago, and Lake Superior iron used, we could make a success in this business.
>
> ABRAM HEWITT, May 9, 1867

CONCLUSION

American Iron

Had Peter Lesley toured the American iron industry after the Civil War, he would have seen significant changes since he, Joseph Lesley, and Benjamin Smith Lyman conducted their survey. Technologies that Lesley and others in the American Iron Association had touted as best practice before the war were more evident in the industrial landscape. Anthracite furnaces such as those he had so admired in Catasauqua in 1855 were thicker on the ground in southeastern Pennsylvania and the Hudson Valley and were now operating as far west as Buffalo, New York. New rolling mills were more likely to be part of integrated works with one or more blast furnaces on site. The Bethlehem Iron Company complex, for example, included an anthracite furnace to supply its wrought-iron rail mill, and a second furnace went into blast in 1867.[1] Integrated iron production was changing the landscape of Pittsburgh as well. The long, low roofs of rolling mills that had lined the Monongahela and Allegheny Rivers for more than three decades were now punctuated by the stacks of coal-burning blast furnaces (fig. 68). Clinton Furnace, the first in Pittsburgh to run on coke, had been erected in 1859. A year later the furnace began burning coke from the Connellsville district. Other Pittsburgh firms built successively larger coke furnaces in 1861, 1863, and 1865, spurring development of the Connellsville mines and coke ovens.[2] Steel production was still minute, but it was growing. Pittsburgh had four crucible steel producers before the war, seven by 1865. Several firms achieved technical breakthroughs in making steel of uniform quality. Bessemer steel was slowly catching on, first in 1864 at an experimental works in Wyandotte, Michigan, then with commercial success in 1865 at the Albany Iron Works in Troy, New York. The Pennsylvania Steel Company in Harrisburg blew in its first converter in 1867.[3]

Outside the Mid-Atlantic region and Pittsburgh, Lesley mostly would have seen isolated ironworks on the same scale as in the 1850s, though ironmaking technologies, ownership, and markets were changing in some places. In the Hanging Rock iron district, Jefferson Furnace was one of a number of charcoal iron companies whose stockholders reinvested their wartime earnings in new coal-fired blast furnaces that burned locally mined bituminous coal. Many old charcoal furnaces in the district continued to operate—Jefferson was the last to go out of business, in 1916—though their share of the region's pig iron production was beginning to decline.[4] The ore-rich Juniata district in central Pennsyl-

Figure 68. Jones and Laughlins American Iron Works, Pittsburgh. Historical Society of Western Pennsylvania. This print, published ca. 1875, shows the new density of rolling mills along the Monongahela River. The tallest chimney is probably venting smoke from the central power generator for the works. The Eliza Furnaces are visible on the opposite shore.

vania continued to produce charcoal pig iron and blooms, but the number of independent producers was dwindling as larger iron companies in Pittsburgh and Johnstown bought furnaces and forges to secure supplies of known quality.[5] Counties north of the Juniata district that had raised hopes so high in the 1830s were virtually abandoned as sites for ironmaking because of their low-grade ores, mediocre coal, and remoteness from markets. New England's long-lived Salisbury district also lost steam after the war. Managers' loyalty to traditional methods of smelting and forging was increasingly out of step. Salisbury was stagnating even as its high-quality charcoal pig fetched a premium of up to 60 percent in eastern markets.[6] The vigorously growing industry in Pittsburgh included more companies of every size, producing more kinds of iron and steel, than the city had at any point before the war.[7] The growth of rail and inland water transportation networks in the North had also softened the boundaries of regional iron markets, which formerly had chiefly been defined by major topographic barriers.[8]

The iron region most profoundly changed by the Civil War was, of course, the Appalachian South. Some ironworks that federal troops had damaged or destroyed were rebuilt in the late 1860s, but the devastation of the southern economy, the collapse of the region's markets, the continuing scarcity of skilled labor, and the ruinous state of transportation inhibited would-be iron manufacturers from investing in the region. Though some tried. In Jefferson County, Alabama, a southerner with financial backing from Cincinnati investors put Irondale Furnace back into blast in 1866. Despite technological improvements to the charcoal furnace, including enlarging the furnace stack and applying the hot blast, the owner gave up in the mid-1870s "after seven years of heartache." Josiah Gorgas, former chief of ordnance for the Confederacy, ran Bibb (later Brierfield) Furnace from 1866 to 1869. Historian W. David Lewis writes that Gorgas "had a hard time recruiting workers and was constantly short of cash. Chronic squabbling with the owners of the Alabama and Tennessee Rivers Railroad weighed him down; eventually, bitter and disillusioned, he decided to sell out."[9]

Northerners thought they could do a better job. Many agreed with Joseph Lesley's assessment of the South. He declared in 1856 that all that was needed to realize the industrial potential of southern resources was for "Yankees or Pennsylvanians to come & show them how."[10] Abram S. Hewitt was one of the first to do so after the war (fig. 69). With typical alacrity, he bought the Chattanooga Rolling Mill at government auction in 1865. Hewitt meant to demonstrate how "to help set the stricken section on its feet." He was an absentee owner in this case, letting others run the mill that John Fritz had refitted for the Union Army. By 1868 the company (renamed the Roane Iron Company) was turning a profit by rerolling rails for southern railroads. Hewitt may also have encouraged the company to erect a large coke blast furnace in 1867, the first such furnace built in the South.[11]

Figure 69. Abram S. Hewitt, ca. 1860. Negative 79116, Collection of the New-York Historical Society.

Generally, however, northern capitalists were wary of investing in the southern iron industry. In addition to the region's many social problems and poor infrastructure, bituminous coal deposits in the South had relatively high levels of impurities, and in most of the southern Appalachians coal and iron ore deposits were not close together.[12] David Thomas's gradual investment in southern iron suggests the wisdom of waiting until the region's infrastructure was improved and its mineral resources were thoroughly evaluated. In 1866, Giles Edwards invited Thomas to visit Alabama. Thomas obliged his old acquaintance and made a tour of central Alabama in 1866 and 1867. Impressed by what he saw, Thomas formed the Pioneer Mining and Manufacturing Company in 1869 with his sons, Samuel and Edwin, and a Pennsylvania associate. Over the next eighteen years, the company acquired Tannehill Furnace and its mines, a former plantation of nearly two thousand acres, and land in four counties at rock-bottom prices. During part of this period, they employed Giles Edwards as a prospector to locate mineral deposits. Edwards may have been the one who found an ore vein and coal seam very near each other in Jones Valley, four miles southwest of the tiny settlement of Birmingham. David Thomas died in 1882. His son Samuel continued to bide his time. Not until 1887 did he resign as president of the Thomas Iron Company and move south to build the first of several coke-fired blast furnaces in the new company town of Thomas, Alabama. With Hokendauqua as its template, Thomas became something of a model town for the New South, as its furnaces were soon leading producers in the new Birmingham iron district.[13]

Another story that connects the antebellum northern industry to the postbellum South involves Daniel Tyler, the former manager at Farrandsville Furnace. In 1872, when Tyler was the seventy-year-old president of the Mobile and Montgomery Railroad, he formed a partnership with English immigrant Samuel Noble to make iron in Alabama. Noble had been tutored in iron manufacturing by his father at foundries, rolling mills, and machine shops in Pennsylvania before the family moved to Georgia. Tyler reputedly told Noble, "I have had the iron business burned into me and have not forgotten my first experience, but if I can find a property that has on it everything for making iron, without buying any raw material, or bringing any to it, I might be tempted to go into the business again." The property at Oxford Furnace in Calhoun County, Alabama, possessed considerably better minerals than Farrandsville had; it had made good charcoal iron during the war. Tyler took an active interest in the venture, which soon included a company town called Anniston in honor of Tyler's daughter-in-law, Annie Tyler. He was buried at Anniston in 1882.[14]

Immediately after the war, Pittsburgh and its surrounding region were the main story in American iron circles. The city that journalist James Parton called "hell with the lid taken off" in 1868 had become the country's leading producer of finished iron and steel and the focal point for a major region of heavy industry.[15] Abram Hewitt worried about the shift of the industry's center of gravity west of the Alleghenies. As rail orders slackened and the US economy went into a general slump after the war, Hewitt tried to persuade his old business partner, Peter Cooper, to relocate the Trenton Iron Works west to be within range of "good coke and cheap iron ore." "I have no doubt," Hewitt wrote his friend Edward Cooper, "that if our mill were moved to Chicago, and Lake Superior iron used, we could make a success in this business."[16] Two large coal-fired rolling mills had already been built in Chicago, the first in a wide arc of coal-fired ironworks and steelworks that would develop around the southern end of Lake Michigan.[17] Any place with lake, canal, and rail connections to the Marquette, Menominee, and northern Minnesota iron ranges (fig. 70) had the potential to produce iron. Not only did the Lake Superior ore beds hold a seemingly inexhaustible supply of iron, but many deposits contained only traces of sulfur and phosphorus. American manufacturers finally had realistic hopes for making Bessemer steel on a truly industrial scale.[18] Blocked by Cooper's unwillingness to leave New York, Hewitt found another means of improving their company's competitiveness. An alternative technology for producing large amounts of steel, the Siemens-Martin open-hearth process, which was perfected in England and France in the mid-1860s, proved more suitable for American ores east of the Alleghenies and in Alabama. Hewitt's decision to build open-hearth furnaces at Trenton in 1867–68 was one of his last contributions to the iron industry before he took up the even more unpredictable business of politics as mayor of New York City.[19]

Hewitt also left an enduring assessment of the American iron industry in his report on the iron exhibits at the Paris Universal Exhibition in 1867. Among his fellow delegates to the Exhibition was Peter Lesley, who was sent to assess the state of European mining technology. By this time Lesley was feeling his age. He worried about his wife, Susan's, fragile health and complained about the heat. His

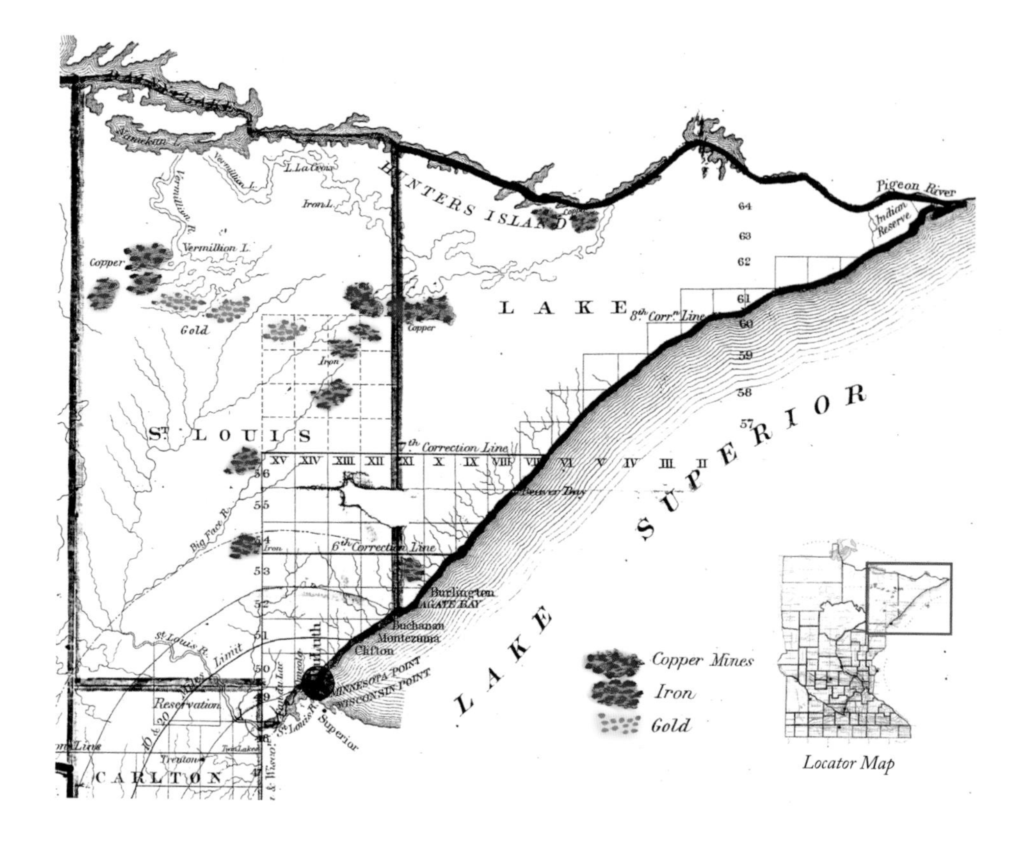

Figure 70. Iron, copper, and gold mines in northern Minnesota, 1866. The figure picks out newly charted metal-mining regions that are highlighted in softer tones on Joseph S. Wilson's original map, "Sketch of the Public Surveys in the State of Minnesota." Minnesota Historical Society.

slim journals from the trip say little about the Exhibition.[20] Hewitt, on the other hand, closely studied the displays of European competitors, and he extended his stay to visit a number of major ironworks on the Continent and in Britain. He concluded that French, German, and British manufacturers held a commanding lead over the United States in all branches of iron and steel manufacturing. Hewitt praised the superiority of European firms' rolled iron and cast steel. He marveled at the enormous showpiece items in their displays, such as a wrought-iron beam that was one foot wide and thirty-two feet long, a steel rail fifty feet long (which had been bent double without a fracture), and a "monster" rifled steel cannon from the Krupp works in Essen that weighed fifty tons and fired shells weighing nearly half a ton. Knowing how difficult it was to roll iron evenly to great lengths, Hewitt admired specimens of coiled wire as long as nine hundred yards from a company in Manchester.[21] The American ironmaster was most impressed—and rather shaken—by his visits to the Krupp works and Le Creusôt in Burgundy. He was astonished by the size and output of these places. "The establishment of Krupp," Hewitt reported, "occupies about 450 acres, of which one-fourth are under roof." The ironworks and steelworks employed 8,000 men, the coal mines, blast furnaces, and ore mines another 2,000. Krupp produced 61,000 tons of cast steel from 412 smaller furnaces and ran 195 steam engines. Le Creusôt rivaled Dowlais, with 15 coke-fired blast furnaces and a workforce of 9,950 men. Its 1,400-foot rolling mill housed 150 puddling and 85 heating furnaces. The company mined all its own coal, and the extensive works were linked by telegraph. The settlement at Le Creusôt, Lesley observed, included 650 houses around a central square. It was a planned industrial city even larger than Dowlais but considerably more attractive.[22]

What mattered most in Hewitt's eyes was the quality of the iron and steel that Krupp and Le Creusôt produced and the sophistication of their manufacturing methods. He figured they led American manufacturers by a decade or more. Where American machine-tool companies welded or bolted sections of iron or steel together to make large machine parts, the Europeans produced apparently stronger beams and pistons without a single seam. Furthermore, they achieved superior quality using the mass-production techniques of the Bessemer process. Americans seemed to have no regard for quality in their manufacturing, Hewitt complained, and the world knew it. "In the Welsh iron works it is notorious that the quality of the article produced is directly proportioned to the price paid for it, and in my visits to those gigantic establishments which have grown up in the mountains of South Wales it was humiliating to find that the vilest trash which could be dignified by the name of iron went universally by the name of the American rail."[23]

The epithet from Wales was a comment on both the poor quality of American iron and the fact that American railroad companies were willing to purchase the lowest grade of rails. Hewitt's frustration with America's lagging development was understandable. His own firm's innovations before and during the Civil War had proved that US iron companies could compete with high-grade European imports. His criticism of American ironmakers was misplaced, however—a rare moment when this astute man somewhat missed the point. Statistics that Hewitt collected at the Paris Exhibition showed that the United States had pulled even with France as the world's second largest iron producer (table 8). It had done so not by building gargantuan, sophisticated facilities at state expense, such as the Krupp works and Le Creusôt, or by replicating the British model of coke iron refined at massive rolling mills and foundries. Although the US industry benefited from government assistance in the form of federal import tariffs and state-funded infrastructure improvements, American ironworks were built entirely by private capital. The US iron industry had grown mainly by making do with whatever resources were available in a given region, though in many places American entrepreneurs and the immigrant workers they hired pushed to implement technologies for larger-scale production. Only in the Lehigh Valley did those technologies succeed on the first attempt, thanks to the similarity of anthracite in Pennsylvania and South Wales, the proximity of the

Table 8. Leading iron producers, 1866 (in tons)

	Pig iron	Wrought iron	Total
England	4,530,051	3,500,000	8,030,051
United States	1,175,000	882,000	2,057,000
France	1,200,320	844,734	2,045,054
Prussia	800,000	400,000	1,200,000
Belgium	500,000	400,000	900,000
Russia	408,000	350,000	758,000
Austria	312,000	200,000	512,000
Zollverein	250,000	200,000	450,000
Sweden	226,676	148,292	374,968

Source: Abram S. Hewitt, "The Production of Iron and Steel in Its Economic and Social Relations," in *Selected Writings of Abram S. Hewitt*, ed. Allan Nevins (New York: Columbia University Press, 1937), 64.

anthracite district to urban markets, the extent of existing transportation infrastructure, and the liberal attitude of the capitalists who recruited David Thomas to implement the processes he knew so well. It took considerably longer to implement British methods in other American iron regions.

There was no single, distinctly American model of ironmaking in the antebellum and Civil War periods. The US industry had, however, developed a number of broad characteristics that differed to varying degrees from the British model and from other European nations' iron industries. First, American ironmaking was remarkably diverse. The value of charcoal iron for making agricultural implements and machinery of all kinds stimulated charcoal furnace construction throughout the period. At the same time, territorial expansion and the dynamic growth of the US population and the industrializing economy inspired entrepreneurs to attempt larger-scale production, which required mineral-fuel, hot-blast technology, larger blowing engines, and coal mining as well as the bundle of technologies used at large rolling mills and foundries. Although the impurities in much Appalachian coal and the distance between coal and ore deposits slowed the development of coke iron production, the presence of coal from Pennsylvania to Alabama and in parts of the Middle West (notably Indiana and Illinois) supported the use of the hot blast and steam power even at rural charcoal furnaces. By 1868, US iron companies were using every known smelting and refining technology, in a variety of hybrid combinations as well as in the classic formations we associate with the extremes of rural plantation ironworks and integrated, large-scale manufacturing. The workforce at American companies was also diverse. Industrial wages remained high enough in the United States to lure English, Welsh, Scottish, Irish, German, and French iron artisans throughout the antebellum period and the Civil War. Diversity could be problematic, for mixing ethnic groups or races proved volatile at ironworks in both the North and the South. Labor relations were generally smoothest where managers maintained racial and ethnic divisions along lines of occupation and skill.

Second, the American industry was far more spatially dispersed, over greater distances, than was the British industry. This dispersion had several facets. For one thing, ironworks were ubiquitous because demand was ubiquitous. Iron was smelted and refined in almost every state east of the Mississippi River, and in some states, in most counties. The United States also had a great many small ironworks in this period because ground transportation was poor in many areas where ironmaking resources were located. The same topographic obstacles that slowed construction of canals and railroads before the Civil War somewhat insulated small, relatively inefficient ironworks from competition from imports and larger domestic producers. The antebellum United States was a nation of small farmers and small manufacturers. The major sectors of iron production were dispersed (blast furnaces distant from rolling mills, for example) because large deposits of iron ore, coal, and limestone rarely existed in the same locale. The few vertically integrated firms in the antebellum period, such as the Trenton and Lehigh Crane iron companies, had to transport raw materials over greater distances than did their British counterparts. In addition, most major deposits of coking coal, metallurgical bituminous coal, and iron ore were in remote areas far from urban settlements. Rolling mills' location along navigable rivers and early railroad lines pointed the way to solving that problem: knitting together dispersed resources and manufacturing centers by building far-reaching transportation networks.

Human connections were crucial to modernizing this American industry as well. David Thomas was the Samuel Slater of the iron industry. Like Slater, the English immigrant who first established British textile factory technology in the United States, Thomas could not implement British technologies simply by replicating machinery at an appropriate site using local labor. Making iron, like weaving textiles, required appropriately skilled labor, ample capital, and markets.[24] Iron artisans were scarcer than textile operatives on both sides of the Atlantic. The teamwork required at rolling mills and foundries put a particularly high premium on replicating, or somehow creating, an efficient work culture, which led many firms to import experienced British workers. Some sent their managers across the Atlantic for a firsthand education in British production processes. But as the case studies here suggest, managerial awareness and immigrant workers did not guarantee success. British methods had developed in situ over time, just as Swedish charcoal ironmaking methods had.[25] In this sense, the gradual development of the American industry repeated the common story of industrial adaptation to geographical circumstances. The international movement and adjustment of people, work cultures, industrial processes, and machinery, as well as capital, patents, and the extensive trade in iron back and forth across the Atlantic, represent an early stage in the globalization of heavy industry.

Third, the American industry emphasized producing iron of adequate quality rather than either high-quality iron or massive amounts of the metal. As in the American railroad and textile industries, this tendency was driven both by the nature of American markets and by the necessity of underpricing better-quality imports.[26] Hewitt may have wanted US companies to produce the world's finest steel, but Americans needed stoves and plows. Pittsburgh's first surge of growth in the antebellum period came from providing foundry iron for such staples of the American market.[27] Although the Civil War exposed the problems of unreliable iron quality, it took many more years for metallurgical science to solve the technical problems of producing iron and steel of consistent quality on a large scale. Antebellum and wartime manufacturers did try to match the quality of specialty iron imported from Europe and to compete with the low cost of British pig iron and rails.[28] The low-cost revolution in American heavy industry, however, did not come until the end of the nineteenth century, and then it came in manufacturing steel, not iron, which remained a more labor-intensive, smaller-scale business.[29]

Fourth, the case studies presented here suggest that American iron companies were most likely to succeed if owners and managers genuinely valued specialized skills and knowledge. Not only were experienced skilled workers essential for implementing scale-changing technologies; knowledge of local conditions, such as the quality of resources or the seasonality of a river's flow, was key to avoiding costly mistakes. Providing scope for considerable personal freedom of action by free and enslaved artisans appears to have been a more effective management strategy than enforcing company rules. Conflicts tended to arise where inexperienced American managers tried to assert more authority than skilled workers thought they deserved, or in urban iron towns when large numbers of workers were threatened with unlivable wages or unemployment.

Last, the American iron industry was embedded in many kinds of places, from the remote pine forests of central Maine, the Adirondacks, and the Cumberland Valley of western Tennessee to the crowded riverfronts of Philadelphia, Pittsburgh, Cincinnati, and St. Louis. These varied landscapes of production were the cumulative expression of the many factors that shaped the industry's development over the course of seven decades. Those places in turn influenced the daily lives of residents in ironmaking communities. No one landscape or building, any more than a single product or machine, could encapsulate this foundational industry.

The period from 1800 to 1868 was more than a transitional phase from the antique world of colonial ironmaking to the age of steel. The industry during this time had its own complex character, which was unlike ironmaking in the eras that came before and after it. Colonial ironworks were uniformly small, burned only charcoal, operated almost exclusively in rural or village settings, and primarily served local markets, though some coastal producers were enmeshed in the Atlantic market.[30] Although ironworks and steelworks remained diverse throughout the nineteenth century and up to the present day, the signal change after the Civil War was the establishment of giant steelworks, such as Carnegie's Edgar Thompson Works in North Braddock, Pennsylvania. The scale of production, size of the workforce, and global market penetration of America's largest iron and steel companies dwarfed all other nations' producers by the first decade of the twentieth century.[31]

That transformation began with the introduction of the Bessemer process, which a handful of US firms were using by 1868. This new smelting technology made a mighty impression (fig. 71).[32] Alexander Holley, the engineer who implemented Bessemer technology at the Albany Iron Works, narrated the process for the *Troy (NY) Daily Times* in 1865:

> As the combustion progresses, the surging mass grows hotter, throwing out splashes of liquid slag; and the discharge from its mouth changes from sparks and streaks of red and yellow gas to thick, full, white, howling, dazzling flame. But such battles cannot last long. In a quarter of an hour the iron is stripped of every combustible alloy, and hangs out the white flag. The converter is then turned upon its side, the blast shut off, and the recarburizer run in. Then for a moment the war of the elements rages again; the mass boils and flames with higher intensity, and with a rapidity of chemical reaction, sometimes throwing it violently out of the converter mouth; then all is quiet, and the product is steel,—liquid, milky steel, that pours out into the ladle from under its roof of slag, smooth, shining, and almost transparent.[33]

Holley's awed, sensuous description echoes Rebecca Blaine Harding's evocation of hell at Wheeling rolling mills, where she saw "fire in every horrible form." But now there were no "bent ghastly wretches stirring the strange brewing."[34] The sublime scene focused on the machine that had replaced laboring men, as indeed the Bessemer process was meant to do. Where an experienced puddler could refine at most 1,500 pounds of wrought iron a day, a Bessemer converter could produce five tons of steel with every blast. Neither Bessemer converters nor open-hearth furnaces were implemented swiftly in the United States. By 1870, only five Bessemer steel plants had been built, two of which lay idle; ten were operating in 1873.[35] Because steel did not directly replace wrought iron for

Figure 71. Bessemer converters blazing at a Pittsburgh steel-works, 1886. Wood engraving in *Harper's Weekly*, drawn by Charles Graham.

most products, however, the number of puddlers, heaters, and rollers actually increased through the 1870s. Historian John William Bennett points out that the new technology eliminated "the puddling bottleneck . . . not by mechanizing the process but by replacing the finished product, wrought iron, with another product, Bessemer steel." The replacement was very gradual. Puddling remained the only way to produce wrought iron on a commercial scale until 1930. Allegheny County, Pennsylvania, which includes Pittsburgh, still had approximately 2,000 puddlers at the turn of the twentieth century.[36]

Technological change was not the only threat to skilled ironworkers. During the Civil War, the general manager of the Cambria Iron Works, Daniel J. Morrell, began a sustained campaign to weaken artisans' control over production. Refusing to allow Cambria puddlers and heaters the conventional prerogative of hiring their own helpers, Morrell hired inexperienced men and then forced the skilled workers to pay the "green hands" the same wages as experienced helpers. When disputes arose in 1866 between helpers and the heaters and puddlers who supervised them, Morrell sided with the helpers, arguing publicly that he wanted to defend the rights of "poor men who are glad to get any kind of employment at any price" against the "secret organization of high paid workmen" who "set a crushing foot upon their necks." By driving a wedge between mostly immigrant artisans and American-born helpers, Morrell stifled the heaters' union and the Sons of Vulcan, both of which deserted Cambria for seven years. Through tough anti-union tactics and paternalistic support of workers' families and social institutions, Cambria made Johnstown a company town whose workers never developed a strong network of mutual support or resistance.[37]

The Sons of Vulcan built a much stronger union in Greater Pittsburgh, where organizers such as Miles Humphreys resisted company demands and firms were smaller and more numerous than in Johnstown. The puddlers' union was founded in Pittsburgh in 1858, not long before workers at the Sligo Rolling Mill had their pictures taken with the tools of their trade (fig. 72). The tintypes of Adam and "Salty" Hart capture the "individuality and the essential dignity of work" that were typical of such working-class portraits. The same look in dozens of other tintypes taken of Sligo workers during the Civil War suggests the pride of ironworkers, whether or not they joined the union.[38] The Sons of Vulcan remained a small, underground organization until September 1862, when Humphreys was elected president and mounted a campaign to establish local "forges" across the country. In 1867, after an eight-month strike, Pittsburgh mill owners recognized the union and agreed to a sliding scale of wages based on the market price of bar iron. By 1870, most puddlers in the city belonged to the union.[39]

Between 1873 and 1876, the Sons of Vulcan led nearly eighty strikes, most of them in Pennsylvania. The combined threats of anti-unionism, technological change, and unemployment, which hit iron and steel communities hard during the 1873–74 depression, galvanized craftsmen and semiskilled workers to find common cause. In 1876 the Amalgamated Association of Iron and Steel Workers was organized. The Amalgamated merged the Vulcans, the Brotherhood of Heaters, Rollers, and Roughers, and the Roll Hands Union.[40] A labor anthem published in Pittsburgh's *National Labor Tribune* in 1875 expressed their hopeful determination. It was written by Reese E. Lewis, who set the words to the tune of the Welsh anthem "Men of Harlech":

Figure 72. Sligo Rolling Mill workers, ca. 1860–64. Adam Hart (*left*) worked at Sligo as a heater. F. "Salty" Hart (*right*) was an assistant roller at the mill. Heinz History Center.

Rouse, ye noble sons of Labor, and protect your country's honor,
Who with bone, and brain, and fibre, make the nation's wealth.
Lusty lads, with souls of fire, gallant sons of noble sire,
Lend your voice and raise your banner, battle for the right.
Heater, roller, rougher, catcher, puddler, helper,
All unite and join the fight, and might for right encounter,
In the name of truth and justice, stem the tide of evil practice,
Mammon's sordid might and av'rice, our land from ruin save.[41]

This stirring song and the union meetings where it was sung marked a turning point in labor-management relations in the American iron industry. Rather than migrating away from low pay or undesirable working conditions, iron artisans and their assistants were choosing to "stay in place and fight," to borrow a phrase from geographer David Harvey.[42] As iron manufacturing became increasingly concentrated in industrial cities and towns such as Johnstown and Pittsburgh, the number of workers and families who stayed in such places also grew. Working-class communities took shape within networks strengthened by numbers and the common experiences of emigration, occupation, unemployment, and politicized resistance. New landscapes were also taking shape, including iron and steel cities that would become even larger and more dominant than Merthyr Tydfil had been at its height, as well as new sites of resource extraction and coke production that created intense pollution and human suffering.[43] These places would be battlegrounds for another generation of struggle in American industry.

ACKNOWLEDGMENTS

While researching and writing this book I have accumulated many debts to individuals and institutions whose support made it possible. Among the friends and professional acquaintances who encouraged me to carry on, I owe deepest thanks to Michael P. Conzen, Robert B. Gordon, Christopher T. Baer, Richard G. Healey, and Edward K. Muller. They believed in this project from the start and provided good advice whenever I came back with more questions. Michael gave me invaluable help with my research on Farrandsville Furnace and provided a key source of supplemental information for the Lesley historical GIS that undergirds much of this book. His enthusiasm for combining fieldwork with archival research has long influenced my approach to historical geography. Bob Gordon patiently corrected my misunderstandings of ironmaking technology. Chris Baer opened my eyes to the significance of the ironworks at Farrandsville, Pennsylvania, and he was never stumped by my detailed questions about the history of Mid-Atlantic railroads and canals. Richard Healey taught me most of what I know about relational databases while helping me design and build the Lesley historical GIS, and he was a stimulating collaborator on our article about the Pennsylvania iron industry. I will never forget the evening he and his wife, Bettina, drove me, blindfolded, to the place outside Portsmouth, England, where Henry Cort invented puddling. Ted Muller has always been willing to offer frank, sensible editorial advice and to write letters for grant and fellowship applications that kept the research afloat. These friends' collegial interest and good humor, as well as the excellence of their own scholarship, has inspired and buoyed me throughout this project. Bob, Chris, and Ted read the entire manuscript, as did Chris Evans, who generously shared his knowledge of continental European ironmaking. Stephen J. Hornsby's critical reading of multiple drafts helped me find the necessary shape of this book. I am grateful to him in more ways than I can say.

Many other colleagues answered questions, suggested sources, or influenced my thinking, including Jeremy Atack, John Bezís-Selfa, David Brody, Andrew J. Croll, Charles B. Dew, Kenneth Fones-Wolf, Gregory Galer, Laurence Glasco, William Hunter, John M. Ingham, Daniel Johnson, William D. Jones, Gregg D. Kimball, Annie Laurant, Ronald L. Lewis, Lee R. Maddex, Patrick M. Malone, Rolf Manne, David R. Meyer, David Montgomery, John R. Moravek, Gerald Morgan, Anne E. Mosher, Michael Moss, William H. Roberts, John Staudenmaier, SJ, Robert Steinfeld, David West, and James Wilson. While working in the emerging field of historical GIS as this book was taking shape, I met scores of scholars who shared my excitement for what geographical questions, sources, and methods bring to historical scholarship. I appreciate all their questions and suggestions. The friendship of Amy Hillier, Geoff Cunfer, Ian N. Gregory, David Bodenhamer, Trevor and Sylvia Harris, John H. Long, James R. Akerman, and the Holocaust Geographies team, especially Alberto Giordano, Paul B. Jaskot, and Tim Cole, sustained me over the long haul. I also appreciate the encouragement of my students at Middlebury College, whose sharp questions and creative work in historical geography classes and research seminars influenced my thinking in many ways.

Excellent archivists and generous private individuals helped me locate, and permitted me to consult, rare materials in their collections. In roughly chronological order, heartfelt thanks to Christopher S. Duckworth, Ohio Historical Society; Marty Everse, Tannehill Iron Works State Park, McCalla, Alabama; Lance E. Metz, National Canal Museum, Easton, Pennsylvania; Margarette Evans, Pittsburgh; Sallie Sypher, Putnam County Historical Department and Records Center, Brewster, New York; Rob Cox, Charles B. Greifenstein, Scott Ziegler, and Valerie-Anne Lutz van Ammers, American Philosophical Society, Philadelphia; Edward Rogers, Poncha Springs, Colorado; Yves Randeynes, Decazeville, France; Jim Dershem and Bill Simcox, Farrandsville, Pennsylvania; Anne McCloskey, Lou Bernard, and Katharine A. Paulhamus, Clinton County Historical Society, Lock Haven, Pennsylvania; Jim Gaffney and the Rev. Douglas Cronce, Catasauqua, Pennsylvania; Chester Kulesa and Connie Richards, Anthracite

Heritage Museum and Iron Furnaces, Scranton, Pennsylvania; John R. Hébert and Edward Redmond, Geography and Maps Division, and Jeffrey M. Flannery and Patrick Kerwin, Manuscript Division, Library of Congress, Washington, DC; Susan Edwards and Charlotte Hodgson, Glamorgan Archives, Cardiff, Wales; Sandra Wheatley, Institute of Geography and Earth Studies, University of Wales, Aberystwyth; David Cobb, Joseph Garvey, C. Scott Walker, and Bonnie Burns, Harvard Map Collection, Harvard University; Christopher Lee, Columbus Chapel and Boal Mansion Museum, Boalsburg, Pennsylvania; Rebecca Ross, Hopewell Furnace National Historic Site, Elverson, Pennsylvania; Laura Linard, Tim Mahoney, and Katherine Fox, Baker Library Historical Collections, Harvard Business School; Art Louderback, Richard Price, and Lauren Zabelsky, Senator John Heinz History Center, Pittsburgh; Bob Kreps of Lonaconing, Maryland; MaryJo Price, Special Collections, Lewis J. Ort Library, Frostburg State University, Frostburg, Maryland; Barbara Tuttle Kent, Lexington, Massachusetts; Richard T. Colton, Springfield Armory National Historic Site; Rachel Plant and Carolyn Tallen, Bixbie Memorial Free Library, Vergennes, Vermont; Steven Engelhart, Adirondack Architectural Heritage, Keeseville, New York; Peter Smithurst, Royal Armouries Museum, Leeds; Sis Hause, Danville, Pennsylvania; Carol Salomon, Cooper Union Library, New York, New York; Carolyn Jacob, Merthyr Tydfil Public Library, Merthyr Tydfil, Wales; Rachel Manning, Davis Library, Middlebury College; Martha Bace, W. S. Hoole Special Collections Library, University of Alabama; Jim Bennett, Alabama Historic Ironworks Commission; and Thomas M. Whitehead, Special Collections Department, Temple University. I learned invaluable lessons about the crafts of puddling and rolling from Kenneth Hall and Graham Collis, Blists Hill Victorian Village, Ironbridge, Shropshire. Linda Cowan and Mary Wolff, AmesINK, Boulder, Colorado, extended great kindness and hospitality in allowing me to work in the J. Peter Lesley papers in the Ames Family Historical Collection.

I was fortunate to collaborate with two talented cartographers and many gifted student research assistants during this project. Michael Hermann worked tirelessly to design the first set of maps, which appeared in print and online in the paper I coauthored with Richard G. Healey, published in the *Journal of Economic History*. Chester Harvey immensely improved my crude compilations for the book's maps and frequently broke the tension of sustained effort with his humor and friendship. Miriam Neirick, my first undergraduate research assistant at Wellesley College, entered most of the attribute data in the Lesley historical GIS. In addition to Chester, my Middlebury College research assistants were Nicholas Emery, who extracted data on labor at ironworks from the McLane Report; Brooke Medley, who designed and built the Pennsylvania transportation GIS; Garrott Kuzzy, who compiled and designed figures 9 and 11 and assisted with the Lesley historical GIS and transportation historical GIS; Christopher Lizotte, whose digital work with the 1851 Ordnance Survey map of Merthyr Tydfil enabled me to analyze the microgeography of the Dowlais and Cyfarthfa ironworks; and Charlie Hofmann, who helped with permissions. Staff at the University of Portsmouth who assisted Richard G. Healey in locating Mid-Atlantic ironworks include Trem Stamp, Paul Carter, and David Kidd. Several skilled translators helped as well. Althea Tyndale translated articles on the Canadian iron industry from the French. Matthieu Sanders translated French documents at the Centre Pompidou and the Bibliothèque Nationale, Paris. Maria Alessandra (Sasha) Woolson's translations of Spanish letters related to investment in Farrandsville Furnace were key to understanding the final phase of that venture.

Mastering Iron was generously supported by research fellowships from the American Council of Learned Societies and the National Endowment for the Humanities, which named this a "We the People" project. Research and travel grants from the British Academy and the Learned Societies Fund and College Research Fund, University of Wales, Aberystwyth, launched my research in American archives. Grants from Wellesley College and Middlebury College paid for research assistants, archival research, and travel to conferences to present the work in progress. Cartography was supported by grants from the National Welsh American Foundation and the Association of American Geographers. Publishing the figures in full color was possible thanks to support from Middlebury College, Columbia University Seminars, Ruth K. Morris, and Richard P. Smith Jr. The index was supported by the Digital Humanities Initiative at Hamilton College, thanks to angel David Nieves. As visiting professor at the École des Hautes Études en Sciences Sociale, Paris, in May 2002 and international visiting scholar at the Centre for Data Digitisation and Analysis at Queen's University, Belfast, in September 2005, I benefited from presenting parts of this work to European audiences. I thank my hosts, Paul-André Rosental and Nancy Green in Paris and Paul S. Ell and Ian N. Gregory in Belfast, for their hospitality and encouragement.

My daughter Kate has lived with this book her whole life; it was conceived the same year she was. One day when she was about ten, I told Kate much more than she wanted to know about my Iron Book. Sensing my anxiety about the complexity of it all, she tried to reassure me that it would all come together in the end. Later that day I found a note next to my computer: "It's just a book." I thank Kate and her dad, Larry Knowles, for their common sense and constant support. I also thank my editors at the University of Chicago Press, Christie Henry and Abby Collier, for their cheerful faith and professionalism and for their choice of perceptive scholars as the anonymous reviewers of the book proposal and manuscript. Alice Bennett's scrupulous editing caught errors I was blind to (thank goodness). Thanks also to the rest of the sterling production team, including senior manuscript editor Erin DeWitt, designer Ryan Li, production manager Joan Davies, indexer Susan Hernandez, and in marketing, Micah Fehrenbacher.

My peculiar interest in the iron industry may have begun when I was about ten years old and became fascinated by pictures of the *Monitor* and the *Virginia* in an American Heritage book about Civil War ironclads. My father read that book aloud at the dinner table after supper, one of many he read to entertain and educate his four children. He loved history, and books. He and our mother organized vacation trips around historic places they wanted us to see—a copper mine in northern Michigan; Monticello and historical museums in Washington, DC; the battlefields of Gettysburg and Custer's last stand; Devils Tower and Mount Rushmore. I remember Dad reading to us from *The Book of Mormon* so that we would understand the significance of the Angel Moroni when we saw him blowing his golden trumpet atop the Mormon temple in Salt Lake City. Our family trips gave me a lifelong curiosity about places where American history happened. What gifts those trips were. Thank you, Papa.

APPENDIX

A Note on Historical GIS

Historical GIS (HGIS) is an interdisciplinary field that focuses on using geographic information systems and related geospatial technologies to study the past. Constructing a database of geographically located historical information constitutes a major part of the research in many HGIS projects, as it did in this one. I have explained elsewhere what kinds of historical subjects and sources have generally proved most amenable to an HGIS approach.[1] Although many historical sources have been digitally scanned for online access in recent years, and numerous infrastructure projects have created digital historical datasets to make HGIS research easier, scholars still often have to create their own databases from the particular sources necessary to answer their questions. In my case, I was very fortunate to find an exceptionally well-suited source of longitudinal data on the American iron industry in J. Peter Lesley's *The Iron Manufacturer's Guide* (1859). Not only does this compendium include the great majority of iron producers in operation from 1785 to 1858, it is also systematically organized, which made it relatively easy to extract the historical and geographical information the *Guide* contains. Lesley captured details of ironworks' construction, location, management, and production that were crucial for mapping and analyzing the historical development of the antebellum iron industry. Few print sources provide such a wealth of information for a major industry in the nineteenth century. I could not have written this book without Lesley's *Guide*, nor could I have made extensive use of that source without converting its contents into GIS format.

That process involved two basic tasks that go into building many historical GIS databases. To create the "attribute" database, Miriam Neirick and I distributed the information contained in each entry into its corresponding field, or slot, in a tabular relational database. This information included names of owners; dates of construction, improvement, and abandonment; amount of iron produced; kinds of technology employed at the works, and so forth (fig. 73). The second process was to create the companion spatial database that located each ironworks.[2] Richard G. Healey and his assistants Paul Carter, Trem Stamp, and David Kidd designed this part of the HGIS and established our basic method for locating the works Lesley identified. For blast furnaces and rolling mills, each locational description in the *Guide* was checked against one or more maps or atlases before the corresponding point was placed on the digital map, drawn in a GIS program. (Forges and foundries outside Pennsylvania, whose locations are far less certain, were mapped only by county.) We relied mainly on the large-scale maps in the DeLorme state atlas and gazetteer series and the US Geological Survey's topographic map series, available online through the Survey's Geonames website.[3] We also consulted historical maps for some cities and other locations that required additional verification, making use of the high-resolution digital

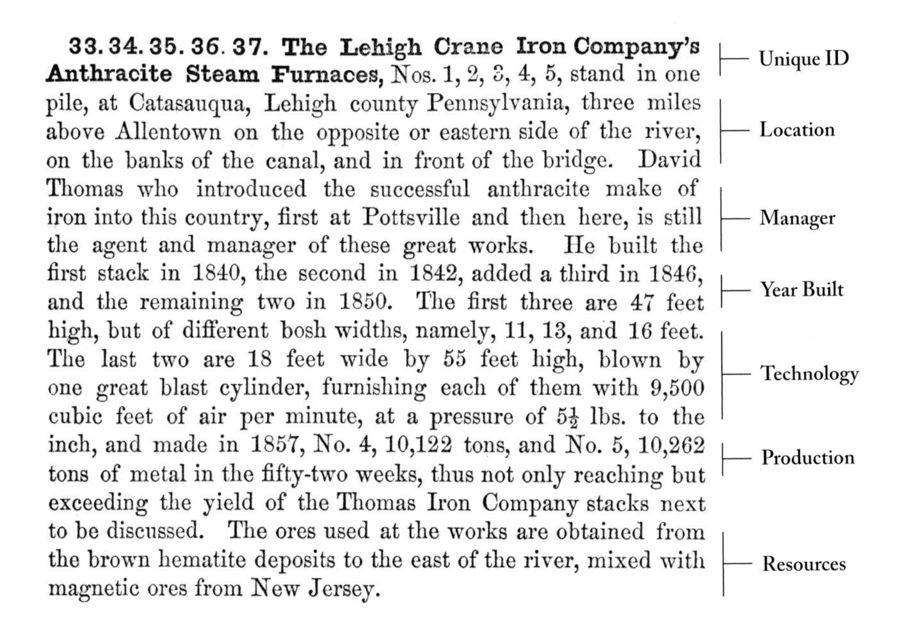

33. 34. 35. 36. 37. The Lehigh Crane Iron Company's Anthracite Steam Furnaces, Nos. 1, 2, 3, 4, 5, stand in one pile, at Catasauqua, Lehigh county Pennsylvania, three miles above Allentown on the opposite or eastern side of the river, on the banks of the canal, and in front of the bridge. David Thomas who introduced the successful anthracite make of iron into this country, first at Pottsville and then here, is still the agent and manager of these great works. He built the first stack in 1840, the second in 1842, added a third in 1846, and the remaining two in 1850. The first three are 47 feet high, but of different bosh widths, namely, 11, 13, and 16 feet. The last two are 18 feet wide by 55 feet high, blown by one great blast cylinder, furnishing each of them with 9,500 cubic feet of air per minute, at a pressure of 5½ lbs. to the inch, and made in 1857, No. 4, 10,122 tons, and No. 5, 10,262 tons of metal in the fifty-two weeks, thus not only reaching but exceeding the yield of the Thomas Iron Company stacks next to be discussed. The ores used at the works are obtained from the brown hematite deposits to the east of the river, mixed with magnetic ores from New Jersey.

Figure 73. Entry in Lesley's *The Iron Manufacturer's Guide* (1859), 8. Lesley's entry for the Lehigh Crane Iron Company, which he much admired, is one of the longest in the *Guide*. This verbal blueprint for how to build a modern ironworks illustrates his aim to provide key details of best practice to ironmasters across the country. The topics picked out on the right became the framework for the relational database of attributes for the Lesley HGIS.

scans of historical maps available through the Library of Congress's American Memory website.[4] Once the two halves of the HGIS were complete, we joined them, linking each uniquely numbered ironworks to its unique location. The HGIS could then return a map in answer to queries. The iterative process of asking hundreds of questions gradually revealed the geographical patterns embedded in Lesley's *Guide*, which are summarized in the maps in chapter 1.

As we built the Lesley HGIS, problems cropped up that required additional research. The most important issue was the need to augment Lesley's incomplete information on the years of construction and abandonment of ironworks. Healey, who located many Mid-Atlantic ironworks, and I found most of the missing dates in county and local histories. A complete bibliography of these sources comes at the end of this appendix, along with other sources used to create reference features in the maps that display Lesley's data. As the work continued, we recorded our sense of the relative accuracy of point locations in the database. In cities and most villages, we believe locations are spatially accurate to within several hundred meters for most ironworks; in rural areas, to within one to three miles. As chapter 1 explains, Lesley and his survey assistants gathered the most complete, and most accurate, information in the parts of the country they knew best, namely Pennsylvania, the Mid-Atlantic region, and New England. Data quality declined as they moved farther from home and deeper into the terra incognita of the trans-Appalachian West and especially the South.

This project also led to the creation of a more specialized historical GIS of Pennsylvania's iron industry in relation to river, canal, and railroad transportation. Then Middlebury College student Brooke Medley was the chief architect of the Pennsylvania transportation HGIS, with substantial assistance from fellow student Garrott Kuzzy. This HGIS was built mainly to answer questions of distance and proximity, which Healey and I explore in detail in our article on Pennsylvania's iron industry.[5] Our method for building this HGIS was somewhat different from that for the Lesley HGIS. Medley scanned and georectified the large sheet maps in Christopher T. Baer's *Canals and Railroads of the Mid-Atlantic States, 1800–1860*, which record the extent of transportation improvements at fifteen-year intervals. She then made a digital mosaic from several score of DOQQs (digital orthographic quarter-quadrangles: high-resolution, georeferenced digital aerial images, on which one can pick out the routes of railroads and canals). Using the Baer maps as visual references, she digitally traced those routes from the DOQQs, adding attributes for each line segment from Baer's tables of the routes' construction history such as year incorporated, year completed, type of canal or railroad, and owner. Lines representing channel improvements along rivers came from hydrographic data provided, like the DOQQs, by the State of Pennsylvania's spatial data clearinghouse.[6] The DeLorme *Pennsylvania Atlas and Gazetteer* and historical maps resolved locational questions. The final HGIS also included thirty-meter digital elevation data for the state of Pennsylvania. Because this database was constructed to answer questions of distance, we took pains to be locationally precise. Fortunately, the quality of Lesley's information on the whereabouts of Pennsylvania blast furnaces and rolling mills made that subset of the Lesley HGIS a good match for the transportation HGIS. We estimate that our distance measurements combining these two datasets are accurate to within five hundred meters.

Those interested in further details of the construction or content of the Lesley HGIS or the Pennsylvania transportation HGIS may consult the metadata documents available on my website (http://www.middlebury.edu/academics/geog/faculty/knowles).

Sources Used to Build and Map the Lesley Historical GIS

Africa, J. Simpson. *History of Huntingdon and Blair Counties, Pennsylvania*. Philadelphia: Louis H. Everts, 1883.

Alexander, John H. *Report on the Manufacture of Iron*. Baltimore: Fielding Lucas Jr., 1840.

American Iron and Steel Association. *The Ironworks of the United States: A Directory of the Furnaces, Rolling Mills, Steel Works, Forges and Bloomaries in Every State*. Philadelphia: American Iron and Steel Association, 1876.

Bayles, Richard M. *History of Providence County, Rhode Island*. Vol. 1. New York: W. W. Preston, 1891.

Baer, Christopher T. *Canals and Railroads of the Mid-Atlantic States, 1800–1860*. Edited by Glenn Porter and William H. Mulligan Jr., cartographers Marley E. Amstutz and Anne E. Webster. Wilmington, DE: Regional Economic History Research Center, Eleutherian Mills–Hagley Foundation, 1981.

Beach, Ursula Smith. *Along the Warioto, or A History of Montgomery County, Tennessee*. Nashville, TN: McQuiddy Press, 1964.

Biggs, Nina Mitchell, and Mabel L. Mackoy. *History of Greenup County, Kentucky*. Louisville, KY: Franklin Press, 1951.

Boyer, Charles Shimer. *Early Forges and Furnaces in New Jersey*. Philadelphia: University of Pennsylvania Press, 1931.

Camplin, Paul. *A New History of Muhlenberg County [Kentucky]*. Nashville, TN: Williams, 1984.

Canby, Henry Seidel. *The Brandywine*. New York: Farrar and Reinhart, 1941.

"City of St. Louis." Map, no scale. N.p., 1885. Perry-Castañeda Library Map Collection, University of Texas at Austin. Accessed online at www.lib.utexas.edu/maps/historical/st_louis_1885.jpg.

Corlew, Robert Ewing. *A History of Dickson County, Tennessee*. Nashville: Tennessee Historical Commission and Dickson County Historical Society, 1956.

DeLorme. *Atlas and Gazetteer* series. Various scales. Yarmouth, ME: DeLorme, various dates.

Dew, Charles B. "David Ross and the Oxford Iron Works: A Study of Industrial Slavery in the Early Nineteenth-Century South." *William and Mary Quarterly* 31, no. 2 (1974): 189–224.

Elmer, Lucius Q. C. *History of the Early Settlement and Progress of Cumberland County, New Jersey*. Bridgeton, NJ: George F. Nixon, 1869.

Fernandez, Kathleen M. *A Singular People: Images of Zoar*. Kent, OH: Kent State University Press, 2003.

Galer, Gregory, Robert Gordon, and Frances Kemmish. "Connecticut's Ames Iron Works: Family, Community, Nature, and Innovation in an Enterprise of the Early American Republic." *Transactions [of the] Connecticut Academy of Arts and Sciences* 54 (December 1998): 83–194.

Hall, Ruby Franklin. *History of Hardin County, Illinois*. Carbondale, IL, 1970.

Historical and Biographical Annals of Columbia and Montour Counties, Pennsylvania. Vol. 1. Chicago: J. H. Beers, 1915.

History of Franklin County, Pennsylvania. Chicago: Warner, Beers, 1887.

History of Litchfield County, Connecticut. Philadelphia: J. W. Lewis, 1881.

History of Oneida County, New York. Philadelphia: Everts and Fariss, 1878.

The History of Tuscarawas County, Ohio. Chicago: Warner, Beers, 1884.

Hurd, Duane Hamilton. *History of Bristol County, Massachusetts*. Philadelphia: J. W. Lewis, 1883.

Jackson County: A History. Dallas, TX: Taylor Publishing for Jackson County [Wisconsin] Historical Society, 1984.

Johnstone, Hallie Tipton. *History of Estill County [Kentucky]*. N.p., 1974.

Landis, George B. "The Society of Separatists of Zoar, Ohio." *Annual Report of the American Historical Association for the Year 1898*. Washington, DC: Government Printing Office, 1899.

Lesley, J. Peter. *The Iron Manufacturer's Guide to the Furnaces, Forges, and Rolling Mills of the United States*. New York: John Wiley, 1859.

[Lesley, J. Peter, ed.]. *Bulletin of the American Iron Association*. Philadelphia: American Iron Association, 1856–59.

Lloyd, Col. Thomas W. *History of Lycoming County, Pennsylvania*. Vol. 1. Topeka, KS: Historical Publishing, 1929.

Lytle, Milton Scott. *History of Huntingdon County, . . . Pennsylvania*. Lancaster, PA: William H. Roy, 1876.

Maddex, Lee. Untitled list of West Virginia blast furnaces. Typescript, courtesy of the author.

Massie, Larry B., and Peter J. Schmitt. *Kalamazoo: The Place Behind the Products*. Sun Valley, CA: American Historical Press, 1998.

Mathews, Alfred, and Austin N. Hungerford. *History of the Counties of Lehigh and Carbon. . . .* Philadelphia: Everts and Richards, 1884.

McClain, Iris Hopkins. "A History of Stewart County, Tennessee." Bound typescript. [Columbia?], TN, 1965.

McGrain, John W. *From Pig Iron to Cotton Duck: A History of Manufacturing Villages in Baltimore County*. Towson, MD: Baltimore County Public Library, 1985.

Meginness, John F., ed. *History of Lycoming County, Pennsylvania*. 1892. Reprint, Baltimore: Gateway Press, 1990.

National Geographic Society. *Road Atlas 1998*. Washington, DC: National Geographic Society, 1998.

National Park Service. *Cumberland Gap*. Map, scale indeterminate. Accessed online at http://www.nps.gov/cuga/planyourvisit/upload/CUGAmap1-3.pdf.

Neilson, William G., comp. *The Charcoal Blast Furnaces, Rolling Mills, Forges and Steel Works of New York in 1867*. Philadelphia: American Iron and Steel Association, 1867.

Patterson, Tom. *Natural Earth I*. 500-meter world land cover map data for the coterminous United States, landcover and shaded relief. Accessed via http://www.shadedrelief.com/natural/pages/download.html.

Perrin, William Henry, ed. *History of Summit County [Ohio]*. Chicago: Baskin and Battey, 1881.

Pesgraves, James S., ed. *Wythe County Chapters*. Wytheville, VA: privately published, 1972.

"Philadelphia." Map, scale indeterminate. In H. S. Tanner, *The American Traveller, or Guide through the United State*, 8th ed. New York, 1842. Accessed online at www.lib.utexas.edu/maps/historical/philadelphia_1842.jpg.

[Providence, Rhode Island]. Untitled map, scale 2,600 feet to one inch. Buffalo, NY: J. N. Matthews, 1904. Accessed online at www.lib.utexas.edu/maps/historical/providence_ri_1904.jpg.

Ransom, James M. *Vanishing Ironworks of the Ramopos*. New Brunswick, NJ: Rutgers University Press, 1966.

Richards, John Adair. *A History of Bath County, Kentucky*. Yuma, AZ: Southwest Printers, 1961.

Rockey, John L. *History of New Haven County, Connecticut*. 2 vols. New York: W. W. Preston, 1892.

Ruttenber, E. M., and K. H. Clark, comps. *History of Orange County, New York*. Philadelphia: Everts and Peck, 1881.

Scharf, Thomas J. *History of Delaware, 1609–1888*. Vol. 2. Philadelphia: L. J. Richards, 1888.

Schneider, Norris F. *Y Bridge City: The Story of Zanesville and Muskingum County, Ohio*. Cleveland: World, 1950.

Shaefer, Peter W. *Official Coal, Iron, Railroad, and Canal Map of Pennsylvania*. Map, no scale. [Harrisburgh]: by authority of the Legislature of Pennsylvania, 1864.

Sharp, Myron B., and William H. Thomas. "A Guide to the Old Stone Blast Furnaces in Western Pennsylvania," part 4. *Western Pennsylvania Historical Magazine* 48, no. 4 (1965): 365–88.

Stevens, Lewis Townsend. *The History of Cape May County, New Jersey*. Cape May City, NJ: L. T. Stevens, 1897.

Thomas, William H. Papers. MS 31. Historical Society of Western Pennsylvania Archives, Pittsburgh.

US Geological Survey. Geographic Names Information System (GNIS). Accessed online at http://geonames.usgs.gov.

———. 7.5 minute digital elevation model (DEM) for Pennsylvania, 30 meter resolution. Reston, VA: US Geological Survey, 2000. Retrieved from http://www.pasda.psu.edu/data/dem24k/.

———. Digital Orthophoto Quarter-Quadrangles (DOQQs) for Pennsylvania. Reston, VA: US Geological Survey, various years. Retrieved from http://www.pasda.psu.edu/data/doq99/.

———. National Hydrography Dataset (NHD) for Pennsylvania. Reston, VA: US Geological Survey, 1999. Retrieved from ftp://www.pasda.psu.edu/pub/pasda/nhd/.

Virginia: A Guide to the Old Dominion. American Guide Series. New York: Oxford University Press, 1940.

Wayland, John W. *A History of Shenandoah County, Virginia*. Strasburg, VA: Shenandoah, 1927.

Welch, Sarah N. *A History of Franconia, New Hampshire, 1772–1972*. Littleton, NH: Courier, 1972.

West Virginia Archives and History. *Place Names in West Virginia*. Accessed online at www.wvculture.org/hiStory/placnamp.html on September 30, 2005.

Wilson, Harold Fisher. *The Jersey Shore: A Social and Economic History of the Counties of Atlantic, Cape May, Monmouth, and Ocean*. Vol. 1. New York: Lewis Historical Publishing, 1953.

NOTES

Introduction

1. "Iron," *Encyclopaedia Britannica*, vol. 12 (Chicago: William Benton, 1971), 598; Robert Raymond, *Out of the Fiery Furnace: The Impact of Metals on the History of Mankind* (University Park: Pennsylvania State University Press, 1986), 50–94.

2. Chris Evans and Göran Rydén, *Baltic Iron in the Atlantic World in the Eighteenth Century* (Leiden: Brill, 2007), chap. 2; Eugenia W. Herbert, *Iron, Gender, and Power: Rituals of Transformation in African Societies* (Bloomington: Indiana University Press, 1993).

3. Chris Evans and Göran Rydén, eds., *The Industrial Revolution in Iron: The Impact of British Coal Technology in Nineteenth-Century Europe* (Aldershot, UK: Ashgate, 2005).

4. Thomas Southcliffe Ashton, *Iron and Steel in the Industrial Revolution* (Manchester: Manchester University Press, 1963; reprint, Augustus M. Kelley, 1968); David S. Landes, *The Unbound Prometheus: Technological Change and Industrial Development in Western Europe from 1750 to the Present* (Cambridge: Cambridge University Press, 1969).

5. Charles B. Dew, *Ironmaker to the Confederacy: Joseph R. Anderson and the Tredegar Iron Works* (New Haven, CT: Yale University Press, 1966), Dew, *Bond of Iron: Master and Slave at Buffalo Forge* (New York: Norton, 1994), and Dew, "David Ross and the Oxford Iron Works: A Study of Industrial Slavery in the Early Nineteenth-Century South," *William and Mary Quarterly* 31 (1974): 189–224; Joseph E. Walker, *Hopewell Village: The Dynamics of a Nineteenth Century Iron-Making Community* (Philadelphia: University of Pennsylvania Press, 1966); James D. Norris,

Frontier Iron: The Story of the Maramec Iron Works, 1826–1876 (Madison: State Historical Society of Wisconsin, 1964); Gerald G. Eggert, *Making Iron on the Bald Eagle: Roland Curtin's Ironworks and Workers' Community* (University Park: Pennsylvania State University Press, 2000). Unpublished works include Julian C. Skaggs, "Lukens, 1850–1870: A Case Study in the Mid-Nineteenth Century American Iron Industry" (PhD diss., University of Delaware, 1975); Linda McCurdy, "The Potts Family Iron Industry in the Schuylkill Valley" (PhD diss., Pennsylvania State University, 1974); and Robert Hamlet McKenzie, "A History of the Shelby Iron Company, 1865–1881" (PhD diss., University of Alabama, 1971).

6. Peter Temin, *Iron and Steel in Nineteenth-Century America: An Economic Inquiry* (Cambridge, MA: MIT Press, 1964); Kenneth Warren, *The American Steel Industry, 1850–1970: A Geographical Interpretation* (Pittsburgh: University of Pittsburgh Press, 1973), Warren, *Wealth, Waste, and Alienation: Growth and Decline in the Connellsville Coke Industry* (Pittsburgh: University of Pittsburgh Press, 2001), and Warren, *Bethlehem Steel: Builder and Arsenal of America* (Pittsburgh: University of Pittsburgh Press, 2008); Paul Paskoff, *Industrial Evolution: Organization, Structure, and Growth of the Pennsylvania Iron Industry, 1750–1860* (Baltimore: Johns Hopkins University Press, 1983); John N. Ingham, *Making Iron and Steel: Independent Mills in Pittsburgh, 1820–1920* (Columbus: Ohio State University Press, 1991); David Brody, *Steelworkers in America: The Nonunion Era* (Cambridge, MA: Harvard University Press, 1960); Daniel J. Walkowitz, *Worker City, Company Town: Iron and Cotton-Worker Protest in Troy and Cohoes, New York, 1855–84* (Urbana: University of Illinois Press, 1978); John Bennett, "Iron Workers in Woods Run and Johnstown: The Union Era" (PhD diss., University of Pittsburgh, 1977); Robert B. Gordon, *American Iron, 1607–1900* (Baltimore: Johns Hopkins University Press, 1996). John Bezís-Selfa, *Forging America: Ironworkers, Adventurers, and the Industrious Revolution* (Ithaca, NY: Cornell University Press, 2004), analyzes labor and management in the colonial period and the early republic. Unpublished geographical treatments of the antebellum industry at the regional scale include John Richard Moravek, "The Iron Industry as a Geographic Force in the Adirondack-Champlain Region of New York State, 1800–1971" (PhD diss., University of Tennessee, 1976), and James Larry Smith, "Historical Geography of the Southern Charcoal Iron Industry, 1800–1860" (PhD diss., University of Tennessee, Knoxville, 1982).

7. Leading surveys of industrialization with this general focus include Walter Licht, *Industrializing America: The Nineteenth Century* (Baltimore: Johns Hopkins University Press, 1995), and David R. Meyer, *The Roots of American Industrialization* (Baltimore: Johns Hopkins University Press, 2003). The enduring historiographical interest in textile mills, railroads, and small arms dates to outstanding early studies in social history, econometric history, and the history of technology, including Thomas Dublin, *Women at Work: The Transformation of Work and Community in Lowell, Massachusetts, 1826–1860* (New York: Columbia University Press, 1979); David J. Jeremy, *Transatlantic Industrial Revolution: The Diffusion of Textile Technologies between Britain and America, 1790–1830s* (Cambridge, MA: MIT Press, 1981); Robert F. Dalzell Jr., *Enterprising Elite: The Boston Associates and the World They Made* (Cambridge, MA: Harvard University Press, 1987); Albert Fishlow, *American Railroads and the Transformation of the Ante-bellum Economy* (Cambridge, MA: Harvard University Press, 1965); Robert William Fogel, *Railroads and American Economic Growth: Essays in Econometric History* (Baltimore: Johns Hopkins University Press, 1964); and Merritt Roe Smith, *Harpers Ferry Armory and the New Technology: The Challenge of Change* (Ithaca, NY: Cornell University Press, 1977).

8. Philip Scranton, *Proprietary Capitalism: The Textile Manufacture at Philadelphia, 1800–1885* (Cambridge: Cambridge University Press, 1983).

9. The process was also prolonged and varied regionally in Britain; see Evans and Rydén, *Industrial Revolution in Iron.*

10. David E. Nye, *America as Second Creation: Technology and Narratives of New Beginnings* (Cambridge, MA: MIT Press, 2003), quotation on 3. See also John Stilgoe, *Metropolitan Corridor: Railroads and the American Scene* (New Haven, CT: Yale University Press, 1985).

11. Tench Coxe, "Digest of Manufactures, Communicated to the Senate, on the 5th of January, 1814," *American State Papers, Finance*, vol. 2, 13th Cong., 2nd sess. (doc. 407), 671.

12. Kim M. Gruenwald, *River of Enterprise: The Commercial Origins of Regional Identity in the Ohio Valley, 1790–1850* (Bloomington: Indiana University Press, 2002), 77–79; Neil Dahlstrom and Jeremy Dahlstrom, *The John Deere Story: A Biography of Plowmakers John and Charles Deere* (De Kalb: Northern Illinois University Press, 2005); Peter Way, *Common Labour: Workers and the Digging of North American Canals, 1780–1860* (Cambridge: Cambridge University Press, 1993), 32, 135–39; Angela Lakwete, *Inventing the Cotton Gin: Machine and Myth in Antebellum America* (Baltimore: Johns Hopkins University Press, 2003), 49–61, 86–87; Jeremy, *Transatlantic Industrial Revolution*, 13; Evans and Rydén, *Baltic Iron*, 139–40.

13. Robert B. Gordon, *American Iron, 1607–1900* (Baltimore: Johns Hopkins University Press, 1996), 202, 212–14; Louis C. Hunter, *Steamboats on the Western Rivers: An Economic and Technological History* (1949; reprint, New York: Dover, 1993), 121–80; Robert B. Gordon, personal communication, February 27 and May 19, 2010; Allan Nevins, *Abram S. Hewitt, with Some Account of Peter Cooper* (New York: Harper, 1935), 113–17; Warren Ripley, *Artillery and Ammunition of the Civil War* (New York: Promontory Press, 1970); William H. Roberts, *Civil War Ironclads: The U.S. Navy and Industrial Mobilization* (Baltimore: Johns Hopkins University Press, 2002).

14. Walter H. Crockett, *How Vermont Maple Sugar Is Made*, Bulletin 21 (Vermont Department of Agriculture, 1915), 10–13; Donald A. Hutslar, *Log Construction in the Ohio Country, 1750–1850* (Athens: Ohio University Press, 1992), 206–7, 211, 215; Fred W. Peterson, *Homes in the Heartland: Balloon Frame Farmhouses of the Upper Midwest, 1850–1920* (Lawrence: University Press of Kansas, 1992), 1, 5–8, 12–13; Stephanie Ellis, "The Piano Industry in America: 1775–1940," in *Atlas of Industrial America*, ed. Anne Kelly Knowles (Middlebury, VT: privately published at Middlebury College, 2007), 46; Charles Edward Goodrich, "Story of the Washburn and Moen Manufacturing Company, 1831–1899," typescript (1935), 5, 8, American Steel and Wire Collection, Baker Library Historical Collections, Harvard Business School.

15. Licht, *Industrializing America*.

16. See Rowland Tappan Berthoff, *British Immigrants in Industrial America, 1790–1950* (Cambridge, MA: Harvard University Press, 1953).

17. For overviews of these fields, see Anne Kelly Knowles, "GIS and History," in *Placing History: How Maps, Spatial Data, and GIS Are Changing Historical Scholarship*, ed. Anne Kelly Knowles (Redlands, CA: ESRI Press, 2008), 1–25; Ian N. Gregory and Paul S. Ell, *Historical GIS: Technologies, Methodologies, and Scholarship* (Cambridge: Cambridge University Press, 2008); and David J. Bodenhamer, John Corrigan, and Trevor M. Harris, eds., *The Spatial Humanities: GIS and the Future of Humanities Scholarship* (Indianapolis: Indiana University Press, 2010).

18. Bruno Latour, *Science in Action: How to Follow Scientists and Engineers through Society* (Cambridge, MA: Harvard University Press, 1987), 215–50.

19. David Harvey, *The Limits to Capital* (Chicago: University of Chicago Press, 1982; Midway reprint, 1989); David R. Meyer, *Networked Machinists: High-Technology Industries in Antebellum America* (Baltimore: Johns Hopkins University Press, 2006).

20. Anne Kelly Knowles, *Calvinists Incorporated: Welsh Immigrants on Ohio's Industrial Frontier* (Chicago: University of Chicago Press, 1997).

Chapter 1

1. Alan Birch, *The Economic History of the British Iron and Steel Industry, 1784–1879* (London: Frank Cass, 1967; reprint, New York: Augustus M. Kelley, 1968), 164–68, 222–24; Peter Temin, *Iron and Steel in Nineteenth-Century America: An Economic Inquiry* (Cambridge, MA: MIT Press, 1964), 21–22; Frederick W. Taussig, *The Tariff History of the United States*, 8th ed. (New York: G. P. Putnam's Sons, 1931; reprint, New York: Augustus M. Kelley, 1967), 123–31; *Memorial from Pennsylvania on the Manufacture of Iron* (Philadelphia: Convention of Iron Masters, 1850); Albert Fishlow, *American Railroads and the Transformation of the Ante-bellum Economy* (Cambridge, MA: Harvard University Press, 1965), 103–18, 137–38.

2. [J. Peter Lesley, ed.], *Bulletin of the American Iron Association* (Philadelphia: American Iron Association, ca. 1857), 1. The *Bulletin* was published serially but without clear dates or volume numbers. Pages in the bound volume I have used are numbered consecutively. Subsequent citations provide dates where Lesley included them.

3. Mary Lesley Ames, ed., *Life and Letters of Peter and Susan Lesley*, 2 vols. (New York: G. P. Putnam's Sons, 1909); John Howard Brown, "Peter Lesley," in *The Cyclopaedia of American Biography*, vol. 5 (Boston: Federal Book Company, 1903), 41; J. P. Lesley, *Manual of Coal and Its Topography* (Philadelphia: J. P. Lippincott, 1856); Benjamin Smith Lyman, "Biographical Notice of J. Peter Lesley," *Transactions of the American Institute of Mining Engineers*, reprinted in Ames, *Life and Letters*, 2:473.

4. Ames, *Life and Letters*, 2:6–23; Lyman, "Biographical Notice," 2:459.

5. Arthur A. Solocow, ed., "Pennsylvania," in *The State Geological Surveys: A History* (n.p.: Association of American State Geologists, 1988), 374; Lyman, "Biographical Notice," 462.

6. Lyman, "Biographical Notice," 463, 474; journal of J. Peter Lesley, March 21–October 4, 1862, box 6-37, Ames Family Historical Collection, AmesINK, Boulder, Colorado (hereafter JPL journal). I have been unable to locate any of Lesley's "great maps."

7. Lyman, "Biographical Notice," 475–78; Clifford H. Dodge, "The Second Geological Survey of Pennsylvania: The Golden Years," *Pennsylvania Geology* 18 (February 1987): 9–15.

8. J. Peter Lesley to Susan Inches Lesley (hereafter JPL to SIL), September 3, 1856, Ames Family Historical Collection.

9. JPL journal, January 7, 1856.

10. [Louis McLane], Secretary of the Treasury, *Documents relative to the Manufactures in the United States,* HR Doc. 308, McLane Report, 2 vols. (Washington, DC: Duff Green, 1833); Tench Coxe, "Digest of Manufactures, Communicated to the Senate, on the 5th of January, 1814," American State Papers, Finance, vol. 2, 13th Cong., 2nd sess. (Doc. 407). The states covered in McLane's report are Maine, New Hampshire, Vermont, Massachusetts, Rhode Island, Connecticut, New York, New Jersey, Pennsylvania, Ohio, and Maryland.

11. See, for example, Benjamin F. French, *History of the Rise and Progress of the Iron Trade of the United States from 1621 to 1857* (New York: Wiley and Halsted, 1858). An early, influential European example is Léon Coste and Auguste Perdonnet, *Mémoires métallurgiques sur le traitement minérals de fer, d'étain et de plomb en Angleterre* (Paris: Bachelier, 1830). The expanded second edition of this work is widely quoted in other iron industry reports: M. Dufrénoy, Élie de Beaumont, Léon Coste, and Auguste Perdonnet, *Voyage métallurgique en Angleterre* [A metallurgical journey in England], 2nd ed., 2 vols. (Paris: Bachelier, 1837).

12. JPL journal, December 2, 1857.

13. Lyman, "Biographical Notice," 459–60.

14. JPL journal, October 3, 20, 1856; January 2, 16, 1857. Someone named "Brown" had previously made lists of technical data for the association. This may have been William R. Brown of Pittsburgh, who joined the AIA by February 15, 1856, or another member of the Brown family who owned the Wayne Rolling Mill and other iron ventures in the Pittsburgh region. J. Peter Lesley, *The Iron Manufacturer's Guide to the Furnaces, Forges, and Rolling Mills of the United States* (New York: John Wiley, 1859), 250, 252, 253; John N. Ingham, *Making Iron and Steel: Independent Mills in Pittsburgh, 1820–1920* (Columbus: Ohio State University Press, 1991), 145, 146, 211, 229n65.

15. JPL to SIL, July 9, 1858; Lesley, *Bulletin*, 170.

16. "The Benjamin Smith Lyman Collection," online finding aid, University of Massachusetts, Amherst Library, accessed at www.library.umass.edu/subject/easian/lyman.htm on March 26, 2006; JPL journal, February 2, 1858; JPL to SIL, July 6 and July 17, 1856.

17. JPL to SIL, July 6, 1856.

18. JPL journals, 1858–65; JPL to SIL, September 13, 1858; "Lyman Collection."

19. JPL journal, December 2, 1856; Joseph Lesley obituary, *Philadelphia Enquirer*, February 12, 1889; Ames, *Life and Letters*, 1:10. Thanks to Charles Greifenstein and Valerie-Anne Lutz van

Ammers, American Philosophical Society Library, for their help in documenting Lesley's date of death.

20. Ames, *Life and Letters*, 1:8–9.

21. Ibid., 9; J. Peter Lesley to Mr. Rood, autobiographical letter cited in ibid., 10.

22. Obituary of Margaret White Lesley Bush-Brown, *New York Times*, November 18, 1944, 13.

23. On preparations for Joseph's survey of eastern Kentucky, see JPL journal, beginning March 17, 1858.

24. Linda Cowan, personal communication, February 5, 2006; JPL to SIL, April 23 and May 4, 1855.

25. Kirk Johnson, "In 200 Years of Family Letters, a Nation's Story," *New York Times*, January 29, 2006, sec. A, 1, 20. Significant portions of the letters are archived at the Massachusetts Historical Society and the American Philosophical Society. The letters and journals I consulted are in the Ames Family Historical Collection.

26. J. Peter Lesley to Joseph Lesley Jr., November 17, 1856, J. Peter Lesley Papers, J. Peter and Joseph Lesley Correspondence (hereafter JPL to JL or JL to JPL), American Philosophical Society.

27. Lesley, *Bulletin*, 31–32.

28. All calculations are based on the Lesley historical GIS (hereafter Lesley HGIS). For more on this database, see the appendix.

29. Ames, *Life and Letters*, 1:20–24; JPL to SIL, November 15, 1856.

30. JPL to SIL, November 12–December 9, 1856; JPL journal, November 9–December 8, 1856; Pennsylvania transport GIS, constructed by Brooke Medley, Anne Kelly Knowles, and Garrott Kuzzy, based on Christopher T. Baer, *Canals and Railroads of the Mid-Atlantic States, 1800–1860*, gen. eds. Glenn Porter and William H. Mulligan Jr., cartographers Marley E. Amstutz and Anne E. Webster (Wilmington, DE: Regional Economic History Research Center, Eleutherian Mills–Hagley Foundation, 1981), hereafter Pennsylvania transport GIS; JPL to SIL, November 12–December 8, 1856. For more on the Pennsylvania transport GIS, see the appendix.

31. JPL to SIL, November 14, 1856.

32. JPL to SIL, November 12, 1856.

33. JPL to SIL, November 17, 1856.

34. JPL to SIL, November 19, 21, 27, 1856.

35. J. P. Lesley, "Anthracite Blast Furnaces in Pennsylvania, 1856–7," *Bulletin of the American Iron Association*, 50; JPL journal, October 15, 1856; July 1, 1857. The table in the *Bulletin*, 50, is the only place where Lesley systematically noted the number of mules and horses used. He may have stopped recording labor because it was of little interest to members of the association or because the number of workers fluctuated so much during the year at many works that no single number could represent the workforce.

36. JPL to SIL, March 19, 1857.

37. Lesley, *Bulletin*, 73. Lesley's journal during and shortly after his tour of the anthracite district includes notes on several machine shops and foundries. See, e.g., November 17, 18, 19, and 21, 1856, and January 4, 9, 1857.

38. Lesley, *Bulletin*, 73; Lesley, *Guide*, 147–218.

39. JPL to SIL, June 2–30, 1857, quotation in June 3, 1857; "John Kintzing Kane" and "Thomas Kane," *American National Biography* online, accessed at www.anb.org on March 27, 2006.

40. JPL to SIL, December 24, 1857.

41. Ibid.

42. JPL to SIL, January 3, 23, 1858; Ben Lyman journal, 1857; JPL journal, 1858.

43. JPL to SIL, January 5, 17, 1858.

44. JPL to SIL, January 16, 1858.

45. JPL to SIL, January 5, 1858.

46. JPL journal, January 25, 1858.

47. The locations marked on Lesley's sketch map compare very favorably to a map compiled

in the early twentieth century by then state geologist Wilbur Stout, in "The Charcoal Iron Industry of the Hanging Rock Iron District: Its Influence on the Early Development of the Ohio Valley," *Ohio Archaeological and Historical Quarterly* 42, no. 1 (1933): 73. Stout's map was based on his own field investigations, which he recorded on US Geological Survey topographic quadrangles, with site photographs and verbal descriptions in a large scrapbook; see the Wilbur Stout Collection, vol. 408, Ohio Historical Society.

48. JPL to JL, August 28, 1857.

49. JPL to JL, February 16–26, 1857; JPL journal, January 25–29, 1858.

50. JL to JPL, May 27, 1857.

51. JL to JPL, June 26, 1857. On Weaver, see Charles B. Dew, *Bond of Iron: Master and Slave at Buffalo Forge* (New York: W. W. Norton, 1994).

52. JL to JPL, July 5, June 28, 1857; Lesley, *Guide*, 75–76, 245. Peter Lesley noted the tendency to overstate production in *Bulletin*, 52.

53. JL to JPL, June 26, 28, 1857; JL to Will [his brother William W. Lesley of Philadelphia], July 1, 1857; JPL to SIL, January 27, 1858.

54. JL to Will, July 1, 1857.

55. Dew, *Bond of Iron*, 15–18; "Flourtown, Pennsylvania," entry and maps, US Geological Survey Geographic Names Information System, accessed online at http://geonames.usgs.gov/gnispublic on March 25, 2006.

56. Ann Anderson, "Horace Ware," Calhoun-Shelby County, Alabama Archives Biographies, USGenWeb Archives accessed at http://www.rootsweb.com/~usgenweb/al/alfiles.htm on March 24, 2006; Joyce Jackson, *History of the Shelby Iron Company, 1862–1868* (Shelby, AL: Historic Shelby Association, 1990), 3–4.

57. JPL to JL, August 28, 1857; JL to JPL, September 10, 12, 21, 28, 1857; JPL journal, September 15–17, 1857; JPL to SIL, September 11, 1857. On the geography of the iron mines and works around Port Henry, see Jerry Jenkins with Andy Keal, *The Adirondack Atlas: A Geographic Portrait of the Adirondack Park* (Syracuse, NY: Syracuse University Press and Adirondack Museum, 2004), 224.

58. JL to JPL, September 28, 1857; E. C. Harder, "Iron Ores, Pig Iron, and Steel," in *Mineral Resources of the United States*, part 1, Metallic Products, Calendar Year 1908, Department of the Interior, US Geological Survey (Washington, DC: Government Printing Office, 1909), 89–90.

59. JPL journal, February 11, 13, 17, 24, 1858; JL to JPL, February 28, 1858.

60. JL to JPL, February 28, March 4, 8, 1858.

61. JL to JPL, March 8, 1858; Ames, *Life and Letters*, 1:283.

62. JL to JPL, March 18, 19, 1858.

63. Charles B. Dew, "Black Ironworkers and the Slave Insurrection Panic of 1856," *Journal of Southern History* 41, no. 3 (1975): 321–38. Peter Lesley mentions the insurrection in the *Bulletin*, 170, in a discussion of Tennessee ironworks based on Joseph's summary; JL to JPL, March 26 [1858].

64. JL to JPL, March 26 [1858].

65. JL to JPL, March 13, 18, 26, 1858; JPL journal, March 1, 8, 10, 17, 1858; "David Dale Owen," *American National Biography* online, accessed on March 27, 2006. Owen originally offered the survey to Peter, but Peter persuaded him to give the job to Joseph, retaining an advisory role himself.

66. James D. Norris, *Frontier Iron: The Story of the Maramec Iron Works, 1826–1876* (Madison: State Historical Society of Wisconsin, 1964); William R. Edgar, "History of Iron Mountain," undated article downloaded from www.rootsweb.com/~mostfran/mine_history/iron_mountain.htm on March 26, 2006.

67. JL to JPL, April 19, 20, 1858; Lesley, *Guide*, 262.

68. Benjamin Smith Lyman to J. Peter Lesley, April 25, 27, 1857, quotation in April 25, Benjamin Smith Lyman Collection, ser. 1, Correspondence, 1856–92 (Lesleys), University of Massachu-

setts at Amherst, W. E. B. DuBois Library, Special Collections and Archives (hereafter BSL to JPL); B. S. Lyman journal, July 11, 1857, Ames Family Historical Collection (hereafter BSL journal).

69. BSL to JPL, May 5–June 28, 1857.

70. JPL to SIL, May 9, 1857.

71. BSL to JPL, July 25, 1857.

72. BSL to JPL, July 26, 1857.

73. BSL to JPL, August 3, 5, 21, 28, 1857.

74. BSL to JPL, September 13, 20, 1857; 1850 Census, Cass County, Georgia, M432, reel 63, household 745.

75. BSL to JPL, October 20, 1857.

76. JPL to SIL, September 22, 1857; JPL journal, November 26 and December 2, 3, 1857; Lesley, *Bulletin*, 167, 174.

77. JPL journal, June 8, September 9, 1858, and January 11, February 3, 25, 26, March 11, 1859; JPL to SIL, August 16, September 25, 1858; quotation in JPL to SIL, June 26, 1858; Lesley, *Guide*.

78. JPL to SIL, July 10, 1859; JPL journal, June 8, 1860.

79. Virtually every history of the industry from James Moore Swank's *History of the Manufacture of Iron in All Ages* (1892) to Temin's *Iron and Steel in Nineteenth-Century America* (1964) has drawn on *The Iron Manufacturer's Guide*. Charles O. Paullin, "Iron and Steel Works, 1858," plate 135D, in *Atlas of the Historical Geography of the United States*, ed. John K. Wright, pub. 401 (Washington, DC: Carnegie Institute of Washington and the American Geographical Society of New York, 1932), maps the distribution of blast furnaces and rolling mills but does not account for the dynamics of works' opening or abandonment. In Kenneth Warren, *The American Steel Industry, 1850–1970: A Geographical Interpretation* (Oxford: Clarendon Press, 1973), figs. 1.1–1.6 map only the Mid-Atlantic and Northeast. See also Michael Williams, *Americans and Their Forests: A Historical Geography* (Cambridge: Cambridge University Press, 1989), 149, 151.

80. For a full explanation of the construction of the Lesley historical GIS and the sources used to build it, see the appendix, "A Note on Historical GIS."

81. Although Lesley's geographical descriptions vary in quality and detail, they proved remarkably accurate for blast furnaces and rolling mills when compared with other sources.

82. Bruno Latour, *Science in Action: How to Follow Scientists and Engineers through Society* (Cambridge, MA: Harvard University Press, 1987), 210–13.

83. Lesley HGIS compared with unpublished list of West Virginia blast furnaces compiled by Lee R. Maddex, historian with Christine Davis Consultants in archaeology and history, Verona, Pennsylvania.

84. Temin, *Iron and Steel*, appendix C; Paullin, *Atlas of Historical Geography*, plate 135; Lester J. Cappon, Barbara Bartz Petchenik, and John Hamilton Long, eds., *Atlas of Early American History: The Revolutionary Era, 1760–1790* (Princeton, NJ: Princeton University Press, 1976), maps on 29, text on 105–6.

85. The notion that spatiotemporal relationships resemble a musical score is indebted to John Krygier, "Sound and Geographic Visualization," in *Visualization in Modern Cartography*, ed. Alan MacEachren and D. R. F. Taylor (New York: Pergamon, 1994), 149–66, available online at http://go.owu.edu/~jbkrygier/krygier_html/krysound.html, and to Richard Garrett's use of computer algorithms to transform weather data into musical compositions in *Weathersongs*, vol. 1, *Days in Wales* (n.p.: Sunday Dance Music, 2006). Thanks to Henry Lamb for introducing me to Garrett's music.

86. Estimate derived by comparing Lesley HGIS maps of ironworks built by 1810, 1820, 1830 with Paullin's maps of population density by county, 1800–1830 in *Atlas of Historical Geography*, plate 76.

87. The data may omit some new furnaces in this last period, since Lesley and his assistants had completed data gathering for some districts by early 1857.

88. David H. Mould, *Dividing Lines: Canals, Railroads, and Urban Rivalry in Ohio's Hocking Valley, 1825–1875* (Dayton, OH: Wright State University Press, 1994), 174.

89. Taussig, *Tariff History*, 126.

90. Lesley HGIS; Warren, *American Steel*, 13.

91. Lesley HGIS; Jay D. Allen, "The Mount Savage Iron Works, Mount Savage, Maryland: A Case Study in Pre–Civil War Industrial Development" (MA thesis, University of Maryland 1970), 21–29; JPL journal, March 29–April 1, 1858.

92. Buffer analysis applied to the Lesley HGIS.

93. Paullin, *Atlas of Historical Geography*, plates 135C (iron and steel, 1810), 6B (iron ore deposits), 76D (population density, 1810), and 138A (rates of travel, 1800); Lesley HGIS; E. C. Harder, "Distribution of Iron Ore in the United States" and "Iron Ores, Pig Iron, and Steel," in *Mineral Resources of the United States*, part 1, Metallic Products, Calendar Year 1908, Department of the Interior, US Geological Survey (Washington, DC: Government Printing Office, 1909), plate 1 and pp. 61–126, respectively.

94. Lesley HGIS; Lesley, *Guide*, 139, 111, 112; William J. Wayne, *Native Indiana Iron Ores and 19th Century Ironworks*, Department of Natural Resources Geological Survey Bulletin 42-E (Bloomington: State of Indiana, 1970); Edward Neal Hartley, *Ironworks on the Saugus: The Lynn and Braintree Ventures of the Company of Undertakers of the Ironworks in New England* (Norman: University of Oklahoma Press, 1957), 28–30, 100–101; Robert B. Gordon, *American Iron, 1607–1900* (Baltimore: Johns Hopkins University Press, 1996), 28.

95. Gordon, *American Iron*, 103; Joseph E. Walker, *Hopewell Village: The Dynamics of a Nineteenth Century Iron-Making Community* (Philadelphia: University of Pennsylvania Press, 1966), 141; Norris, *Frontier Iron*, 42–43; Anne Kelly Knowles, *Calvinists Incorporated: Welsh Immigrants on Ohio's Industrial Frontier* (Chicago: University of Chicago Press, 1997), 166–68. In the heat of combustion, calcium in the limestone chemically combined with phosphorus, sulfur, and other unwanted minerals in the ore to form a glassy substance that floated above the molten iron, which could then be tapped in a nearly pure state.

96. Lesley, *Guide*, 4; Lesley, *Bulletin*, passim.

97. Lesley, *Bulletin*, passim; Harder, "Iron Ores," 63–66, 74–75, 81–107.

98. Robert Hunt, "Map of Great Britain and Ireland, Shewing the Districts in Which Coal and Iron Are Respectively Found," scale approx. 27 miles to the inch, in *Plans Referred to in the Report of the Commissioners Appointed to Inquire into the Application of Iron to Railway Structures* (London: William Clowes and Sons, for Her Majesty's Stationery Office, 1849); Chris Evans and Göran Rydén, eds., *The Industrial Revolution in Iron: The Impact of British Coal Technology in Nineteenth-Century Europe* (Aldershot, UK: Ashgate, 2005).

99. Harder, "Iron Ores," 92–93; Lesley, *Guide*.

100. Harder, "Iron Ores," 89–90; Baer, *Canals and Railroads*; Lesley HGIS. On the development of the anthracite coal industry, see Richard G. Healey, *The Pennsylvania Anthracite Coal Industry, 1860–1902: Economic Cycles, Business Decision-Making, and Regional Dynamics* (Scranton, PA: University of Scranton Press, 2007).

101. Cappon, Petchenik, and Long, *Atlas of Early American History*, 29; Gordon, *American Iron*, 186.

102. Norris, *Frontier Iron*; Eugene L. Hardy, "A Topographical Map of the Etowah Property, Cass County, Georgia," scale 2 inches to the mile (n.p., 1856); Lesley, *Guide*, 138, 217, 260, and 77, 246, 193, respectively. The first integrated ironworks in North America was built at Lynn, Massachusetts Bay Colony, in the 1640s; Hartley, *Ironworks on the Saugus*.

103. See Robert C. Allen, "The Peculiar Productivity History of American Blast Furnaces, 1840–1913," *Journal of Economic History* 37 (September 1977): 605–33.

104. Lesley, *Guide*, 96, 133, 126; James E. Fell Jr., "Iron from 'the Bend': The Great Western and Brady's Bend Iron Companies," *Western Pennsylvania Historical Magazine* 67 (October 1984): 328; John Adair Richards, *A History of Bath County, Kentucky* (Yuma, AZ: Southwest Printers, 1961), 98–99.

105. Ursula Smith Beach, *Along the Warioto, or A History of Montgomery County, Tennessee* (Nashville, TN: McQuiddy Press, 1964), 123; Lesley, *Guide*, 60, 100, 120, 123, 108, 128, 93, 138, 5; Knowles, *Calvinists*, 171–74.

106. Allan Nevins, *Abram S. Hewitt, with Some Account of Peter Cooper* (New York: Harper, 1935; reprint, New York: Octagon Books, 1967), 102.

107. Excepting years of construction and abandonment, the data analyzed in this section are limited to Lesley's *Guide* and the *Bulletin* reports of 1856–58 unless otherwise noted.

108. Exploratory or iterative mapping means using GIS to produce maps that display subsets of the data, display the results of GIS analytical techniques, and/or combine two or more kinds of data for the sake of visualizing distributions that help one formulate and test hypotheses and descriptions. Early essays on this aspect of geovisualization include E. B. MacDougal, "Exploratory Analysis, Dynamic Statistical Visualization, and Geographic Information Systems," *Cartography and Geographic Information Systems* 19 (1992): 237–46; and Alan M. MacEachren and Menno-Jan Kraak, "Exploratory Cartographic Visualization: Advancing the Agenda," *Computers and Geosciences* 23 (1997): 335–43. A pioneering work is John Wilder Tukey, *Exploratory Data Analysis* (Reading, PA: Addison-Wesley, 1997).

109. George H. Cook and John C. Smock, *Northern New Jersey, Showing the Iron-Ore and Limestone Districts*, horizontal scale 2 miles to the inch, map, vertical scale 2,000 feet to the inch (n.p., 1874); Lesley, *Guide*, 62.

110. See Weymouth Furnace, Lesley, *Guide*, 62.

111. Longevity was calculated as the difference between construction year and abandonment year where both are known, Lesley HGIS.

112. Charles B. Dew, *Ironmaker to the Confederacy: Joseph R. Anderson and the Tredegar Iron Works* (New Haven, CT: Yale University Press, 1966), 32–33; "Sketch Illustrating the Positions of the Commercial Cities and Towns of the Eastern, Middle, and Western States," scale 58.5 miles to the inch (n.p., ca. 1850), accessed at Library of Congress, American Memory, Maps, http://memory.loc.gov, on May 13, 2006; Christopher Baer, personal communication, June 18, 2010. The US manufacturing census recorded twenty-nine blast furnaces operating in Virginia in 1850 versus sixteen in 1860. Lesley counted fifty-four furnaces operating in Virginia (including the future territory of West Virginia) in 1847 versus thirty-four in 1857–58.

113. Lesley, *Bulletin*, 124, 125, 88–89; Aaron J. Davis, *History of Clarion County* (Syracuse, NY: D. Mason, 1887), 115–21; Williams, *Americans and Their Forests*, 104–10.

114. Lesley HGIS and Pennsylvania transport GIS, analyzed with the "Near" function in ArcGIS 9.0, which calculates the shortest linear distance from each point to the nearest feature in the specified feature class—here, the distance from each blast furnace to the nearest river or canal.

115. Temin, *Iron and Steel*, 77; Louis C. Hunter, "A Study of the Iron Industry at Pittsburgh before 1860" (PhD diss., Harvard University, 1928), 426.

116. Hunter, "Study of the Iron Industry," 407–33.

117. Ibid.; Lesley HGIS.

118. Because of the scarcity of production data, it is not possible to form a precise estimate of the net loss of pig iron owing to the closure of Allegheny and Shenango Valley furnaces. Applying average production of the region's charcoal furnaces with known output (603 tons) to all that closed from 1848 to 1858 yields a rough estimate of 31,356 tons, plus 5,400 tons from closed raw-coal furnaces, for a total of approximately 36,756 tons lost by abandonment. Subtracting the estimated added production of furnaces built in the district in 1848–58 (16,017 tons) leaves a net loss of about 20,739 tons.

119. Hunter, "Study of the Iron Industry," 393–433; Knowles, *Calvinists*, chap. 4.

120. Lesley, *Guide* and *Bulletin*. Catherine Elizabeth Reiser notes that "much of the iron" sent to Pittsburgh in the antebellum period "was transported from forges east of Johnstown"; *Pittsburgh's Commercial Development, 1800–1850* (Harrisburg: Pennsylvania Historical and Museum Commission, 1951), 106. An early assessment of the quality of iron ore in Essex and Clinton

Counties, New York, is Ebenezer Emmons, "Survey of the Second Geological District," in *Geology of New York*, part 2 (Albany: Carroll and Cook, 1842), 231–63, 291–308.

121. Technically, ore was not a source of iron but was used as a reagent in puddling furnaces, which Lesley noted as "used in lining" furnaces. See, for example, *Bulletin*, 149–51.

122. William Cronon, *Nature's Metropolis: Chicago and the Great West* (New York: Norton, 1992).

123. Gordon, *American Iron*, 58–59.

124. John J. Toomey and Edward P. B. Rankin, *History of South Boston* (Boston: printed by authors, 1901), 232–34; Arthur B. Cohn, *Lake Champlain's Sailing Canal Boats: An Illustrated Journey from Burlington Bay to the Hudson River* (Basin Harbor, VT: Lake Champlain Maritime Museum, 2003), 49, 114, 122. This and following paragraphs discussing fig. 19 and table 3 also draw on the Lesley HGIS.

125. John K. Brown, *The Baldwin Locomotive Works, 1831–1915* (Baltimore: Johns Hopkins University Press, 1995); Healey, *Pennsylvania Anthracite*; Susan Dieffenbach and Craig A. Benner, *Cornwall Iron Furnace: Pennsylvania Trail of History Guide* (Mechanicsburg, PA: Stackpole Books, for the Pennsylvania Historical and Museum Commission, 2003).

126. Anne Kelly Knowles and Richard G. Healey, "Geography, Timing, and Technology: A GIS-Based Analysis of Pennsylvania's Iron Industry, 1825–1875," *Journal of Economic History* 66 (September 2006): 608–34.

127. John Bezís-Selfa, *Forging America: Ironworkers, Adventurers, and the Industrious Revolution* (Ithaca, NY: Cornell University Press, 2004); Kathleen Bruce, *Virginia Iron Manufacture in the Slave Era* (1930; reprint, New York: Augustus M. Kelley, 1968); Charles B. Dew, "David Ross and the Oxford Iron Works: A Study of Industrial Slavery in the Early Nineteenth-Century South," *William and Mary Quarterly* 31, no. 2 (1974): 189–224; Dew, *Bond of Iron*; and Dew, *Ironmaker to the Confederacy*.

128. W. David Lewis, *Sloss Furnaces and the Rise of the Birmingham District: An Industrial Epic* (Tuscaloosa: University of Alabama Press, 1994); Marjorie Longenecker White, *The Birmingham District: An Industrial History and Guide* (Birmingham, AL: Birmingham Historical Society, 1981).

129. John Fritz, *The Autobiography of John Fritz* (New York: John Wiley, 1912), 70–74; JL to JPL, April 19, 20, 1858. Joseph probably meant Iron Mountain, Michigan, on the Wisconsin border.

Chapter 2

1. Robert Burns, quoted in H. R. Campbell, *Carron Company* (Edinburgh: Oliver and Boyd, 1961), 39.

2. J. G. Wood, *The Principal Rivers of Wales Illustrated* (n.p., 1813), 59, and Wood, "Account of Myrther-tedvel," *Monthly Magazine* 7 (1799): 357, quoted in Chris Evans, *The Labyrinth of Flames: Work and Social Conflict in Early Industrial Merthyr Tydfil* (Cardiff: University of Wales Press, 1993), 31. See also George Borrow, *Wild Wales: The People, Language, and Scenery* (London: J. M. Dent, 1906), 588, and *Leigh's Guide to Wales and Monmouthshire* (London: M. A. Leigh, 1833), 299.

3. Rebecca Harding Davis, *Life in the Iron Mills*, ed. Cecelia Tichi (Boston: Bedford Books, 1998), 3–25, 39–74; Lee R. Maddex, "La Belle Iron Works," Historic American Engineering Record WV-47, typescript, August 1990, 8; J. Peter Lesley, *The Iron Manufacturer's Guide to the Furnaces, Forges, and Rolling Mills of the United States* (New York: John Wiley, 1859), 254–55; Earl Chapin May, *Principio to Wheeling, 1715–1945: A Pageant of Iron and Steel* (New York: Harper, 1945), 106–9. Harding became a prolific author of popular fiction under her married name, Rebecca Harding Davis. She published *Life in the Iron Mills* anonymously.

4. Davis, *Life in the Iron-Mills*, 45.

5. John R. Stilgoe, *Common Landscape of America, 1580–1845* (New Haven, CT: Yale University Press, 1982), 267.

6. Hywel Teifi Edwards, *Arwr Glew Erwau'r Glo: Delwedd y Glowr yn Llenyddiaeth y Gymraeg, 1850–1950* [Brave Hero of the Coal Fields: The Image of the Miner in Welsh Literature, 1850–1950] (Llandysul: Gwasg Gomer, 1994), xx, quoting from an Eisteddfod essay of 1865.

7. Kay Parkhurst Easson and Roger R. Easson, eds., William Blake, *Milton* (Boulder, CO: Shambhala, 1978).

8. David E. Nye, *American Technological Sublime* (Cambridge, MA: MIT Press, 1994), 54–55.

9. Alan Birch, *The Economic History of the British Iron and Steel Industry, 1784–1879* (London: Frank Cass, 1967; reprint, New York: Augustus M. Kelley, 1968), 244.

10. William S. Rodner, *J. M. W. Turner: Romantic Painter of the Industrial Revolution* (Berkeley: University of California Press, 1997), 86.

11. Derrick Pritchard Webley, *East to the Winds: The Life and Work of Penry Williams (1802–1885)* (Aberystwyth: National Library of Wales, [ca. 1995]).

12. My discussion of Welsh artists' depictions of industrial works owes much to the pioneering works of art historian Peter Lord, including *Arlunwyr Gwlad/Artisan Painters* (Aberystwyth: National Library of Wales, 1993), Lord, *Hugh Hughes: Arlunydd Gwlad* [artisan painter], *1790–1863* (Llandysul: Gwasg Gomer, 1995), and particularly Lord, *The Visual Culture of Wales: Industrial Wales* (Cardiff: University of Wales Press, 1998).

13. Webley, *East to the Winds,* 11–17.

14. Walt Whitman, "A Song of Occupations," stanza 5, and "A Song of Joys," in *Leaves of Grass* (New York: Signet Classic, 1958), 188 and 160, respectively. See also "Crossing Brooklyn Ferry," *Leaves of Grass,* 149. Olson notes the difference between Whitman's descriptions and Harding's but does not explore it; Tillie Olson, afterword in *Life in the Iron Mills* by Rebecca Harding Davis, 164n11.

15. Betsy Fahlman, *John Ferguson Weir: The Labor of Art* (Newark: University of Delaware Press, 1997), 41, 79.

16. Ibid., 81.

17. Charles Russell Lowell to [Daniel] Tyler, May 11, 1835, box C.1, Charles Russell Lowell correspondence, Lycoming Coal Company Records, Baker Library Historical Collections, Harvard Business School (hereafter LCC).

18. See Anne Kelly Knowles, "Wheeling Iron and the Welsh: A Geographical Reading of 'Life in the Iron Mills,' " in *Transnational West Virginia: Ethnic Work Communities during the Industrial Era,* ed. Ronald Lewis and Kenneth Fones-Wolf (Morgantown: West Virginia University Press, 2002), 228–32.

19. Davis, *Life in the Iron Mills,* 39–40.

20. James E. Reeves, *The Physical and Medical Topography . . . of the City of Wheeling* (Wheeling, WV: Daily Register, 1870), 18. Reeves's descriptions of Wheeling's atmosphere are very similar to those in Anna Egan Smucker's autobiographical story about growing up in the steel town of Weirton, West Virginia, in the mid-twentieth century: Smucker, *No Star Nights* (New York: Alfred A. Knopf, 1989).

21. Robert B. Gordon and Patrick M. Malone, *The Texture of Industry: An Archaeological View of the Industrialization of North America* (New York: Oxford University Press, 1994), 49–50.

22. Reeves, *Physical and Medical Topography,* 38.

23. Compare Childs's painting with the map of Dowlais diagrammed in figure 41.

24. United Nations, Economic Commission for Europe, *Problems of Air and Water Pollution Arising in the Iron and Steel Industry* (New York: United Nations, 1970); Joel A. Tarr, "Searching for a 'Sink' for an Industrial Waste: Iron-Making Fuels and the Environment," *Environmental History Review* 18 (Spring 1994): 9–34; Gordon and Malone, *Texture of Industry,* 164.

25. Michael Williams, *Americans and Their Forests: A Historical Geography* (Cambridge: Cambridge University Press, 1989), 106. See also Anne Kelly Knowles, *Calvinists Incorporated: Welsh Immigrants on Ohio's Industrial Frontier* (Chicago: University of Chicago Press, 1997), 166–74.

26. Williams, *Americans and Their Forests,* 148–52, 342–44. Williams estimates it was only 0.8

percent if one accounts for cutting of regenerated woodland (344). Both estimates are based on a moderate average consumption of 500 acres of timber per 1,000 tons of iron.

27. George M. Kober, "Iron, Steel, and Allied Industries," in *Industrial Health*, ed. George M. Kober and Emery R. Hayhurst (Philadelphia: P. Blakiston's Son, 1924), 176–84.

28. Interview with Graham Collis at Blists Hill, Ironbridge Gorge Museum, June 23, 2004.

29. Interviews with Kenneth Hall at Blists Hill, Ironbridge Gorge Museum, June 23, 2004, and August 30, 2005.

30. Chris Evans, "Work and Workloads during Industrialization: The Experience of Forgemen in the British Iron Industry, 1750–1850," *International Review of Social History* 44 (1999): 210–11; brackets in original.

31. Gordon and Malone, *Texture of Industry*, 51.

32. Ivor Wilks, *South Wales and the Rising of 1839: Class Struggle as Armed Struggle* (Urbana: University of Illinois Press, 1984); Gwyn Alf Williams, *The Merthyr Rising* (London: Croom Helm, 1978); John William Bennett, "Iron Workers in Woods Run and Johnstown: The Union Era, 1865–1895" (PhD diss., University of Pittsburgh, 1977), 42ff. The Sons of Vulcan was founded by puddlers and boilers, the latter being puddlers who used a slightly different technique that "boiled" the pig iron more rapidly.

33. James J. Davis, *The Iron Puddler: My Life in the Rolling Mills and What Came of It* (Indianapolis: Bobbs-Merrill, 1922), 87, 91, 98–99; interview with Kenneth Hall, August 30, 2005.

34. B. L. Coombes, *These Poor Hands: The Autobiography of a Miner Working in South Wales* (1939), quoted in Bill Jones and Chris Williams, *B. L. Coombes* (Cardiff: University of Wales Press, 1999), 1.

35. Knowles, *Calvinists*, 57–58. David Gatley's compilation from the 1851 British Population Census lists 23,939 men aged twenty or over in ironmaking and metalworking occupations in Merthyr Tydfil registration district.

36. [Thomas Wright] A Journeyman Engineer, *Some Habits and Customs of the Working Classes* (London: Tinsley Brothers, 1867), 83. See also [Wright], *The Great Unwashed* (1868; reprint, London: Frank Cass, 1970).

37. A rigorous categorization of ironworks by population would require research in the US Census to determine the size of each settlement at the time an ironworks was built or to determine an "average" size during the antebellum period. The generalized categories in figure 31 were assigned based on Lesley's descriptions of works' locations, contextual information that I gleaned from US Geological Survey topographic quadrangles and historical maps while plotting locations in the Lesley HGIS, and supplemental research on individual communities in county histories and online community history websites. "Rural" works were not in preexisting settlements and are not known to have developed community or urban functions beyond those required for producing iron and sustaining the workforce. "Village" works were within one mile of a small preexisting settlement. "Urban" works were within two miles of a market town or city of approximately five thousand or more residents. I labeled works "unknown" if I could not confidently relate their location to known settlements.

38. Arthur Cecil Bining, *Pennsylvania Iron Manufacture in the Eighteenth Century* (1938; reprint, New York: Augustus M. Kelley, 1970), 29–48; James Moore Swank, *History of the Manufacture of Iron in All Ages*, 2nd ed. (1884; Philadelphia: American Iron and Steel Association, 1892; reprint, New York: Burt Franklin, 1965), 189, quoted in Peter Temin, *Iron and Steel in Nineteenth-Century America: An Economic Inquiry* (Cambridge, MA: MIT Press, 1964), 84–85. See also Gerald G. Eggert, *Making Iron on the Bald Eagle: Roland Curtin's Ironworks and Workers' Community* (University Park: Pennsylvania State University Press, 2000), 81.

39. Charles S. Aiken, *The Cotton Plantation South since the Civil War* (Baltimore: Johns Hopkins University Press, 1998), 3–28; Gerald G. Eggert, *The Iron Industry in Pennsylvania,* Pennsylvania Historical Studies 25 (University Park: Pennsylvania Historical Association, 1994), 16–39.

40. Joseph E. Walker, *Hopewell Village: The Dynamics of a Nineteenth Century Iron-Making*

Community (Philadelphia: University of Pennsylvania Press, 1966); Eggert, *Making Iron on the Bald Eagle*; Charles B. Dew, "David Ross and the Oxford Iron Works: A Study of Industrial Slavery in the Early Nineteenth-Century South," *William and Mary Quarterly* 31, no. 2 (1974): 189–224, and Dew, *Bond of Iron: Master and Slave at Buffalo Forge* (New York: Norton, 1994); John Bezís-Selfa, "A Tale of Two Ironworks: Slavery, Free Labor, Work, and Resistance in the Early Republic," *William and Mary Quarterly,* 3rd ser., 56, no. 4 (1999): 677–700.

41. William A. Sullivan, *The Industrial Worker in Pennsylvania, 1800–1840* (Harrisburg: Pennsylvania Historical and Museum Commission, 1955), 59–60; Dew, *Bond of Iron,* 111.

42. Dew, *Bond of Iron,* 181.

43. Norris, *Frontier Iron,* 39.

44. Eggert, *Making Iron on the Bald Eagle,* 44–48; Theodore W. Kury, "Labor and the Charcoal Iron Industry: The New Jersey–New York Experience," *Material Culture* 25, no. 3 (1993): 27–28.

45. Robert B. Gordon, *A Landscape Transformed: The Ironmaking District of Salisbury, Connecticut* (New York: Oxford University Press, 2001), 72–73; Paul M. Heberling, "Status Indicators: Another Strategy for Interpretation of Settlement Pattern in a Nineteenth-Century Industrial Village," in *Consumer Choice in Historical Archaeology*, ed. Suzanne M. Spencer-Wood (New York: Plenum Press, 1987), 199–216; Jack Larkin, *Where We Lived: Discovering the Places We Once Called Home* (Taunton, MA: Taunton Press, 2006); Sullivan, *Industrial Worker,* 59; site visits to numerous ironmaking communities, including Irondale, New York; Hopewell, Pennsylvania; and Glendon, Pennsylvania.

46. Sullivan, *Industrial Worker,* 60–62; Knowles, *Calvinists,* 171–74, 214–16.

47. Dew, *Bond of Iron,* 76; Eggert, *Making Iron on the Bald Eagle,* 44; Kury, "Labor and the Charcoal Iron Industry," 22.

48. Eggert, *Making Iron on the Bald Eagle,* 87–102; Kury, "Labor and the Charcoal Iron Industry," 24–27. The system described here was very similar to Swedish charcoal iron furnace and forge settlements. See Goran Höppe and John Langton, *Flows of Labour in the Early Phase of Capitalist Development: The Time-Geography of Longitudinal Migration Paths in Nineteenth-Century Sweden,* Historical Geography Research Series 29 (n.p.: Institute of British Geographers, Historical Geography Research Group, 1992), and Robert C. Ostergren, *Patterns of Seasonal Industrial Labor Recruitment in a Nineteenth Century Swedish Parish: The Case of Matfors and Tuna, 1846–1873,* Report 5 from the Demographic Data Base (Umeå, Sweden: Umeå University, 1990).

49. See, for example, Dew, *Bond of Iron,* 22.

50. Charles B. Dew, "Slavery and Technology in the Antebellum Southern Iron Industry: The Case of Buffalo Forge," in *Science and Medicine in the Old South,* ed. Ronald L. Numbers and Todd L. Savitt (Baton Rouge: Louisiana State University Press, 1989), 125–26; Dew, *Bond of Iron,* 332–33; Anne Kelly Knowles, "Labor, Race, and Technology in the Confederate Iron Industry," *Technology and Culture* 42, no. 1 (2001): 12–13.

51. Dew, *Bond of Iron,* 333.

52. Eggert, *Making Iron on the Bald Eagle,* 22, 120–21.

53. Anne Kelly Knowles and Richard G. Healey, "Geography, Timing, and Technology: A GIS-Based Analysis of Pennsylvania's Iron Industry, 1825–1875," *Journal of Economic History* 66 (September 2006): 624–25, 627–30.

54. Eggert, *Making Iron on the Bald Eagle,* 83; Dew, *Bond of Iron,* 123, 83–97, 124–29.

55. Dew, *Bond of Iron,* 261–62, 276–77;

56. Ibid., 280; Dew, "David Ross," 189–224.

57. Sullivan, *Industrial Worker,* 59–65; Kury, "Labor and the Charcoal Iron Industry," 29–30; Bezís-Selfa, *Forging America,* 128–31, 198–200, 205–6, quotation on 206.

58. *Webster's New International Dictionary of the English Language,* 2nd ed., s.v. "hamlet."

59. Lesley, *Guide,* 42, 25, 72, 154.

60. James T. Lemon, *The Best Poor Man's Country: A Geographical Study of Early Southeastern Pennsylvania* (Baltimore: Johns Hopkins University Press, 1972); Edward C. Carter II, John C. Van

Horne, and Charles E. Brownell, eds., *Latrobe's View of America, 1795–1820: Selections from the Watercolors and Sketches* (New Haven, CT: Yale University Press, 1985), 238–39; T. J. Kennedy, "Map of Chester Co., Pennsylvania," map, no scale (n.p., 1860), accessed online at Library of Congress, American Memory, http://memory.loc.gov/ammem on July 1, 2007.

61. Warwick Township, Chester County, Pennsylvania, 1850 Manuscript Census, microfilm roll 274 (Chester-Clinton [part of]). Thanks to Hopewell National Historic Site for lending me this microfilm.

62. Lesley, *Guide*, 40.

63. Ibid., 40, 232; Julian C. Skaggs, "Lukens, 1850–1870: A Case Study in the Mid-Nineteenth Century American Iron Industry" (PhD diss., University of Delaware, 1975), 7–11, 17–19, 41–45.

64. John Richard Moravek, "The Iron Industry as a Geographic Force in the Adirondack-Champlain Region of New York State, 1800–1971" (PhD diss., University of Tennessee, Knoxville, 1976), 42–43; Lesley, *Guide*, 141; Greg Smith, "A History of the McIntyre Mine near Newcomb, N.Y.," accessed online at http://www.adirondack-park.net/history/mcintyre.mine.html on July 16, 2007. Smith dates the furnace's construction to 1837 and says that the company first tried to ship out raw ore.

65. Allan Nevins, *Abram S. Hewitt, with Some Account of Peter Cooper* (New York: Harper; reprint, New York: Octagon Books, 1967), 102.

66. Lesley, *Guide*, passim.

67. Sullivan, *Industrial Worker*, 59–71, treats all rural Pennsylvania ironworks as plantations. Applying my typology, most of the ironworks he discusses were small companies in hamlets, not self-sufficient plantations.

68. Gordon and Malone, *Texture of Industry*, 84; G. M. Hopkins, *Clark's Map of Litchfield County, Connecticut*, map, scale 1:50,688 (n.p., 1859), downloaded from Library of Congress, American Memory, http://memory.loc.gov/ammem on July 9, 2007. Gordon and Malone include a similar detail of *Clark's Map of Litchfield County* on 85.

69. Gordon, *Landscape Transformed*, 14, 28–38; Lesley, *Guide*, 2, 28–31.

70. Gordon, *Landscape Transformed*; Christopher Clark, *The Roots of Rural Capitalism: Western Massachusetts, 1780–1860* (Ithaca, NY: Cornell University Press, 1990), 59–117.

71. Gordon, *Landscape Transformed*.

72. Michael S. Raber et al., *Forge of Innovation: An Industrial History of the Springfield Armory, 1794–1968*, ed. Richard Colton (1989; reprint, Fort Washington, PA: Eastern National [Division, National Park Service], 2008), 144–47; Gordon, *Landscape Transformed*, 43–44.

73. Gregory Galer, Robert Gordon, and Frances Kemmish, "Connecticut's Ames Iron Works: Family, Community, Nature, and Innovation in an Enterprise of the Early American Republic," *Transactions [of the] Connecticut Academy of Arts and Sciences* 54 (December 1998): 83–194; Watson and Hillman, Empire Iron Works, Kentucky, to Peter Cooper, January 21, 1846, and James Cox, Lehigh Coal and Navigation Company, Philadelphia, to Abram S. Hewitt, October 28, 1847, Cooper, Hewitt and Company Papers, Manuscript Division, Library of Congress (hereafter Cooper, Hewitt Papers). In the absence of other evidence, Galer et al. concluded that Ames and his partners developed wood-fired puddling; Robert B. Gordon, personal correspondence, May 19, 2010. The correspondence cited here asks Cooper for details of "his" wood-puddling technique, but I have not found evidence confirming that Cooper developed the method.

74. Gordon, *Landscape Transformed*.

75. Knowles, *Calvinists*, 116–224; quotation, 179, from Thomas Ll. Hughes, *Y Cyfaill o'r Hen Wlad* [The Friend from the Old Country] (April 1854), 153, trans. from the Welsh by the author.

76. Knowles, *Calvinists*, 208–19.

77. John W. McGrain, *From Pig Iron to Cotton Duck: A History of Manufacturing Villages in Baltimore County* (Towson, MD: Baltimore County Public Library, 1985), 34–35; Lesley, *Guide*, 24.

78. See John C. Hudson, *Plains Country Towns* (Minneapolis: University of Minnesota Press, 1985), and John W. Reps, *Cities of the American West: A History of Frontier Urban Planning* (Princeton, NJ: Princeton University Press, 1979).

79. McGrain, *From Pig Iron*, 42–43.

80. W. David Lewis, *Sloss Furnaces and the Rise of the Birmingham District: An Industrial Epic* (Tuscaloosa: University of Alabama Press, 1994), 39–69; Henry M. McKiven Jr., *Iron and Steel: Class, Race, and Community in Birmingham, Alabama, 1875–1920* (Chapel Hill: University of North Carolina Press, 1995), 7–22.

81. Lesley, *Guide*, 77, 193, 246; US Geological Survey, "Allatoona Dam," map, original scale 1:24,000 (n.d.), accessed online at USGS Geographic Names Information System, http://geonames.usgs.gov on July 13, 2007; Eugene L. Hardy, "A Topographical Map of the Etowah Property, Cass County, Georgia," map, scale 2 inches to the mile, National Archives and Records Administration, Washington, DC, 1856; *An Exposition of the Property of the Etowah Manufacturing and Mining Co.* (New York: George F. Nesbitt, 1860), 1–5, quotation on 2; Brewster and Alley Fleming, "Birds-Eye Map of the Western and Atlantic R.R., the Great Kennesaw Route; Army Operations, Atlanta Campaign, 1864," map, no scale (New York, 1887), accessed at Library of Congress, American Memory, http://memory.loc.gov on July 13, 2007.

82. Hardy, "Topographical Map."

83. US Manuscript Census, 1850, Cass County, Georgia, M423, reel 63.

84. On the conflicting advantages of single versus married men, see LCC. Particularly pertinent correspondence is box C.2, General Correspondence, C. Bolinger to Dan[iel] Tyler, April 18, 1835; John R. Bowes to Tyler, December 1, 1835; E. Tyler to D. Tyler, June 12, 1837; W. T. Walters to Tyler, August 5, 1837.

85. US Manuscript Census, 1850, Slave Schedule, Cass County, Georgia, M423, reel 88.

86. Lesley, *Guide*, 130–32; US Manuscript Census, 1850 Slave Schedule, Stewart County, Tennessee, M423, reel 907.

87. David B. Parker, "Mark Anthony Cooper (1800–1885)," *The New Georgia Encyclopedia* (article dated September 3, 2002), accessed online at www.georgiaencyclopedia.org on July 13, 2007; Ethel Armes, *The Story of Coal and Iron in Alabama* (Birmingham, AL: Chamber of Commerce, 1910; reprint, New York: Arno Press, 1972), 64, 163; Deposition of Moses Stroup, Chancery Court Records, Loose Papers File, drawer S, Shelby County Museum and Archives, Columbiana, Alabama.

88. McGrain, *From Pig Iron*, 34, 36.

89. Exposition of the Property of Etowah, 4; Carol Bundy, *The Nature of Sacrifice: A Biography of Charles Russell Lowell, Jr., 1835–64* (New York: Farrar, Straus and Giroux, 2005), 38.

90. Moravek, "Iron Industry," 16–19, 107.

91. Ibid., 67–69, 76, 94–98, 104; [Louis McLane], Secretary of the Treasury, *Documents relative to the Manufactures in the United States*, HR Doc. 308, 2 vols. (Washington, DC: Duff Green, 1833), doc. 10, nos. 37, 38, 40.

92. "Clintonville, Town of Ausable," in Frederick W. Beers, *Atlas of Clinton County, New York* (New York: Beers, Ellis, and Soule, 1869), 38.

93. Moravek, "Iron Industry," 94–97; Lesley, *Guide*, 262; Arthur B. Cohn, *Lake Champlain's Sailing Canal Boats: An Illustrated Journey from Burlington Bay to the Hudson River* (Basin Harbor, VT: Lake Champlain Maritime Museum, 2003), 36–40.

94. L. F. Henning, *West View of Catasauqua* (ca. 1852); Lesley, *Guide*, 8; [J. Peter Lesley], ed., *Bulletin of the American Iron Association* (Philadelphia: American Iron Association, ca. 1857), 50; *Population of the United States in 1860* (Washington, DC: Government Printing Office, 1864), 427.

95. Lesley, *Guide*, 19–20, 236–37; Lesley HGIS; F. V. Lahr, after James Queen, *Montour Iron Works, Danville, Pennsylvania*, color lithograph (Philadelphia: P. S. Duval, [ca. 1855], Palmer Museum of Art, O'Connor-Yeager Collection, Pennsylvania State University, accessed online at http://www.psu.edu/dept/palmermuseum on June 17, 2007.

96. 1850 Census, Montour County, Pennsylvania, borough of Danville, M432, reel 801, 577–655.

97. Knowles, *Calvinists*, chap. 1; database of Pittsburgh residents extracted from the 1850 US Manuscript Census by Laurence Glasco, Department of History, University of Pittsburgh.

98. Knowles, *Calvinists*, chap. 1; William D. Jones, *Wales in America: Scranton and the Welsh, 1860–1920* (Cardiff: University of Wales Press, 1993); telephone interview with Fiona Powell, Danville, August 1, 2000; "Joseph Parry (1841–1903), Composer and Musician," *Casglu'r Tlysau/ Gathering the Jewels, Website for Welsh Cultural History*, National Library of Wales, accessed online at http://www.gtj.org.uk on July 27, 2007. Welsh American denominational periodicals document musical, literary, and religious festivals and competitions in many iron towns throughout the antebellum period. See particularly *Y Cyfaill o'r Hen Wlad* [The Friend from the Old Country], 1838–60.

99. Bruno Latour, *Science in Action: How to Follow Scientists and Engineers through Society* (Cambridge, MA: Harvard University Press, 1987), 215–57.

100. Ibid., 249.

101. David R. Meyer, *Networked Machinists High-Technology Industries in Antebellum America* (Baltimore: Johns Hopkins University Press, 2006).

102. Patricia A. Schechter, "Free and Slave Labor in the Old South: The Tredegar Ironworkers' Strike of 1847," *Labor History* 35 (Spring 1994): 165–86; Charles B. Dew, *Ironmaker to the Confederacy: Joseph R. Anderson and the Tredegar Iron Works* (New Haven, CT: Yale University Press, 1966), 23, 26; unsigned note in hand of J. R. Anderson to S. H. Hartman, November 18, 1845, Tredegar Company Supplementary Records 24808 (23), Minutes of the Director and Stockholders, January 12, 1838–January 9, 1850, Tredegar Company Papers, Library of Virginia; 1850 and 1860 federal census of population, Henrico County, Virginia, M432, reel 951 and M653, reels 1352 and 1353, respectively.

103. Murdock Leavitt and Co. to Peter Cooper, November 5, 1846, William Green Jr., Boonton Iron Works to Peter Cooper, February 23, 1847, and March 1, 1847, Cooper, Hewitt Papers.

104. James Linaberger, "The Rolling Mill Riots of 1850," *Western Pennsylvania Historical Magazine* 47 (January 1964): 1–18.

105. Alexander Keyssar, *Out of Work: The First Century of Unemployment in Massachusetts* (Cambridge: Cambridge University Press, 1986), 14–17, 124–30, quotations on 14 and 128.

106. Linaberger, "Rolling Mill Riots," 17–18; John A. Fitch, *The Steel Workers* (Pittsburgh: University of Pittsburgh Press, 1989), 76–77.

107. Jones, *Wales in America*, 1–27; J. Peter Lesley to Susan Inches Lesley, November 15, 1856, Ames Family Historical Collection, AmesINK, in Boulder, Colorado.

108. David Harvey, *The Limits to Capital* (Chicago: University of Chicago Press, 1982; Midway reprint, 1989), 204–38, 380–85; Brinley Thomas, *Migration and Economic Growth: A Study of Great Britain and the Atlantic Economy* (Cambridge: Cambridge University Press, 1954), and Thomas, "Wales and the Atlantic Economy," *Scottish Journal of Political Economy* 6 (1959): 169–92; Anne Kelly Knowles, "Immigrant Trajectories through the Rural-Industrial Transition in Wales and the United States, 1795–1850," *Annals of the Association of American Geographers* 85 (1995): 246–66.

109. David Montgomery, *Citizen Worker: The Experience of Workers in the United States with Democracy and the Free Market during the Nineteenth Century* (New York: Cambridge University Press, 1993); Robert J. Steinfeld, *The Invention of Free Labor: The Employment Relation in English and American Law and Culture, 1350–1870* (Chapel Hill: University of North Carolina Press, 1991), and Steinfeld, *Coercion, Contract, and Free Labor in the Nineteenth Century* (Cambridge: Cambridge University Press, 2001); Amy Dru Stanley, *From Bondage to Contract: Wage Labor, Marriage, and the Market in the Age of Slave Emancipation* (Cambridge: Cambridge University Press, 1998), chap. 1.

110. Harvey, *Limits to Capital*, 381–85.

111. Steinfeld, *Coercion*.

112. The evidence of constraints imposed on ironworkers' geographical mobility contributes to the critique of the Whig view of democracy in the United States. See, for example, Alexander Keyssar, *The Right to Vote: The Contested History of Democracy in the United States* (New York: Basic

Books, 2000). The phrase comes from Eric J. Hobsbawm, "Artisan or Labour Aristocrat?" *Economic History Review*, 2nd ser., 37, no. 3 (1984): 355–72.

113. Williams, *Merthyr Rising*; Wilks, *South Wales and the Rising of 1839*.

114. David Brody, "Time and Work during Early American Industrialism," in *In Labor's Cause: Main Themes on the History of the American Worker* (New York: Oxford University Press, 1993), 26–27; Knowles, "Labor, Race, and Technology," 16–17, 25; Wright, *Some Habits and Customs*, and Wright, *The Great Unwashed*; "Lonaconing Iron Works Records, 1837–1840," typescript of George's Creek Coal and Iron Company Journal, December 23, 1837, MS 2292, Maryland Historical Society (hereafter "Lonaconing Records").

115. LCC, box C.2, Daniel Tyler to Thos. S. Williams, January 16, 1837; letters by federal ordnance inspectors complaining of founders' resistance to the inspectors' standards, Record Group 74, Records of the Bureau of Ordnance [Navy], entry 20, letters received from foundries, 1855–1861, National Archives and Research Administration.

116. Bill of Complaint of William Weaver, November 19, 1825, Case Papers, *Weaver v. Mayburry*, Superior Court of Chancery Records, Augusta County Court House, Staunton, Virginia, quoted in Dew, *Bond of Iron*, 22.

117. Ibid., 217.

118. Dew, "Slavery and Technology," 107–26.

119. US Census of Manufactures and Population Census, 1850, Stewart County, Tennessee; Lesley, *Guide*, 130–33, 259. Charles B. Dew provides a much higher estimate in "Black Ironworkers and the Slave Insurrection Panic of 1856," *Journal of Southern History* 41, no. 3 (1975), 324–25.

120. Robert S. Starobin, *Industrial Slavery in the Old South* (New York: Oxford University Press, 1970), 87–90; Dew, "Black Ironworkers," 321–38.

121. Starobin, *Industrial Slavery*, 33–34, 88, 90.

122. See chapter 3.

123. LCC, box C.2, James Hartshorne to Mr. Walters, February 11, 1838.

124. Frances F. Dunwell, *The Hudson River Highlands* (New York: Columbia University Press, 1991), 73; "Lonaconing Records."

125. Philip Scranton, *Proprietary Capitalism* (Cambridge: Cambridge University Press, 1983).

126. See Dew, "David Ross."

127. Harvey, *Limits to Capital*, 382.

Chapter 3

1. Robert F. Dalzell Jr., *Enterprising Elite: The Boston Associates and the World They Made* (Cambridge, MA: Harvard University Press, 1987), 85–91; James E. Vance Jr., *The North American Railroad: Its Origin, Evolution, and Geography* (Baltimore: Johns Hopkins University Press, 1995), 18–21; Christopher T. Baer, *Canals and Railroads of the Mid-Atlantic States, 1800–1860*, ed. Glenn Porter and William H. Mulligan Jr., cartographers Marley E. Amstutz and Anne E. Webster (Wilmington, DE: Regional Economic History Research Center, Eleutherian Mills–Hagley Foundation, 1981), appendix C.

2. W. Ross Yates, "Discovery of the Process for Making Anthracite Iron," *Pennsylvania Magazine of History and Biography* 98 (1974): 207–8.

3. M. [Pierre Armand] Dufrénoy, *On the Use of Hot Air in the Iron Works of England and Scotland*, trans. from a report made to the director general of mines in France, . . . in 1834 (London: J. Murray, 1836).

4. Edgar Jones, *A History of GKN* [Guest, Keen and Nettlefolds], vol. 1, *Innovation and Enterprise, 1759–1918* (Houndsmills, UK: Macmillan, 1987), 6–7, 23; Charles K. Hyde, *Technological Change and the British Iron Industry, 1700–1870* (Princeton, NJ: Princeton University Press, 1977), 60–63, 90–91; Chris Evans, *"The Labyrinth of Flames": Work and Social Conflict in Early Industrial Merthyr Tydfil* (Cardiff: University of Wales Press, 1993), 27–28.

5. Jean-François Belhoste and Denis Woronoff, "The French Iron and Steel Industry during

the Industrial Revolution," trans. Paul Smith, in *The Industrial Revolution in Iron: The Impact of British Coal Technology in Nineteenth-Century Europe*, ed. Chris Evans and Göran Rydén (Aldershot, UK: Ashgate, 2005), 76–80, 83–85; Donald Reid, *The Miners of Decazeville: A Genealogy of Deindustrialization* (Cambridge, MA: Harvard University Press, 1985), 17–20.

6. Belhoste and Woronoff, "French Iron and Steel," 87–89; Annie Laurant, *Des fers de Loire à l'acier Martin: Maîtres de forges en Berry et Nivernais* (Paris: Royer, 1995), 117–19; "Welsh People in France," typescript by Yves Randeynes (n.d.), transcription of births, marriages, and deaths from parish records; electronic correspondence with Joane Jay (Tucson, AZ) and Brian Wagstaffe (Neath, Glamorganshire, UK) regarding the migration history of their relative, Rees Joshua Prosser, and his family.

7. Reid, *Miners of Decazeville*, 18.

8. Evans and Rydén, *Industrial Revolution in Iron*.

9. Gerard Ralston to Daniel Tyler, January 20, 1835, and Daniel Tyler to Thos. S. Williams, January 16, 1837, box C.2, General Correspondence, Lycoming Coal Company Records, Baker Library Historical Collections, Harvard Business School (hereafter LCC).

10. Robert B. Gordon, *American Iron, 1607–1900* (Baltimore: Johns Hopkins University Press, 1996). Gordon notes that the first professor of metallurgy in the United States, George J. Brush, was appointed at Yale University in 1855. He was sent to England and continental Europe to be trained for two years before beginning to teach. *American Iron*, 21, and personal communication, May 19, 2010.

11. Hyde, *Technological Change*, 27–62; Thomas Southcliffe Ashton, *Iron and Steel in the Industrial Revolution* (Manchester: Manchester University Press, 1963; reprint, New York: Augustus M. Kelley, 1968), 13–38.

12. Evans, *"Labyrinth of Flames,"* 28.

13. John Davies, *A History of Wales* (London: Penguin Books, 1990), 319–35.

14. Evans, *"Labyrinth of Flames"*; Anne Kelly Knowles, *Calvinists Incorporated: Welsh Immigrants on Ohio's Industrial Frontier* (Chicago: University of Chicago Press, 1997), chap. 2; Brinley Thomas, "Migration of Labour into the Glamorganshire Coalfield, 1861–1911," *Economica* 30 (1930): 275–94; W. T. R. Pryce, "The Welsh Language, 1750–1961," in *National Atlas of Wales*, ed. Harold Carter (Cardiff: University of Wales Press, 1980), plate 3.1.

15. The clannishness of skilled ironworkers (and later steelworkers) features in John Bezís-Selfa, *Forging America: Ironworkers, Adventurers, and the Industrious Revolution* (Ithaca, NY: Cornell University Press, 2004), and Francis G. Couvares, *The Remaking of Pittsburgh: Class and Culture in an Industrializing City, 1877–1919* (Albany: State University of New York Press, 1984).

16. Ordnance Survey, "Merthyr Tydfil," map, 42 sheets, scale 1:528 (1851), Glamorgan Archives; Harold Carter and Sandra Wheatley, *Merthyr Tydfil in 1851: A Study of the Spatial Structure of a Welsh Industrial Town*, Social Science Monograph 7 (Aberystwyth: University of Wales, Board of Celtic Studies, 1982), 8–13.

17. Hyde, *Technological Change*, 203–4; *Reader's Digest Complete Atlas of the British Isles* (London: Reader's Digest Association, 1965).

18. Calculated in ArcGIS by intersecting the point layers for blast furnaces and rolling mills with a hundred-mile buffer of the Atlantic coast.

19. *Reader's Digest Atlas of the British Isles*, 222–23.

20. Christopher T. Baer, personal communication, June 1, 2010.

21. William Cothren, "Farrand Family," in *History of Ancient Woodbury, Connecticut, from the First Indian Deed in 1659 to 1854 . . .* (Waterbury, CT: Bronson Brothers, 1854), 544; Franklin Bowditch Dexter, *Biographical Sketches of the Graduates of Yale College with Annals of the College History*, vol. 5, June 1792–September 1805 (New York: Henry Holt, 1911), 325; Guy R. Woodall, "The American Review of History and Politics," *Pennsylvania Magazine of History and Biography* 93, no. 3 (1969): 392–93, 407–8; Ellis Paxson Oberholtzer, *The Literary History of Philadelphia* (Philadelphia: George W. Jacobs, 1906), 189–92; John A. Paxton, *The Philadelphia Directory and Register*

for 1819 (Philadelphia, [ca. 1819]).

22. Frank Willing Leach, "Old Philadelphia Families," in *The North American* (Philadelphia, 1912), transcribed by Vince Summers as "Ralston Family History: Philadelphia and Chester Counties, Pennsylvania," accessed online at http://ftp.rootsweb.com/pub/usgenweb/pa/philadelphia/history/family/ralston.txt on September 22, 2007. Gerard's given name was Gerardus.

23. LCC, *A Brief Description of the Property Belonging to the Lycoming Coal Company* (Poughkeepsie, NY: P. Potter, 1828), 1–6; LCC, box C.2, Contract between William P. Farrands [*sic*] for Mathew [*sic*] C. Ralston and others and David McCloskey, February 4, 1832; box C.1, William Lyman correspondence, Wm. Lyman to Wm. P. Farrand, May 29, 1832.

24. LCC, box C.1, vol. 1, copy of the articles of copartnership, indenture dated June 1, 1832; Dalzell, *Enterprising Elite.*

25. Mary Malloy, *"Boston Men" on the Northwest Coast: The American Maritime Fur Trade, 1788–1844* (Anchorage: University of Alaska Press, 1998), 19, 42, 112, 123, 128, 129, 169.

26. "Thomas Handasyd Perkins," *National Cyclopedia of American Biography,* vol. 5 (New York: J. T. White, 1898), 245; Dalzell, *Enterprising Elite,* 60, 62, 85–89, appendix; *The First Railroad in America: A History of the Origin and Development of the Granite Railway at Quincy, Massachusetts* (Quincy, MA: privately published for the Granite Railway Company, 1926), 21, quoted in Vance, *North American Railroad,* 327n10.

27. Carl Seaburg and Stanley Paterson, *Merchant Prince of Boston: Colonel T. H. Perkins, 1764–1854* (Cambridge, MA: Harvard University Press, 1971), 387–404.

28. Monkton Iron Works records, Bixby Memorial Free Library, Vergennes, Vermont. The most salient records for this discussion are *Cash Book of the Monkton Iron Company, Second Agency, No. 2* (November 1814–September 1825), and the folder "Letters from Thomas H. Perkins, Boston to Benjamin Wells, Supt. Monkton Iron Co., 1813–1814." See also Peter H. Templeton, "'Energy, Enterprise, and the Sinews of Success': The Monkton Iron Company, 1807–1816," student research paper, March 10, 1980, typescript at the Bixby Memorial Free Library, Vergennes, Vermont.

29. Alan Seaburg, "Thomas Handasyd Perkins," *Dictionary of Unitarian and Universalist Biography,* accessed online at http://www25.uua.org/uuhs/duub on August 21, 2007; Kenneth A. Degree, *Vergennes in the Age of Jackson* (Vergennes, VT, 1996), 8.

30. Monkton Iron Works records.

31. J. D. Van Slyck, *Representatives of New England,* vol. 2, *Manufactures* (Boston: Van Slyck, 1879), 300–303; Dalzell, *Enterprising Elite,* 9, 30.

32. Dalzell, *Enterprising Elite,* 36–37.

33. Ibid., 45, 60–61, 235.

34. Ibid., 95, appendix, 88, 60; quotation on 88.

35. Ibid., 60–61.

36. Ibid., 80, 88–90, appendix.

37. LCC, Administrative Records, vol. 1, Stockholders and Directors Records, 1832–48.

38. Walter Isard and John H. Cumberland, "New England as a Possible Location for an Integrated Iron and Steel Works," *Economic Geography* 26, no. 4 (1950): 245–59; George Sweet Gibb, *The Saco-Lowell Shops: Textile Machinery Building in New England, 1813–1949* (Cambridge, MA: Harvard University Press, 1950), 50, 93–94.

39. LCC, box C.1, William Lyman to William P. Farrand, May 29, 1832; Edmund Dwight to W. P. Farrand, July 3, 1832; Dwight to Farrand, July 20, 1832; Lyman to Farrand, August 13, 1832.

40. Baer, *Canals and Railroads,* 52–53; Pennsylvania transportation GIS; "Map of the Several Canals and Rail Roads by Which the Lycoming Coal Can Be Sent to Market," in LCC, *Brief Description of the Property Belonging to the Lycoming Coal Company,* 6–9; "Distribution of Pennsylvania Coals," Commonwealth of Pennsylvania, Department of Conservation and Natural Resources, Bureau of Topographic and Geologic Survey, 2000, accessed online at www.dcnr.state.pa.us/topogeo on June 8, 2008.

41. *Journal of the Senate of the Commonwealth of Pennsylvania*, vol. 37 (Harrisburg, 1826–27), 475–80; *Journal of the Thirty-Seventh House of Representatives of the Commonwealth of Pennsylvania*, vol. 2 (Harrisburg, 1826–27), 402–17, 614–15; *Journal of the Thirty-Eighth House of Representatives of the Commonwealth of Pennsylvania*, vol. 2 (Harrisburg, 1827), 103–4; *Journal of the Thirty-Ninth House of Representatives of the Commonwealth of Pennsylvania*, vol. 2 (Harrisburg, 1828–29), 67–71; *Journal of the Forty-Second House of Representatives of the Commonwealth of Pennsylvania*, vol. 2 (Harrisburg, 1831–32), 116–17, 261–64; *Appendix to Vol. 2 of the Journal of the House of Representatives, 1835–36, Containing the Canal Commissioner's Report and Accompanying Documents* (Harrisburg, 1835), 100–101; *Appendix to Vol. 2 of the Journal of the House of Representatives, 1836–7, Containing the Canal Commissioner's Report and Accompanying Documents* (Harrisburg, 1836), 21–22, 176–82; *Appendix to Vol. 2 of the Journal of the House of Representatives, 1838–39, Containing the Canal Commissioner's Report and Accompanying Documents* (Harrisburg, 1838), 22; *Appendix to Vol. 2 of the Journal of the House of Representatives, Session of 1840, Containing the Canal Commissioner's Report and Accompanying Documents* (Harrisburg, 1840), 30; *Journal of the Senate of the Commonwealth of Pennsylvania, Session 1842, Containing the Canal Commissioner's Report and Accompanying Documents*, vol. 3 (Harrisburg, 1842), 27. See also Benjamin Aycrigg's summary report, *Communication from B. Aycrigg, Civil Engineer, relative to the West Branch and Allegheny Canal . . .* (Harrisburg, 1839).

42. Lesley HGIS; Gerald G. Eggert, *Making Iron on the Bald Eagle: Roland Curtin's Ironworks and Workers' Community* (University Park: Pennsylvania State University Press, 2000).

43. LCC, box C.1, Lyman to Farrand, Boston, June 16, 1832; T. H. Perkins to Edmund Dwight, July 5, 1832.

44. LCC, box C.1, T. H. Perkins to Edmund Dwight, July 5, 1832; Wm. Lyman to Wm. P. Farrand, May 29, 1832.

45. LCC, box C.1, Dwight to Farrand, August 20, 1832; Dwight to Farrand, December 17, 1832.

46. LCC, box C.1, Dwight to Farrand, March 28, 1833; Dwight to Farrand, June 2, 1833; emphasis in the original.

47. LCC, box C.1, Dwight to Farrand, July 6, 1833; March 28, 1833; June 2, 1833; June 19, 1833; emphasis in the original.

48. LCC, box C.1, Dwight to Farrand, July 29, 1833; emphasis in the original.

49. LCC, box C.1, Lyman to Farrand, August 13, 1832; Dwight to Farrand, September 5, 1833; emphasis in the original.

50. "Daniel Tyler," *Dictionary of American Biography*, vol. 19 (New York: C. Scribner's Sons, 1936), 86–87; George W. Cullum, *Notices of the Biographical Register of Officers and Graduates of the U.S. Military Academy at West Point*, vol. 1 (New York: J. Miller, 1879), 183–96; [Daniel Tyler], *Daniel Tyler: A Memorial Volume . . .* (New Haven, CT: privately published, 1883), 20–25, quotation on 24.

51. LCC, box C.1, vol. 1, board's contract with Daniel Tyler, January 1, 1834; William C. Davis, *Battle at Bull Run: A History of the First Major Campaign of the Civil War* (Garden City, NY: Doubleday, 1977), 39.

52. LCC, box C.1, vol. 1, minutes of the board of trustees, November 24, 1834; Charles Russell Lowell to Daniel Tyler, December 20, 1834; Carol Bundy, *The Nature of Sacrifice: A Biography of Charles Russell Lowell, Jr., 1835–64* (New York: Farrar, Straus and Giroux, 2005), 22–28.

53. Dalzell, *Enterprising Elite*, 15–25; LCC, box C.1, vol. 1, Charles Russell Lowell to Daniel Tyler, January 8, 1835; J. R. Hume and J. Butt, "Muirkirk, 1786–1802: The Creation of a Scottish Industrial Community," *Scottish Historical Review* 45 (1966): 160–83.

54. Robert and Sarah Ralston Family, Rootsweb, accessed online at http://ftp.rootsweb.com/pub/usgenweb/pa/ philadelphia/history/family/ralston.txt on May 6, 2004; Gerard Ralston to Dowlais Iron Co., December 9, 1835, and A. and G. Ralston to Guest Lewis and Co., April 2, 1836, in Madeleine Elsas, ed., *Iron in the Making: Dowlais Iron Company Letters, 1782–1860* (Cardiff:

County Records Committee of the Glamorgan Quarter Sessions and County Council and Guest Keen Iron and Steel Company, 1960), 190–91; General Registry Office, Marriages, September Quarter 1838, Cardiff 26 475.

55. LCC, box C.2, Gerard Ralston to Daniel Tyler, January 20, 1835; Jones, *History of GKN*, 1:116.

56. LCC, box C.2, Antes Snyder to Gerard Ralston, April 4, 1835.

57. LCC, box C.2, Gerard Ralston and Daniel Treadwell to Edward Thomas, August 4, 1835; Gerard Ralston and Daniel Treadwell to Directors of the Lycoming Coal Company, August 4, 1835; A. and G. Ralston to Daniel Tyler, August 20, 1835; Ynysgau [Independent] Chapel, Merthyr Tydfil, baptismal records, researched by Carolyn Jacob, personal communication, March 21, 2008.

58. Edward Thomas to Bess [Thomas], January 11, 1836; Edward Thomas to William [Thomas], February 2, 1836; Edward Thomas to Bess [Thomas], March 7, 1836; typescripts of original manuscript letters on microfilm at the Heinz History Center, Pittsburgh, MFF 2181, AB/A-80:62/Edward Thomas. Handwritten transcriptions of these letters are also held at the Heisey Museum, Clinton County Historical Society, Lock Haven, Pennsylvania, in the folder Farrandsville/Letters of Edward and Bess Thomas 1836.

59. LCC, box C.2, W. E. McDonald to Tyler, February 17, 1835; C. Bolinger to Tyler, April 6, 1835; Edwin Tyler to Daniel Tyler, April 26, 1835.

60. LCC, box C.7, General Accounts, "Lyman, Forbes and Co. Accounts, 1835–1838"; Edward Thomas to William [Thomas], February 2, 1836.

61. LCC, box C.2, contracts and agreements, memorandum of agreement between James Pitfields and Daniel Tyler, September 15, 1835.

62. LCC, box C.2, John Carruthers to Barnebus Shipley, February 3, 1835.

63. Edward Thomas to Bess [Thomas], January 11, 1836.

64. Ibid.; Edward Thomas to William [Thomas], February 2, 1836.

65. LCC, box C.2, W. P. Farrand to [Daniel] Tyler, November 14, 1835; Edward Thomas to William [Thomas], February 2, 1836.

66. The first reference to the Hemlock property is LCC, box C.2, Daniel Tyler to T. S. Williams, November 26, 1836. Location and distance from Pennsylvania transportation GIS and USGS Geographic Names Information System, accessed online at http://geonames.usgs.gov on December 12, 2008.

67. LCC, box C.2, Edward Thomas to the Lycoming Coal Company, November 18, 1835.

68. Edward Thomas to Bess [Thomas], January 11, 1836.

69. LCC, box C.2, Edward Thomas to D[aniel] Tyler, June 28, 1836.

70. LCC, box C.2, E[dward] Thomas to "My dear Sister" [Bess Thomas], March 7, 1836; E[dward] Tyler to D[aniel] Tyler, August 6, 1836.

71. LCC, box C.2, Edward Thomas to D[aniel] Tyler, July 26, 1836, and July 27, 1836; LCC, box C.1, vol. 1, minutes of board meeting, August 29, 1836.

72. LCC, box C.3, Contract with James Ralston, January 11, 1837; Contract with Furnace Keeper, January 23, 1837; LCC, box C.2, memorandum by David Napier, Camlachin Foundry, January 13, 1837.

73. LCC, box C.2, Daniel Tyler to Thomas S. Williams, January 16, 1837.

74. LCC, box C.3, Agreement to arbitration between James Ralston and Daniel Tyler, February 15, 1839. Ralston's obituary claims that he had put Farrandsville Furnace in blast, but company records do not support this. It also claims that Ralston was the first to implement the hot blast in Scotland. He may have stretched the truth to land the job at Farrandsville as well. Garretson, "James Ralston—the Origin of the Hot Blast," *United States Railroad and Mining Register*, June 18, 1864.

75. LCC, box C.3, Contract with Benjamin Perry, July 23, 1838, with addendum of December 3, 1838; Craig L. Bartholomew and Lance E. Metz, *The Anthracite Iron Industry of the Lehigh Val-*

ley, ed. Ann Bartholomew (Easton, PA: Center for Canal History and Technology, 1988), 31–32, 35–36.

76. LCC, box C.2, Charles Mieg to [Daniel] Tyler, June 1, 1838.

77. LCC, box C.2, P. Flanagan to D[aniel] Tyler, May 26, 1837; W. T. Walters to D[aniel] Tyler, August 4, 1837; Walters to Tyler, August 5, 1837; Walters to Tyler, August 31, 1837; LCC, Importation of Labor, 1837–38. The boilerplate contracts in the company files specify rate of pay in shillings. They say nothing about accommodation. See LCC, box C.3, Contracts and Agreements, 1837.

78. Knowles, *Calvinists*, chap. 1.

79. LCC, Contracts and Agreements, 1837; box C.2, A. V. [?] Parsons to D. Tyler, October 14, 1837. If the worker stayed to the end of the year, the cost of passage would be reimbursed. LCC, box C.3, contract with John Derby, June 12, 1837.

80. LCC, box C.2, Walters to Tyler, October 14, 1837; D. Tyler to Josiah W. Smith, October 17, 1837.

81. LCC, box C.2, E. Greenough to D. Tyler, January 20 and 22, 1838; Josiah W. Smith to D. Tyler, January 27, 1838.

82. Robert J. Steinfeld, *Coercion, Contract, and Free Labor in the Nineteenth Century* (Cambridge: Cambridge University Press, 2001), 62, 108.

83. LCC, box C.2, Josiah W. Smith to Daniel Tyler, February 14, 1838; E. Greenough to Col. Tyler, February 28, 1838.

84. Robert J. Steinfeld, *The Invention of Free Labor: The Employment Relation in English and American Law and Culture, 1350–1870* (Chapel Hill: University of North Carolina Press, 1991), 143–46.

85. Baer, *Canals and Railroads*, 52–54, and Christopher T. Baer, personal communications, June 1, 2010, and January 25, 2011.

86. LCC, box C.9, Rules and Regulations for Working, 1833; box C.1, miners' contract, November 1, 1835; box C.2, lease for double log house rented as boardinghouse to Alexander Hamilton and James Innes, June 14, 1836.

87. LCC, box C.1, proposal to the board by P. T. Jackson, February 9, 1838.

88. LCC, box C.7, Lyman to P. T. Jackson, June 16, 1838.

89. Bundy, *Nature of Sacrifice*, 36–39, quotation on 36.

90. Cullum, *Notices of the Biographical Register*, 190–91; Davis, *Battle at Bull Run*, 39, 112–24, 154–57, 234–37, 254–59; Richard Sewell, personal communication, October 2002.

91. LCC, box C.8, General Accounts, Inventories, sales and disposition of furniture and effects, including inventories dated April 17, 1839; May 29–31, 1840; July 11 and 18, 1840; June 3 and 4, 1840.

92. Seaburg and Patterson, *Merchant Prince of Boston*; Bundy, *Nature of Sacrifice*, 45–46, quotation on 46.

93. Lesley, *Guide*, 23, 93.

94. *Advantages of Ralston, Lycoming Co., Pennsylvania, for the Manufacture of Iron* (New York: Tribune Job Printing Establishment, 1843), 4; J. Peter Lesley, *The Iron Manufacturer's Guide to the Furnaces, Forges, and Rolling Mills of the United States* (New York: John Wiley, 1859), 93; J. Peter Lesley to Susan Inches Lesley, June 3, 1857, Ames Family Historical Collection, AmesINK, in Boulder, Colorado.

95. Thomas W. Lloyd, *History of Lycoming County, Pennsylvania*, vol. 1 (Topeka, KS: Historical Publishing, 1929), 183–84, 139, 185; Lesley HGIS; Leach, "Old Philadelphia Families."

96. Farrandsville Company corporate annual reports, LCC; deed books A, C, D, G, H, I, MD, Register and Recorder's Office, Lycoming County Courthouse, Lock Haven, Pennsylvania.

97. Christopher Fallon, "Report to the Stockholders" (1856), 4–9, Historic Corporate Reports Collection, Farrandsville Company annual report, Baker Library Historical Collections, Harvard Business School.

98. John H. Campbell, *History of the Friendly Sons of St. Patrick and of the Hibernian Society for the Relief of Emigrants from Ireland* (Philadelphia: Hibernian Society, 1892), 403–4; Anonymous [A Merchant of Philadelphia], *Memoirs and Auto-biography of Some of the Wealthy Citizens of Philadelphia* (Philadelphia: Booksellers, 1846), 22; Christopher Fallon Papers, MS 1673, Historical Society of Pennsylvania, Philadelphia (hereafter Fallon Papers); Pennsylvania transportation GIS; Christopher T. Baer, "PRR [Pennsylvania Railroad] Chronology: A General Chronology of the Pennsylvania Railroad Company, Predecessors and Successors and Its Historical Context," 1856, March 2005 edition, 2, accessed online at http://www.prrths.com/Hagley/PRR1856%20Mar%2005.pdf on January 22, 2011. Fallon declined reelection on the grounds that the line had "refused his offer to rescue it" (4).

99. Campbell, *Friendly Sons of St. Patrick*, 403–4; Dean R. Wagner, ed., *Historic Lock Haven: An Architectural Survey* (Lock Haven, PA: Clinton County Historical Society, 1979), 39.

100. Fallon Papers, manuscript correspondence in Spanish from June 9, 1853, to October 13, 1859.

101. William Baynton to Peter Cooper, March 28, April 8, 12, 1865, Cooper, Hewitt Papers.

102. Pat Dooley, "Traces of Queen's Mansion Remain at Farrandsville," *Grit*, November 15, 1981, news section, 53, in Farrandsville/Queen of Spain folder, Heisey Museum, Clinton County Historical Society, Lock Haven, Pennsylvania; United States Geological Survey, "Farrandsville," map, scale 1:24,000 (Washington, DC: Government Printing Office, 1979).

103. Photograph documenting the hotel sign's inscription by Lou Bernard, curator of the Clinton County Historical Society, Lock Haven, Pennsylvania; personal correspondence, Lou Bernard, February 22, 2010.

104. D. S. Maynard, *Historical View of Clinton County* (Lock Haven, PA: Enterprise Printing, 1875), 131–32; "Farrandsville, Rich in History, Observes 150th Anniversary," *Express* (Lock Haven, PA), July 22, 1982, 11.

105. Katherine A. Harvey, ed., *The Lonaconing Journals: The Founding of a Coal and Iron Community, 1837–1840*, Transactions of the American Philosophical Society, n.s., 67, part 2 (Philadelphia: American Philosophical Society, 1977), 7–10; "J. H. Alexander's Statement for the Arbitrators in the matter referred between himself and the George's Creek Coal and Iron Company," November 5, 1850, in John Henry Alexander Papers, 1800–83, MS 10, box 1, Maryland Historical Society.

106. *George's Creek Coal and Iron Company* (1836), Rare PAM 10047, Maryland Historical Society; Baer, *Canals and Railroads*, "Railroad Companies: Basic Organizational Information" (table), 60, 52; US Geological Survey, 1:24,000 topographic quadrangles for the area from Cumberland to Lonaconing, Maryland: "Cumberland, Md.–Pa.–W.Va." (1993); "Frostburg, Md.–Pa." (1998); "Lonaconing, Md.–W.Va." (1998); Google Earth, accessed online on June 10, 2008. Alexander wrote in the company journal that a site had been laid out for "the four furnaces" in September 1837. Only one furnace was built. "Lonaconing Iron Works Records," September 4, 1837; site visit, February 2002.

107. J. T. Ducatel and J. H. Alexander, "Report on the Projected Survey of the State of Maryland" (Philadelphia, 1834), 34, quoted in James D. Dilts, *The Great Road: The Building of the Baltimore and Ohio, the Nation's First Railroad, 1828–1853* (Stanford, CA: Stanford University Press, 1993), 286n13.

108. Katherine A. Harvey, *The Best-Dressed Miners: Life and Labor in the Maryland Coal Region, 1835–1910* (Ithaca, NY: Cornell University Press, 1969), 3–8.

109. Harvey, *Lonaconing Journals*, 69.

110. Maryland *Laws* 1835, chaps. 328, 382, quoted in Harvey, *Lonaconing Journals*, 9nn19, 20.

111. Harvey, *Lonaconing Journals*, 12–13.

112. The following narrative is drawn from "Lonaconing Iron Works Records" and Alexander, "Statement." Additional sources are noted as they apply.

113. LCC, box C.2, Thomas Davies to D. Tyler, July 8, 1837.

114. Passenger list of the bark *Tiberias*, reprinted in Harvey, *Lonaconing Journals*, 70.

115. Harvey, *Best-Dressed Miners*, 377–78.

116. Although Alexander claimed to have done both translations, it is highly unlikely he could have translated the punctilious rules into Welsh. *The Lonaconing Journals* refers elsewhere to his using David Hopkins as a translator. Alexander, "Statement," MS 2292, "Lonaconing Iron Works Records," January 19, 1839.

117. "Lonaconing Iron Works Records," July 10, 1839.

118. Bureau of Topographic and Geologic Survey, *Distribution of Pennsylvania Coals*, 3rd ed., 2nd printing, rev. (Middletown, PA: Commonwealth of Pennsylvania, Department of Conservation and Natural Resources, 2000); Kenneth Warren, *Wealth, Waste, and Alienation: Growth and Decline in the Connellsville Coke Industry* (Pittsburgh: University of Pittsburgh Press, 2001).

119. Harvey, *Lonaconing Journals*, 51–57.

120. Alexander, "Statement." Alexander identifies the miners as English in his statement, but the evidence in the journal suggests they were Welsh.

121. Baer, *Canals and Railroads*, "Railroad Companies: Basic Organizational Information" (table), 60, 52.

122. William Pamplin, "Cyfarthfa Works and Water Wheel," drawing, pencil on paper, item no. 823.992, Cyfarthfa Castle Museum and Gallery, Merthyr Tydfil, Wales, 1791–1800; W. David Lewis, "The Early History of the Lackawanna Iron and Coal Company: A Study in Technological Adaptation," *Pennsylvania Magazine of History and Biography* 96 (1972): 424–68; Lesley, *Guide*, 8, 21; Chester Kulesa, site administrator, Anthracite Heritage Museum and Iron Furnaces, Scranton, Pennsylvania, personal communication, June 12, 2008.

123. Harvey, *Lonaconing Journals*, 67–69.

124. Anne Kelly Knowles and Richard G. Healey, "Geography, Timing, and Technology: A GIS-Based Analysis of Pennsylvania's Iron Industry, 1825–1875," *Journal of Economic History* 66 (September 2006): 608–34.

Chapter 4

1. Robert B. Gordon, *American Iron, 1607–1900* (Baltimore: Johns Hopkins University Press, 1996), 78, 162; J. Peter Lesley, *The Iron Manufacturer's Guide to the Furnaces, Forges, and Rolling Mills of the United States* (New York: John Wiley, 1859), 237, 244; Jay D. Allen, "The Mount Savage Iron Works, Mount Savage, Maryland: A Case Study in Pre–Civil War Industrial Development" (MA thesis, University of Maryland, 1970); Michael D. Thompson, "The Iron Industry in Western Maryland," unpublished paper, Morgantown, WV, 1976, 128–32, National Canal Museum.

2. J. M. Forbes to Paul S. Forbes, December 21, 1859; emphasis in the original, Forbes Family Business Records, Baker Library Historical Collections, Harvard Business School; Thompson, "Iron Industry in Western Pennsylvania," 132–33, cited in John Lauritz Larson, *Bonds of Enterprise: John Murray Forbes and Western Development in America's Railway Age*, rev. ed. (Iowa City: University of Iowa Press, 2001), 211–12n47.

3. Darwin H. Stapleton, *The Transfer of Early Industrial Technologies to America* (Philadelphia: American Philosophical Society, 1987), 169–201; Alfred D. Chandler Jr., "Anthracite Coal and the Beginnings of the Industrial Revolution in the United States," *Business History Review* 46 (Summer 1972): 141–81; Peter Temin, *Iron and Steel in Nineteenth-Century America: An Economic Inquiry* (Cambridge, MA: MIT Press, 1964), 63–76.

4. Albert Fishlow, *American Railroads and the Transformation of the Ante-bellum Economy* (Cambridge, MA: Harvard University Press, 1965), 136–45. For a counterargument about the importance of railroads in the antebellum economy, see Robert William Fogel, *Railroads and American Economic Growth: Essays in Econometric History* (Baltimore: Johns Hopkins University Press, 1964).

5. Temin, *Iron and Steel*, table C.2, 266.

6. Alan Birch, *The Economic History of the British Iron and Steel Industry, 1784–1879* (London:

Frank Cass, 1967; reprint, New York: Augustus M. Kelley, 1968), 170; Temin, *Iron and Steel*, table C.2, 266–67.

7. Philip Scranton, *Proprietary Capitalism* (Cambridge: Cambridge University Press, 1983), 39–41, quotation on 39.

8. David R. Meyer, *Networked Machinists: High-Technology Industries in Antebellum America* (Baltimore: Johns Hopkins University Press, 2006).

9. Chandler, "Anthracite Coal," 152–58.

10. *A History of the Lehigh Coal and Navigation Company*, published by order of the Board of Managers (Philadelphia: William S. Young, 1840), 1–35, quotation on 5; Christopher T. Baer, *Canals and Railroads of the Mid-Atlantic States, 1800–1860,* ed. Glenn Porter and William H. Mulligan Jr., cartographers Marley E. Amstutz and Anne E. Webster (Wilmington, DE: Regional Economic History Research Center, Eleutherian Mills–Hagley Foundation, 1981), appendix A; personal communication, Christopher T. Baer, June 1, 2010.

11. Stephen S. Witte and Marsha V. Gallagher, eds., *The North American Journals of Prince Maximilian of Wied*, vol. 1, *May 1832–April 1833*, trans. William J. Orr, Paul Schach, and Dieter Karch (Norman: University of Oklahoma Press, 2008), 85.

12. Ibid., 136; Christopher T. Baer, personal communication, June 1, 2010.

13. Witte and Gallagher, *North American Journals*, 137–43; John F. Sears, "Tourism and the Industrial Age: Niagara Falls and Mauch Chunk," in *Sacred Places: American Tourist Attractions in the Nineteenth Century,* by John F. Sears (New York: Oxford University Press, 1989), 191–92.

14. Stapleton, "David Thomas," 179–80; Craig L. Bartholomew and Lance E. Metz, *The Anthracite Iron Industry of the Lehigh Valley*, ed. Ann Bartholomew (Easton, PA: Center for Canal History and Technology, 1988), 23–24; Alfred Mathews and Austin N. Hungerford, *History of the Counties of Lehigh and Carbon, in the Commonwealth of Pennsylvania* (Philadelphia: Everts and Richards, 1884), 674–75, quotation on 675.

15. *History of the Lehigh . . . Company*, 66.

16. Stapleton, "David Thomas," 173; Bartholomew and Metz, *Anthracite Iron Industry*, 12–19, 31–33.

17. Solomon W[hite] Roberts, "Gwneyd Haiarn â Maenlo" [Made Iron with Stone Coal], obituary of George Crane, trans. into Welsh, *Y Cyfaill o'r Hen Wlad* [The Friend from the Old Country], 1846, 170–71. "Ynyscedwyn" is the current standard Welsh spelling; it is sometimes spelled "Yniscedwyn."

18. Stapleton, "David Thomas," 173–76; Samuel Thomas, "Reminiscences of the Early Anthracite-Iron Industry," *Transactions of the American Institute of Mining, Metallurgical, and Petroleum Engineers* 29 (1899): 902–4, quoting [Edward Roberts], "David Thomas: The Father of the Anthracite Iron Trade," *Red Dragon,* October 1883, 288–99; [Roberts], "David Thomas," 292. Scholars have not been able to determine whether Thomas or Crane deserves chief credit for inventing the process, a debate that began before Thomas's death. See Thomas, "Reminiscences," 927–28. Crane claimed to be using only anthracite at Ynyscedwyn in April 1837, but in a letter to David Thomas in October 1839 he reported good results using six-sevenths anthracite, presumably mixed with coke. George Crane to Thomas or John Evans, April 14, 1837, in Madeleine Elsas, ed., *Iron in the Making: Dowlais Iron Company Letters, 1782–1860* (Cardiff: County Records Committee of the Glamorgan Quarter Sessions and County Council and Guest Keen Iron and Steel Company, 1960), 202–3; George Crane to David Thomas, October 17, 1839, David Thomas Papers, National Canal Museum (hereafter Thomas Papers).

19. Thomas, "Reminiscences," 903–4; Gordon, *American Iron,* 109; Harry Scrivenor, *History of the Iron Trade* (London: Longman, Brown, Green, and Longmans, 1854), 259–60; Bartholomew and Metz, *Anthracite Iron Industry,* 16; "Crane's Patent for Smelting Iron with Anthracite Coal," *Journal of the Franklin Institute,* June 1838, 405–6.

20. Scrivenor, *Iron Trade,* 260–61.

21. Thomas, "Reminiscences," 902–4, quotation on 903–4; Laurence Ince, "The Neath Abbey Ironworks," *Industrial Archaeology* 11–12 (1977): 25–31.

22. On Crane's patent, see Bartholomew and Metz, *Anthracite Iron Industry,* 17.

23. Crane to Thomas or John Evans, April 14, 1837, in Elsas, *Iron in the Making,* 202–3. Crane may have been unaware that Dowlais had been using the hot blast at some of its coke furnaces since at least 1836; Edgar Jones, *A History of GKN* [Guest, Keen and Nettlefolds], vol. 1, *Innovation and Enterprise, 1759–1918* (London: Macmillan, 1987), 65–68.

24. Stapleton, "David Thomas," 178; Robert and Sarah Ralston Family, Rootsweb, accessed online at http://ftp.rootsweb.com/pub/usgenweb/pa/ philadelphia/history/family/ralston.txt on May 6, 2004; Bartholomew and Metz, *Anthracite Iron Industry,* 18; Matthew S. Henry, *History of the Lehigh Valley* (Easton, PA: Bixler and Corwin, 1860), 415–16; Robert Earp, Report to the Directors of the Lehigh Crane Iron Company, November 9, 1840, in Minutes of Stockholders Meetings, vol. 1a, Lehigh Crane Iron Company Papers, acc. no. 1198, Hagley Museum and Library (hereafter Lehigh Crane Papers).

25. Bartholomew and Metz, *Anthracite Iron Industry,* 20, 16; Stapleton, "David Thomas," 180; George Crane to David Thomas, May 21, 1839, and October 17, 1839, Thomas Papers. According to Solomon White Roberts, Crane was unable to validate his claim to the US patent rights before his death in 1846; "Gwneyd Haiarn â Maenlo" [Made Iron with Stone Coal], 171.

26. Roberts, "David Thomas," 291.

27. "David Thomas," *Dictionary of American Biography,* vol. 18 (New York: Charles Scribner's Sons, 1936), 427; Ince, "Neath Abbey Ironworks," 25–29; Roberts, "David Thomas," 292. Ynyscedwyn had three furnaces by about 1835; S. W. Rider and A. E. Trueman, *South Wales: A Physical and Economic Geography* (London: Methuen, 1929), 92.

28. Carol Siri Johnson, "Prediscursive Technical Communication in the Early American Iron Industry," *Technical Communication Quarterly* 15, no. 2 (2006): 171–89. R. Elwyn Hughes concluded that few if any technical industrial manuals were published in Welsh because "nineteenth-century Welsh writers were reluctant to venture into unfamiliar territory." I think it more likely that no one needed written explanations for industrial processes because skills were transmitted through oral instruction and practice. R. Elwyn Hughes, "The Welsh Language in Technology and Science, 1800–1914," in *The Welsh Language and Its Social Domains, 1801–1911,* ed. Geraint H. Jenkins (Cardiff: University of Wales Press, 2000), 414.

29. Stapleton, "David Thomas," 181; J. H. Alexander to Capt. D. Tyler, July 19, 1838, mentions Lonaconing's using a West Point Foundry–made blowing engine with a blowing cylinder five feet in diameter, the size of the engine Thomas specified for Lehigh Crane; Lycoming Coal Company Records, Baker Library Historical Collections, Harvard Business School (hereafter LCC), box C.2, General Correspondence.

30. Thomas, "Reminiscences," 910–14; Stapleton, "David Thomas," 181–82; Bartholomew and Metz, *Anthracite Iron Industry,* 25–27.

31. Furnace height as given in J. Peter Lesley, *The Iron Manufacturer's Guide to the Furnaces, Forges, and Rolling Mills of the United States* (New York: John Wiley, 1859), 8.

32. 1840 US Manuscript Census, Hanover Township, Lehigh County, Pennsylvania, M704, roll 469, National Archives and Research Administration (hereafter NARA); Bartholomew and Metz, *Anthracite Iron Industry,* 27. On the clustering of Welsh immigrants in urban and rural American communities, see Anne Kelly Knowles, *Calvinists Incorporated: Welsh Immigrants on Ohio's Industrial Frontier* (Chicago: University of Chicago Press, 1997), chap. 1. Place of birth was not recorded in the federal census until 1850. Because census takers usually recorded households in sequence as they were visited, the Manuscript Census captures the relative location of households, door to door; adjoining entries in the manuscript list can be assumed to have been neighbors. See Michael P. Conzen, "Spatial Data from Nineteenth-Century Manuscript Censuses: A Technique for Rural Settlement and Land Use Analysis," *Professional Geographer* 25 (1969): 337–43.

33. 1850 US Manuscript Census, Hanover Township, Lehigh County, Pennsylvania, M432, roll 792, NARA. Hunt's wife, Gwenna, was David and Elizabeth Thomas's daughter; Roberts, "David Thomas," 298.

34. Minutes of Stockholders Meetings, passim, Lehigh Crane Iron Company Papers; Gordon, *American Iron*, 137–38.

35. Anne Kelly Knowles, "Labor, Race, and Technology in the Confederate Iron Industry," *Technology and Culture* 42 (January 2001): 23–24; Sharon A. Brown, *Historic Resource Study: Cambria Iron Company* (Washington, DC: US Department of the Interior, National Park Service, 1989), 264; R. Bruce Council, Nicholas Honerkamp, and M. Elizabeth Will, *Industry and Technology in Antebellum Tennessee: The Archaeology of Bluff Furnace* (Knoxville: University of Tennessee Press, 1992), 75.

36. Minutes of special meeting called to discuss hiring of David Thomas, July 2, 1839, vol. 2a, Lehigh Crane Papers.

37. Minutes of Stockholders Meetings, 1839–71, and Minutes of Board of Directors, 1839–48, Lehigh Crane Papers.

38. Bartholomew and Metz, *Anthracite Iron Industry*, 27; Earp, Report to the Directors, Lehigh Crane Papers.

39. Minutes of Stockholders Meetings, November 22, 1842, November 13, 1843, February 14, 1848, vol. 1a, Lehigh Crane Papers; Thomas, "Reminiscences," 905. The original contract provided for an addition of £50 sterling to Thomas's salary of £250 for each furnace brought into blast. At the standard exchange rate of $6.66 US to the British pound, this means that at some point Thomas's bonus was increased by two-thirds to $500.

40. Minutes of Stockholders Meetings, February 14, 1848, vol. 1a, Lehigh Crane Papers.

41. Minutes of Stockholders Meetings, July 23, 1845; February 14, 1848, vol. 1a, Lehigh Crane Papers.

42. Bruno Latour, *Science in Action: How to Follow Scientists and Engineers through Society* (Cambridge, MA: Harvard University Press, 1987), 249, 250.

43. Josiah White, *Josiah White's History, Given by Himself* (Philadelphia: Lehigh Coal and Navigation Company, ca. 1910), passim. See also Anthony F. C. Wallace, *Rockdale: The Growth of an American Village in the Early Industrial Revolution* (New York: W. W. Norton, 1972), and Scranton, *Proprietary Capitalism*.

44. Baer, *Canals and Railroads*, "Mid-Atlantic Canals and Railroads 1845," map, scale 1:1,000,000, and table of canal segment lengths, 52. Distance from Lonaconing to Cumberland estimated on Google Earth.

45. Minutes of Stockholders Meetings, November 8, 1841, vol. 1a, Lehigh Crane Papers; Lesley HGIS; Baer, *Canals and Railroads*, "Canals and Railroads 1845" and appendix.

46. Fishlow, *American Railroads*, 18–20, 33–36, 243–44.

47. Lesley HGIS and Pennsylvania transportation HGIS. Percentages were calculated using the "select by location" function in ArcGIS (features contained within the Mid-Atlantic region as defined in figure 19).

48. Fishlow, *American Railroads*, 136–45. Fishlow notes that rail mills consumed only a fraction of the mineral-fuel pig iron produced in the antebellum period. As much or more anthracite iron was probably used at the Mid-Atlantic's sheet, plate, and bar mills, which were concentrated in the Philadelphia area.

49. Lesley, *Guide*, 219–63; [J. Peter Lesley, ed.], *Bulletin of the American Iron Association* (Philadelphia, 1856–59). Several southern and western rolling mills were integrated mill-furnace-forge operations before Cambria, but they smelted ore in single charcoal blast furnaces and were much smaller overall. These included Etowah (founded in 1849 in Cass County, Georgia) and Eagle (1831 in Centre County, Pennsylvania). The Safe Harbor works, built in 1848 in Lancaster County, Pennsylvania, included a rolling mill, foundry, and anthracite blast furnace but not, so far as I know, a coal mine. Lesley, *Guide*, 13, 54, 77, 238, 239, 246.

50. Minutes of Stockholders Meetings, November 4, 1844, vol. 1a, and Minutes of Board of Directors, April 7, 1845, vol. 2a, Lehigh Crane Papers; Fr. R. Backus to Peter Cooper, July 17, 1846, Cooper, Hewitt Papers.

51. Glenn Porter and Harold Livesay, *Merchants and Manufacturers* (Baltimore: Johns Hopkins University Press, 1971), 37–61, 79–115.

52. Allan Nevins, *Abram S. Hewitt, with Some Account of Peter Cooper* (New York: Harper, 1935; reprint, New York: Octagon Books, 1967).

53. Ibid., 71–72.

54. Ibid., 71, 82–83; letters to Peter Cooper, 1844–45, Cooper, Hewitt Papers. For a contemporary view of the location of Cooper's New York ironworks and its distance from New York Harbor and other lines of communication, see J. H. Colton, *Topographical Map of the City and County of New-York . . .*, map, scale approx. 1:16,000 (New York: J. H. Colton, 1836).

55. Nevins, *Abram S. Hewitt*, 91–92, 95–110; "Works of the Trenton Iron Company," n.d., n.p., courtesy Cooper Union Library. Estimate of image date based on John K. Brown, *The Baldwin Locomotive Works, 1831–1915* (Baltimore: Johns Hopkins University Press, 1995), 15.

56. Brown, *Baldwin Locomotive Works*, 65–71, 77–78. On the Croton Aqueduct, see Matthew Gandy, *Concrete and Clay: Reworking Nature in New York City* (Cambridge, MA: MIT Press, 2002), 29–37.

57. Nevins, *Abram S. Hewitt*, 40–44.

58. Cooper, Hewitt Papers, passim; Nevins, *Abram S. Hewitt*, 24–30, 147.

59. Letters to Peter Cooper, May–December 1845, Cooper, Hewitt Papers; Sarah A. Dewick, *The Ancestry of John S. Gustin and His Wife Sarah McComb* (Boston: David Clapp, 1900), 34–35.

60. Charles Hewitt to Peter Cooper, January 21, 1846, and June 19, 1846, Cooper, Hewitt Papers. Twenty years later, Charles Hewitt, reflecting on his experience in Trenton, estimated that it would take eighteen to twenty-four months to equip a mill as completely as their mill was at that point. Charles Hewitt to Abram Hewitt, October 11, 1866, Cooper, Hewitt Papers.

61. Livingston and Lyman, Philadelphia, to Peter Cooper, February 11, 1846, Cooper, Hewitt Papers; Gordon, *American Iron*, 216.

62. Charles Hewitt to Peter Cooper, April 16, 1847, Cooper, Hewitt Papers.

63. Benjamin Latrobe to Peter Cooper, March 14, 1846, and Abram S. Hewitt to Peter Cooper, February 19, 1846, Cooper, Hewitt Papers; Lesley HGIS.

64. Cooper, Hewitt Papers, passim, September 1845–November 1847; folder of orders and receipts [for Peter Cooper's rolling mill in New York City], 1844.

65. Charles Hewitt to Peter Cooper, June 1, 1846, and May 14, 1846, Cooper, Hewitt Papers.

66. Benjamin Latrobe to Peter Cooper, January 29, 1846; Charles Hewitt to Abram Hewitt, October 11, 1866, Cooper, Hewitt Papers. For an example of the particularity of specification for rails, see Benjamin Latrobe's letter and diagram for the Baltimore and Ohio Railroad, January 29, 1846, Cooper, Hewitt Papers.

67. Fr. R. Backus to Peter Cooper, March 4, 1846; Abram Hewitt to Peter Cooper, March 30, 1847; James Hall to Peter Cooper, August 4, 1847; Abram S. Hewitt to Peter Cooper, September 17, 1847; G. G. Palmer to A[bram] S. Hewitt, November 14, 1847, Cooper, Hewitt Papers; Lesley, *Guide*, 6, 8; Nevins, *Abram S. Hewitt*, 90.

68. George H. Cook and John C. Smock, *Northern New Jersey, Showing the Iron-Ore and Limestone Districts*, map, horizontal scale 2 miles to the inch, vertical scale 2,000 feet to the inch (n.p., 1874); Baer, *Canals and Railroads*, table, Cumulative Railroad Mileage 1826–62; Lesley HGIS.

69. Nevins, *Abram S. Hewitt*, 90.

70. Abram Hewitt to Peter Cooper, May 7, 1847, September 15, 1847, and November 14, 1847, Cooper, Hewitt Papers.

71. Abram S. Hewitt to Peter Cooper, September 17, 1847, and Charles Hewitt to Peter Cooper, January 22, 1846, Cooper, Hewitt Papers.

72. Edward Cooper to Peter Cooper, November 17, 1846, Cooper, Hewitt Papers; emphasis in the original.

73. Charles Hewitt to Peter Cooper, November 10, 1845, Cooper, Hewitt Papers.

74. Edward Cooper to Peter Cooper, October 13, 1845; Charles Hewitt to Peter Cooper, January 2, 1846, Cooper, Hewitt Papers.

75. Charles Hewitt to Edward Cooper, March 16, 1846, Cooper, Hewitt Papers.

76. Charles Hewitt to Peter Cooper, April 20, 1847, Cooper, Hewitt Papers.

77. A. B. Winder [?] for C[harles] Hewitt, to James Hall, June 16, 1846, Cooper, Hewitt Papers; Lesley, *Guide*, 230; Wallace, *Rockdale*, 98.

78. Murdock Leavitt and Co. to Peter Cooper, November 5, 1846; William Green Jr., Boonton Iron Works, to Peter Cooper, February 23, 1847, and March 1, 1847, Cooper, Hewitt Papers.

79. Edward Cooper to Peter Cooper, February 17, 1847, Cooper, Hewitt Papers.

80. Charles Hewitt to Peter Cooper, February 22, 1847, and March 1, 1847, Cooper, Hewitt Papers.

81. Charles Hewitt to Peter Cooper, February 5, 1846, April 29, 1847, April 20, 1846, and February 25, 1846, Cooper, Hewitt Papers.

82. John Fritz, *The Autobiography of John Fritz* (New York: John Wiley, 1912), 53–54.

83. Lesley, *Guide*; Nevins, *Abram S. Hewitt*, 104.

84. James Moore Swank, *History of the Manufacture of Iron in All Ages*, American Classics of History and Social Science 6 (Philadelphia: American Iron and Steel Association, 1892; reprint, New York: Burt Franklin, 1965), 435; Gordon, *American Iron*, 162.

85. Nevins, *Abram S. Hewitt*, 105–8.

86. Ibid., 106–8.

87. Ibid., 111–18.

88. Ibid., 102; Harold C. Livesay, "Marketing Patterns in the Antebellum American Iron Industry," *Business History Review* 45, no. 3 (1971): 290, 289.

89. John N. Ingham, *Making Iron and Steel: Independent Mills in Pittsburgh, 1820–1920* (Columbus: Ohio State University Press, 1991), 51.

90. Charles B. Dew, *Ironmaker to the Confederacy: Joseph R. Anderson and the Tredegar Iron Works* (New Haven, CT: Yale University Press, 1966), 35–37; Robert C. Black III, *The Railroads of the Confederacy* (Chapel Hill: University of North Carolina Press, 1952), 85–86.

91. Fred Bateman and Thomas Weiss, *A Deplorable Scarcity: The Failure of Industrialization in the Slave Economy* (Chapel Hill: University of North Carolina Press, 1981).

Chapter 5

1. Warren Ripley, *Artillery and Ammunition of the Civil War* (New York: Promontory Press, 1970), 358–61; Ian V. Hogg, *Weapons of the Civil War* (New York: Military Press and Crown Publishers, 1987), 11–27, 34–46; Carl L. Davis, *Arming the Union: Small Arms in the Civil War* (Port Washington, NY: Kennikat Press, 1973), plates 15 and 16; Michael S. Raber et al., *Forge of Innovation: An Industrial History of the Springfield Armory, 1794–1968*, ed. Richard Colton (1989; reprint, Fort Washington, PA: Eastern National [Division, National Park Service], 2008), 80–81, 149; Josiah Gorgas, "Notes on the Ordnance Department of the Confederate Government," *Southern Historical Society Papers* 12 (1884): 70; J. W. Mallet, "Work of the Ordnance Bureau of the War Department of the Confederate States, 1861–1865," *Alumni Bulletin of the University of Virginia* 3 (1910): 162–70; Records of the Bureau of Ordnance [Navy], Record Group 74, entry 20, Letters received from foundries, boxes 1–5, National Archives and Research Administration (hereafter NARA RG 74); Merritt Roe Smith, *Harpers Ferry Armory and the New Technology: The Challenge of Change* (Ithaca, NY: Cornell University Press, 1977), 310–22; Raimondo Luraghi, *A History of the Confederate Navy*, trans. Paolo E. Coletta (Annapolis, MD: Naval Institute Press, 1996), 34–50.

2. James M. McPherson, *Battle Cry of Freedom: The Civil War Era* (New York: Oxford University Press, 1988), 319–20, quoting Gorgas, 320. The standard sources for Gorgas are Frank E. Vandiver, *Ploughshares into Swords: Josiah Gorgas and Confederate Ordnance* (Austin: University of Texas Press, 1952); Frank E. Vandiver, ed., *The Civil War Diary of General Josiah Gorgas* (Tuscaloosa: University of Alabama Press, 1947); and Sarah Woolfolk Wiggins, ed., *The Journals of Josiah Gorgas, 1857–1878* (Tuscaloosa: University of Alabama Press, 1995).

3. J. Gorgas, Chief of Ordnance, to James A. Seddon, Secretary of War, October 13, 1864, reprinted in Fred C. Ainsworth and Joseph W. Kirkley, eds., *The War of the Rebellion: A Compilation of the Official Records of the Union and Confederate Armies,* series 4 (Washington, DC: Government Printing Office, 1900), 3:733–34. See also Gorgas to Seddon, December 31, 1864, ibid., 3:986–87.

4. Raber et al., *Forge of Innovation,* 144–47; William H. Roberts, *Civil War Ironclads: The U.S. Navy and Industrial Mobilization* (Baltimore: Johns Hopkins University Press, 2002), 38–39, 71–72; Luraghi, *Confederate Navy,* 42. See also William H. Thiesen, *Industrializing American Shipbuilding: The Transformation of Ship Design and Construction, 1820–1920* (Gainesville: University of Florida Press, 2006), chap. 5.

5. Robert C. Black III, *The Railroads of the Confederacy* (Chapel Hill: University of North Carolina Press, 1952), 148–63; Thomas Weber, *The Northern Railroads in the Civil War, 1861–1865* (New York: King's Crown Press, 1952), 15–24.

6. Peter Temin, *Iron and Steel in Nineteenth-Century America: An Economic Inquiry* (Cambridge, MA: MIT Press, 1964), table C.15; Anne Kelly Knowles, *Calvinists Incorporated: Welsh Immigrants on Ohio's Industrial Frontier* (Chicago: University of Chicago Press, 1997), 208; Weber, *Northern Railroads,* 15–16.

7. Minutes of the Board of Directors, October 24, 1863, box 3, Brady's Bend Iron Company Records, MS 10, Historical Society of Western Pennsylvania Archives; John B. Jones diary, August 11, 1863, quoted in Black, *Railroads of the Confederacy,* 200.

8. Gorgas, "Notes on the Ordnance Department"; Deposition of Thomas Wildsmith, in Depositions, *Shelby Iron Company and Others v. Horace Ware,* Chancery Court Records, Loose Papers File, drawer S, Shelby County Museum and Archives, Columbiana, Alabama (hereafter Chancery Court Records); Mallet, "Work of the Ordnance Bureau," 162–67; Black, *Railroads of the Confederacy,* 23; Lesley HGIS.

9. The ratio of Southern to Northern blast furnaces was 1:3.57, the ratio of rolling mills 1:4.77.

10. Lesley HGIS. On Bluff Furnace, see R. Bruce Council, Nicholas Honercamp, and M. Elizabeth Will, *Industry and Technology in Antebellum Tennessee: The Archaeology of Bluff Furnace* (Knoxville: University of Tennessee Press, 1992).

11. The extent to which certain technologies demanded particular kinds of labor is not addressed in the classic collection of essays on technological determinism, Merritt Roe Smith and Leo Marx, eds., *Does Technology Drive History? The Dilemma of Technological Determinism* (Cambridge, MA: MIT Press, 1994).

12. Ronald L. Lewis, *Coal, Iron, and Slaves: Industrial Slavery in Maryland and Virginia, 1715–1865* (Westport, CT: Greenwood Press, 1979), 20–35; Charles B. Dew, *Bond of Iron: Master and Slave at Buffalo Forge* (New York: Norton, 1994); Dew, "David Ross and the Oxford Iron Works: A Study of Industrial Slavery in the Early Nineteenth-Century South," *William and Mary Quarterly* 31, no. 2 (1974): 195–97.

13. Robert B. Gordon, *American Iron, 1607–1900* (Baltimore: Johns Hopkins University Press, 1996), 85; J. Peter Lesley, *Iron Manufacturer's Guide to the Furnaces, Forges, and Rolling Mills of the United States* (New York: John Wiley, 1859), 246; US Manuscript Census, 1850, Cass County, Georgia, M432, reels 63 and 88.

14. W. David Lewis, *Sloss Furnaces and the Rise of the Birmingham District: An Industrial Epic* (Tuscaloosa: University of Alabama Press, 1994), 484–89; Population and Slave Schedules of the Seventh Census, 1850, Stewart County, Tennessee, M432, reels 896 and 907.

15. Ira Berlin and Herbert G. Gutman, "Natives and Immigrants, Free Men and Slaves: Urban Workingmen in the Antebellum American South," *American Historical Review* 88 (1983): 1175–1200; Dennis C. Rousey, "Aliens in the WASP Nest: Ethnocultural Diversity in the Antebellum Urban South," *Journal of American History* 79, no. 1 (1992): 152–64.

16. Rhys Davies obituary, *Richmond Enquirer,* September 14, 1838, 3; Kathleen Bruce, *Virginia Iron Manufacture in the Slave Era* (New York: Century, 1930; reprint, New York: Augustus M. Kelley, 1968), 151, 153, 224.

17. Tredegar Company Papers; Richmond City Court Hustings Minutes 17, 1846–1848 (reel 91); Population Schedules of the Seventh Census, 1850, Henrico County, City of Richmond, Virginia, M432, reel 951; J. R. Anderson to the Board of Directors, June 17, 1842, Tredegar Company Supplementary Records 24808 (23), Minutes of the Directors and Stockholders, January 12, 1838, to January 9, 1850, Tredegar Company Papers; Patricia Schechter, "Free and Slave Labor in the Old South: The Tredegar Ironworkers' Strike of 1847," *Labor History* 35, no. 2 (1994): 165–86; Charles B. Dew, *Ironmaker to the Confederacy: Joseph R. Anderson and the Tredegar Iron Works* (New Haven, CT: Yale University Press, 1966), 23–26.

18. J. R. Anderson to Dr. W. E. Daniell, Savannah, October 28, 1850, Tredegar Letter Book, cited in Bruce, *Virginia Iron Manufacture*, 239; Joseph R. Anderson to Major Mark A. Cooper, Altoona Works, Cass County, Georgia, December 15, 1851, and Letterbook, Out, March 20, 1851–December 7, 1852, 394–95, Tredegar Iron Works Records; Joseph R. Anderson to Horace Ware, February 12, 1859, Shelby Iron Company records, W. S. Hoole Special Collections Library, University of Alabama (hereafter Shelby Iron Company records), box 583, unnumbered folder "Horace Ware." After I conducted my research in the Shelby papers, the collection was recataloged. I have included the revised box and volume numbers where possible.

19. Bruce, *Virginia Iron Manufacture*, 232–35, 246, 248–49; Gregg D. Kimball, *American City, Southern Place: A Cultural History of Antebellum Richmond* (Athens: University of Georgia Press, 2000), 166–75, quotation on 173.

20. The problem of white artisans refusing to train slaves dated back to colonial ironmaking. "When the Principio [Maryland] proprietors wanted the British artisans to teach their skills to blacks, they encountered difficulties: 'all the Arguments yet could be used cou'd not prevail with the Gloucestershire finers to admit of a clause to teach Negroes.' " Gordon, *American Iron*, 118.

21. "Tredegar Battalion," NARA RG 109, M324, roll 452; NARA RG 109, chapter 4, vol. 107, Ordnance Department, Record of Enlisted Men Detailed, January–November 1864. The proportion of immigrants at Tredegar was higher than at other Richmond iron companies. For a fuller discussion of these statistics, see Anne Kelly Knowles, "Labor, Race, and Technology in the Confederate Iron Industry," *Technology and Culture* 42 (January 2001): 18–20.

22. Eric J. Hobsbawm, "The Tramping Artisan," *Economic History Review*, 2nd ser., 3 (1951): 299–320; Humphrey R. Southall, "Towards a Geography of Unionization: The Spatial Organization and Distribution of Early British Trade Unions," *Transactions of the Institute of British Geographers* 13, no. 4 (1988): 466–83; Humphrey R. Southall, "The Tramping Artisan Revisits: Labour Mobility and Economic Distress in Early Victorian England," *Economic History Review* 44 (1991): 272–96.

23. Davis to the Confederate House of Representatives, March 13, 1862, reprinted in Ainsworth and Kirkley, *War of the Rebellion*, ser. 4, 1:993.

24. Reports of November 30, 1863, April 30, 1864, and November 5, 1864, folder of reports submitted November 1863–April 1864, Navy Records transferred from RG 45 to RG 109, formerly subject file VN (reports submitted by Confederate secretary of the navy [Stephen Mallory] to the Confederate Congress), 1861–65.

25. Steven G. Collins, "System in the South: John W. Mallet, Josiah Gorgas, and Uniform Production at the Confederate Ordnance Department," *Technology and Culture* 40 (July 1999): 532–34, 540.

26. Gorgas to Seddon, October 25, 1864, M437, roll 148, letter G48, NARA RG 109. Gorgas says much the same thing in Gorgas to Seddon, October 13, 1864, reprinted in Ainsworth and Kirkley, *War of the Rebellion*, ser. 4, 3:734.

27. Depositions of Samuel Clabaugh and Andrew T. Jones, Chancery Court Records; Correspondence, 1860–64, Shelby Iron Company records, box 885, folder 3, and Payroll records, 1865, vol. 429. These rates of increase far exceed those recorded at Springfield Armory, where wages increased by about 60 percent during the war. Felicia Johnson Deyrup, *Arms Makers of the Connecticut Valley: A Regional Study of the Economic Development of the Small Arms Industry, 1798–1870*, Studies in History 33 (Northampton, MA: Smith College, 1948), 200–201.

28. Navy Records transferred to RG 109, Recapitulations of Estimates of Navy Department, 1861–65; Dew, *Ironmaker to the Confederacy*, 239–42.

29. Dew, *Ironmaker to the Confederacy*, 238–39.

30. S. R. Mallory, secretary of the navy, to the president [Jefferson Davis], July 1, 1864, reprinted in Ainsworth and Kirkley, *War of the Rebellion*, ser. 4, 3:520–21.

31. Navy Records transferred to RG 109, folder of reports submitted November 1863–April 1864, letters from John M. Brooke to Mallory, quotation from April 30, 1864.

32. See Brooke's correspondence with the supervisor of the Selma foundry and rolling mill, Capt. Catesby ap R. Jones: RG 109, M1091 (Subject File of the Confederate States Navy, 1861–65), roll 9, file BA, Ammunition (Papers of Catesby ap R. Jones), Brooke to Jones, February 12, 1864; August 17, 1864; August 25, 1864; September 8, 1864; and Jones to Brooke, May 8, 1864, and May 14, 1864, reprinted in Ainsworth and Kirkley, *War of the Rebellion*, ser. 4, 3:523; RG 109, chap. 9, M437, roll 131, J2, December 12, 1863.

33. Luraghi, *Confederate Navy*, 28. On the Confederate Navy's initial difficulties in acquiring men and munitions, see 26–30, 32–54.

34. RG 109, chap. 9, M437, roll 80, A150, J. R. Anderson to Seddon, July 29, 1863. For other Anderson letters requesting details of skilled ironworkers and other favors to assist production at the Tredegar works and its suppliers, see roll 30, A406, A414, A460, A461, A462, A463, A471, A503, A510; roll 80, A66, A107, A149, A152, A264, A351.

35. Ainsworth and Kirkley, *War of the Rebellion*, ser. 4, 2:240; Sixth Battalion Virginia Infantry ("Tredegar Battalion"), Local Defense Troops, RG 109, M324, roll 452.

36. Ella Lonn, *Foreigners in the Confederacy* (Chapel Hill: University of North Carolina Press, 1940), 394–401; William L. Shaw, "The Confederate Conscription and Exemption Acts," *American Journal of Legal History* 6 (1962): 368–405. The first Confederate act of legislation granting exemptions, passed on April 21, 1862, included among exempt occupations "all artisans, mechanics, and employés in the establishments of the Government for the manufacture of arms, ordnance, ordnance stores, and other munitions of war . . . who may be certified by the officer in charge thereof, as necessary for such establishments; also, all artisans, mechanics, and employés in the establishments of such persons as are or may be engaged under contracts with the Government in furnishing arms, ordnance, ordnance stores, and other muntions of war: *Provided*, that the chief of the Ordnance Bureau, or some ordnance officer authorized by him for that purpose, shall approve of the number of the operatives required in such establishments." "An Act to Exempt Certain Persons from Military Duty . . . ," reprinted in Ainsworth and Kirkley, *War of the Rebellion*, ser. 4, 3:160–62.

37. RG 109, chap. 9, M437, roll 148, G27, draft legislation introduced with letter from Gorgas to Seddon, Richmond, October 25, 1864. Brooke proposed a similar plan for naval ordnance workers. S. R. Mallory to the president [Jefferson Davis], July 1, 1864, reprinted in Ainsworth and Kirkley, *War of the Rebellion*, ser. 4, 3:520–21.

38. The Enrollment Act of March 3, 1863, eliminated occupational exemptions in the North. Drafted men could avoid service by paying a $300 commutation fee or providing a substitute. James W. Geary, *We Need Men: The Union Draft in the Civil War* (DeKalb: Northern Illinois University Press, 1991), 66. Northern workers commonly but erroneously believed that men employed by federal arms manufactories were exempt from service. Deyrup, *Arms Makers*, 199–200.

39. RG 109, chap. 9, M437, roll 132, L68, J. W. Lapsley to Seddon, February 15, 1864.

40. Navy Records transferred to RG 109, John K. Mitchell, commander of the Navy Office of Order and Detail, to Stephen Mallory, Richmond, November 16, 1863; RG 109, chap. 9, M437, roll 134 (December 1863–February 1864), M(WD)77, G. A. Myers to Seddon, Richmond, January 27, 1864.

41. Depositions of Samuel Clabaugh, Giles Edwards, Henry R. Erwin, Richard Fell, Robert Hall, Moses Stroup, Robert Thomas, and J. K. Wilson, Chancery Court Records.

42. W. Craig Remington and Thomas J. Kallsen, *Historical Atlas of Alabama*, vol. 1, *Historical Locations by County* (Tuscaloosa: Department of Geography, University of Alabama, 1997), 285,

289; Joyce Jackson, *History of the Shelby Iron Company, 1862–1868* (Shelby, AL: Historic Shelby Association, 1990), 3–5; Ethel Armes, *The Story of Coal and Iron in Alabama* (Birmingham, AL: Chamber of Commerce, 1910; reprint, New York: Arno Press, 1972), 162–63.

43. Benjamin Smith Lyman to J. Peter Lesley, September 28, 1857, Benjamin Smith Lyman Collection, ser. 1, Correspondence, 1856–92 (Lesleys), University of Massachusetts at Amherst, W. E. B. DuBois Library, Special Collections and Archives (hereafter BSL to JPL).

44. John Shropshire and John Musgrove to Horace Ware, November 17, 1861, Shelby Iron Company records, box 885, folder 28.

45. Plan for new construction in the hand of Giles Edwards, no date, and balance sheet for Shelby Iron Company, December 31, 1863, Shelby Iron Company records, box 969, unnumbered folder "Statements (1862–1864)"; Depositions of Clabaugh, Wildsmith, Edwards, Stroup, and Thomas, Chancery Court Records; draft of contract between Shelby Iron Company and the Confederate States of America, ca. July 1862, Shelby Iron Company records, box 885, folder 2; Lesley, *Guide*, 79.

46. Depositions of Hall, Clabaugh, Edwards, and Wildsmith, Chancery Court Records.

47. C. J. Hazard to C. J. McRae, July 3, 1862, Shelby Iron Company records, box 885, folder 25; medical accounts, 1863, box 885, folder 16; Minutes of Board of Directors, box 420; list of boards of directors elected, 1862–70, box 969, unnumbered folder "Legal—Resolutions 1862, 1867, 1885"; Armes, *Story of Coal and Iron*, 64, 163; Deposition of Stroup, Chancery Court Records.

48. BSL to JPL, September 28, 1857; Fred Bateman, James Foust, and Thomas Weiss, "The Participation of Planters in Manufacturing in the Antebellum South," *Agricultural History* 48, no. 2 (1974): 289. On the general problem of underinvestment in Southern manufacturing, see Fred Bateman and Thomas Weiss, *A Deplorable Scarcity: The Failure of Industrialization in the Slave Economy* (Chapel Hill: University of North Carolina Press, 1981), and John Ashworth, *Slavery, Capitalism, and Politics in the Antebellum Republic*, vol. 1, *Commerce and Compromise, 1820–1850* (Cambridge: Cambridge University Press, 1995).

49. Depositions of Ware and Clabaugh, Chancery Court Records; Accounts of Horace Ware, April 1–July 17, 1862, Shelby Iron Company records, box 420; Knowles, *Calvinists Incorporated*, 174–77.

50. Minutes of Board of Directors, February 17, 1863, Shelby Iron Company records, box 420; J. W. Lapsley to Messrs. Ware and Jones, March 12, 1862; A. T. Jones to Col. J. Gorgas, September 27, 1862, box 885, folder 22.

51. Public Record Office, 1841 Manuscript Census, Merthyr Tydfil, Glamorganshire, microfilm ref. no. HO 107/1415, enumeration district 23; Armes, *Story of Iron and Coal*, 172–74; *Plan of Merthyr Tydfil, from Actual Survey*, attributed to John Wood, no scale, 1836, Glamorgan County Record Office.

52. US Manuscript Census, 1850, Lehigh County, Pennsylvania, Hanover Township; Armes, *Story of Iron and Coal*, 174–75. Armes gives the location of the Thomas ironworks as Tamauqua, but in fact there were no major ironworks there. The Thomas works, built in 1855 in Hokendauqua, are probably where Edwards worked and where he may have gained his first experience in designing a rolling mill. Lesley, *Guide*, 8–9.

53. Giles Edwards to John Fritz, Cambria Iron Works, April 20, 21, 25, and 28, 1855, February 27, 1858, John Fritz Papers, National Canal Museum Archives (hereafter Fritz Papers); John Fritz, *Autobiography of John Fritz* (New York: John Wiley, 1912), 108–15, 149–72; Council, Honerkamp, and Will, *Industry and Technology*, 62–65, 67–74; Edwards to Fritz, July 18, 1859; L. R. Speer to John Fritz, July 18, 1860; Speer to Fritz, July 31, September 6 and 15, 1860, Fritz Papers.

54. Giles Edwards expenses for trip to Chattanooga, April 30–May 8, 1862, Shelby Iron Company records, box 885, folder 28; Accounts of Giles Edwards, May–December 1863, June–July 1862, and February–May 1863, box 969, unnumbered folder "Statements (1862–1864)"; Employee and Negro Time Records, March 1862–December 1868, list for March 1863, vol. 372.

55. Richmond City Hustings Court Minutes no. 17, 1846–1848 (reel 91), Tuesday, April 20, 1847; US Manuscript Census, 1850, City of Richmond, Henrico County, Virginia, M432, reel 951; D. J. Davies to Giles Edwards, Montgomery, October 12, 1862, Shelby Iron Company records; Population Schedules of the Seventh Census, 1850, Cass County, Georgia, M432, reels 63 and 88. Davies remained at Shelby through the Civil War; Employees Time Record, 1862–1864, Shelby Iron Company records, vol. 372; Payroll records, 1865, vol. 429.

56. According to a family friend, Edwards and his wife spoke Welsh together, and Edwards subscribed to a Welsh-language newspaper all his life. Armes, *Story of Coal and Iron*, 230.

57. Memorandum from Office of the Shelby [Iron] Co., September 13, 1862, Shelby Iron Company records, box 969, unnumbered folder "Legal—Resolutions 1862, 1867, 1885."

58. Deposition of Hall, Chancery Court Records.

59. Ibid.; Employee and Negro Time Records, March 1862–December 1868, Shelby Iron Company records, vol. 372; Payroll records, 1865, vol. 429; "Negroes belonging to R. T. Ford," box 969, unnumbered folder "Employee records . . . 1863–64(?)"

60. "A List of Clothing given off to Negros," July 28, 1862, and lists of Negroes hired, 1862, Shelby Iron Company records, box 885, folder 28.

61. At the end of the war, the company owned at least twenty-nine adult male slaves; "Time of Hands Belonging to Shelby Iron Co. for Work Done on the Ala. and Tenn. RR Road," April 26–May 12, 1865, Shelby Iron Company records.

62. Sam[uel] Kirkman to A. T. Jones, September 27, 1862, emphasis in original, and J. J. Hutchenson to A. T. Jones, December 16, 1862, Shelby Iron Company records.

63. Lists of Negro "pay role" and extra time, 1864, Shelby Iron Company records, box 885, folder 14, and box 969, unnumbered folder "Employee records . . . 1863–64(?)." The ore bank at Shelby was a surface deposit that was probably worked by scraping or benching, a kind of shallow strip mining in which men used a metal scoop to dig ore directly from the hillside, leaving stepped banks. Knowles, *Calvinists Incorporated*, 172.

64. Various correspondence, unlabeled folder of correspondence, 1865, Shelby Iron Company records.

65. Letter in hand of A. T. Jones to Major Thomas Peters, December 16, 1864; Thomas Peters to A. T. Jones, January 16, 1865; and R. B. G. to A. T. Jones, January 26, 1865, Shelby Iron Company records.

66. A. T. Jones to Gorgas, July 16, 1862, Shelby Iron Company records, box 885, folder 12.

67. J. M. Tillman to C. J. Hazard, July 2, 1863, Shelby Iron Company records, box 969, unnumbered folder "Correspondence, Internal re: supplies 1862"; Report of Work done by the Ore Bank hands from May 1 to December 15, 1863, box 969, unnumbered folder "Employee records . . . 1863–64(?)."

68. J. M. Tillman to A. T. Jones, January 26, 1862, Shelby Iron Company records, box 969, unnumbered folder "Correspondence, Internal re: labor 1862"; "List of Clothing given off to Negros," July 28, 1862, and lists of Negroes hired, 1862, box 885, folder 28; Steven E. Woodworth and Kenneth J. Winkle, *Oxford Atlas of the Civil War* (Oxford: Oxford University Press, 2004), 102–3.

69. Tillman to Jones, July 15, 1862, Shelby Iron Company records, box 969, unnumbered folder "Correspondence, Internal re: labor 1862."

70. Tillman to J. W. Lapsley, December 19, 1862, Shelby Iron Company records.

71. Tillman to Jones, January 13, 1863, Shelby Iron Company records, box 969, unnumbered folder "Correspondence, incoming to A. T. Jones . . . Tillman to Wallace"; Statement of Tillman's expenses, January 12–27, 1863, box 969, unnumbered folder "Statements 1862–64." A. T. Jones sought assistance from a Mr. T. T. Stratton, whom he asked to hire "as many as 100 No. 1 negro men" in Mississippi. The outcome of that request is unknown. Jones to Stratton, January 20, 1863, box 969, unnumbered folder "Incoming Correspondence—alphabetical . . . Adams to Kirkman."

72. Tillman to Jones, January 13, 1863, Shelby Iron Company records, box 969, unnumbered folder "Correspondence, incoming to A. T. Jones . . . Tillman to Wallace."

73. Tillman to Jones, April 11, 1863, and Tillman to Jones, December 5, 1863, Shelby Iron Company records, box 969, unnumbered folder "Correspondence, incoming to A. T. Jones . . . Tillman to Wallace."

74. Tillman to Jones, October 3, 13, 1863, Shelby Iron Company records, box 969, unnumbered folder "Correspondence, incoming to A. T. Jones . . . Tillman to Wallace."

75. Woodward and Winkle, *Oxford Atlas of the Civil War*, 220–21, 230–31.

76. Statement of Tillman's expenses, January 12–27, 1863, Shelby Iron Company records, box 969, unnumbered folder "Statements 1862–64"; Black, *Railroads of the Confederacy*, 6; Tillman to Jones, December 5, 1863, Shelby Iron Company records.

77. William Wallace to Jones, April 4, 1864, and Tillman to Jones, May 31, 1864, Shelby Iron Company records, box 969, unnumbered folder "Correspondence, incoming to A. T. Jones . . . Tillman to Wallace."

78. Woodward and Winkle, *Oxford Atlas of the Civil War*, 230–31, 252–57, 272–73, 336; Tillman to Jones, June 6, 1864, Shelby Iron Company records, box 969, unnumbered folder "Correspondence, incoming to A. T. Jones . . . Tillman to Wallace."

79. P. L. Griffiths, Charges against the Shelby County Iron Co., June 5, 1863, Shelby Iron Company records, box 885, folder 109. See also Griffiths to Jones, October 12, 1863, box 885, folder 9.

80. Unsigned draft of letter in hand of A. T. Jones to Col. John T. Morgan, June 20, 1863, Shelby Iron Company records, box 885, folder 9.

81. RG 109, chap. 9, M437, roll 132, L68, J. W. Lapsley to Seddon, February 15, 1864.

82. General Order 22, Headquarters of the Vol. and Cons. Bureau, Department [of] Tenn., Ala. and Miss., November 5, 1863, Shelby Iron Company records.

83. Unsigned draft, Jones to Morgan, June 20, 1863, Shelby Iron Company records, box 885, folder 9; list of workers, no date, ca. 1863, box 969, unnumbered folder "Employee records, . . . 1863–64(?)"; James T. Pettit to A. T. Jones, November 3, 1863, box 885, folder 16.

84. Giles Edwards to A. T. Jones, November 3, 1862, Shelby Iron Company records, box 583, unnumbered folder "A. T. Jones."

85. R. H. Criswell and others to Giles Edwards, May [no year], Shelby Iron Company records, box 969, unnumbered folder "Employee Records, . . . 1863–1864(?)."

86. See correspondence in Shelby Iron Company records, box 885, folder 9, including J. B. Evans to Horace Ware, September 21, 1861; A. E. Mott [?] to A. T. Jones, July 28, 1862; G. W. Summers to Shelby Iron Company, October 16, 1862; Richard O'Neal to John McClanahan, January 25, 1863; Isaac Ketler to A. T. Jones, January 27, 1863 [?]; J. C. O'Neal to A. T. Jones, April 13, 1863; and box 969, unnumbered folder "Correspondence, incoming to A. T. Jones . . . Tillman to Wallace," John F. Welch to A. T. Jones, December 25, 1863; John Dunlop to Co[l]. Peters, July 15, 1864. See also T. T. Darker to [Robert] Hall, October 18, 1864, and Horace Ware to A. T. Jones, January 20, 1865, Shelby Iron Company records.

87. H. D. Calhoun to Robt. Hall, December 29, 1863, Shelby Iron Company records, box 969, unnumbered folder "Incoming Correspondence—alphabetical . . . Adams to Kirkman"; Geo. H. Harris to A. T. Jones, March 23, 1863, box 583, unnumbered folder "A. T. Jones"; Sam Kirkman to Jno. R. Keenan [Kenan], December 28, 1862. The company dismissed one overseer for being "too severe"; see A. T. Jones to Benja. Grist, August 27, 1862, Shelby Iron Company records, box 885, folder 12.

88. R. H. Criswell and others to Giles Edwards, May [no year], Shelby Iron Company records, box 969, unnumbered folder "Employee Records, . . . 1863–1864(?)."

89. Draft letter in hand of Andrew T. Jones to Col. I. M. St. John, September 8, 1863, Shelby Iron Company records, unnumbered folder "Legal—Contracts 1862–1868."

90. A. T. Jones to Genl. C. Robinson, February 9, 1863, and Shelby Board of Directors to Maj. W. R. Hunt, July 25, 1864, Shelby Iron Company records. The company appears to have resorted to burning wood in the puddling furnaces at times; list of "extra time at ore bank & cutting pudling [*sic*] wood," May 1864, box 885, folder 14.

91. Draft notes for modifying Shelby's contract with the Confederate States of America, in Giles Edwards's hand (ca. 1862–1863), Shelby Iron Company records, box 969, unnumbered folder "Legal—Contracts 1862–68"; A. T. Jones to W. R. Hunt, July 20, 1863; Shelby Board of Directors to Maj. W. R. Hunt, July 25, 1864; William Wallace, estimate for construction of Shelby Iron Co. R[ail] R[oad], December 25, 1863. On the use of iron as a medium of exchange, see Benj. Avarett to Robert Hall, July 9, 1864, box 969, unnumbered folder "Incoming Correspondence, alphabetical . . . Adams to Kirkland" and Isaac Stone to A. T. Jones, June 12, 1864, unnumbered folder "Incoming Correspondence, alphabetical . . . Lapsley to Yates"; Giles Edwards to M. Russell and Co., April 28, 1864, box 969, unnumbered folder "Company correspondence outgoing from A. T. Jones." Company records and depositions taken by the Chancery Court are not conclusive on whether Shelby's officers decided to build a mineral-fuel furnace. Horace Ware, Giles Edwards, and E. G. Walker testified that the company's bituminous coal was adequate for smelting iron.

92. A. T. Jones to Maj. Thomas Peters, March 3, 20, and 28, 1865, and Peters to Jones, March 29, 1865, Shelby Iron Company records.

93. Unsigned, undated penciled note in front of General Ledger no. 1 [former numbering system], Shelby Iron Company records; Woodworth and Winkle, *Oxford Atlas of the Civil War*, 302.

94. J. D. Johns to [Colin J.] McRae, October 29, 1862, and C. J. McRae to A. T. Jones, October 31, 1862, Shelby Iron Company records, box 583, unnumbered folder "A. T. Jones"; James T. Pettit to Major Thomas Peters, February 23, 1865.

95. This may be difficult, if not impossible, because most records of the Confederate Nitre and Mining Bureau were destroyed in the evacuation of Richmond in 1865.

96. William S. Pelletreau, *History of Putnam County, New York* (Philadelphia: W. W. Preston, 1886; commemorative edition Brewster, NY: Landmarks Preservation Committee, 1975), 560; extract of 1850 Putnam County, New York, census, entries for Philipstown, prepared by Sallie Sypher, Putnam County Historical Society, n.d.; West Point Foundry Records, box 2.1, genealogical material on William Young, Putnam County Historical Society. Distance measured on Google Earth.

97. Charles R. Isleib and Jack Chard, *The West Point Foundry and the Parrott Gun: A Short History* (Fleischmanns, NY: Purple Mountain Press, 2000), 10–16; Charles B. Stuart, *The Naval Drydocks of the United States* (New York: Charles B. Norton, 1852), plate 11; Ralph Brill Associates, "West Point Foundry Historic Landscape Report," typescript, 1979, table 1, 6, courtesy of the Putnam County Historical Society; William J. Blake, *The History of Putnam County, N.Y.* (New York: Baker and Scribner, 1849), 239–44.

98. Isleib and Chard, *West Point Foundry*, 17–26; Lesley, *Guide*, 5; Pelletreau, *History of Putnam County*, 561; Ripley, *Artillery and Ammunition*; Robert P. Parrott to Brig. Gen. A. B. Dyer, October 5, 1864, West Point Foundry Records, box 2.5.

99. Arthur M. Alger, *A Genealogical History of . . . Thomas Alger of Tauton and Bridgewater, in Massachusetts, 1665–1875* (Boston: David Clapp, 1876), 24–25; John Leander Bishop, *A History of American Manufactures from 1608 to 1860 . . .* (Philadelphia: Edward Young, 1864), 278–81, quotation on 281; Thomas O'Connor, *Civil War Boston: Home Front and Battlefield* (Boston: Northeastern University Press, 1997), 162.

100. Lesley HGIS; Julian C. Skaggs, "Lukens, 1850–1870: A Case Study in the Mid-Nineteenth Century American Iron Industry" (PhD diss., University of Delaware, 1975), 24–25; John K. Brown, *The Baldwin Locomotive Works, 1831–1915* (Baltimore: Johns Hopkins University Press, 1995), 5–9; John N. Ingham, *Making Iron and Steel: Independent Mills in Pittsburgh, 1820–1920* (Columbus: Ohio State University Press, 1991), 21–46; Louis C. Hunter, *Steamboats on the Western Rivers: An Economic and Technological History* (1949; reprint, New York: Dover, 1993), 5, 108–9.

101. Craig Cowing, "The Katahdin Iron Works: The Early Years, 1846–1863," in *Entrepreneurship in Maine: Essays in Business Enterprise*, ed. Stuart Bruchey (New York: Garland, 1995), 10, 13.

102. James D. Norris, *Frontier Iron: The Story of the Maramec Iron Works, 1826–1876* (Madison: State Historical Society of Wisconsin, 1964), 126–28, 151–52.

103. Samuel Harries Daddow and Benjamin Bannan, *Coal, Iron, and Oil, or The Practical American Miner* (Pottsville, PA: Benjamin Bannan, 1866), 684–95; Fritz, *Autobiography of John Fritz*, 144–46; Weber, *Northern Railroads*, 216, 199–212; James F. Doster, "The Chattanooga Rolling Mill: An Industrial By-Product of the Civil War," *East Tennessee Historical Society's Publications* 36 (1964): 47–48. Daddow and Bannon date the Chattanooga mill to 1863, Fritz to 1864.

104. Roberts, *Civil War Ironclads*, 63.

105. Ibid., 19, 29; Frank R. Donovan, *Ironclads of the Civil War* (New York: American Heritage, 1964), 38–39; David R. Meyer, *Networked Machinists: High-Technology Industries in Antebellum America* (Baltimore: Johns Hopkins University Press, 2006).

106. Roberts, *Civil War Ironclads*, 48–56, 63; see figure 58, above.

107. Roberts, *Civil War Ironclads*, 58–63, 75, 128–33, quotation on 145.

108. Raber et al., *Forge of Innovation*, 145; Robert J. Schneller Jr., *A Quest for Glory: A Biography of Rear Admiral John A. Dahlgren* (Annapolis, MD: Naval Institute Press, 1996), 128, 141, 142–43; Skaggs, "Lukens, 1850–1870," 124.

109. Estimate based on Temin, *Iron and Steel*, table C.2, 266, and Lesley HGIS. To estimate capacity in Union-controlled territory, I took the average production at all cold- and warm-blast charcoal furnaces in Northern states and in border states (Missouri, Kentucky, Tennessee) and multiplied by the total number of furnaces, then subtracted that average for furnaces that Lesley reported as having been abandoned by 1850. Almost no charcoal furnaces returned to production after a decade or more of quiescence.

110. Gordon, *American Iron*, 73; "A Great Cannon Foundry," *Scientific American* 11 (September 10, 1864): 165; R. P. Parrott to Brig. Genl. A. B. Dyer, October 5, 1864, box 2.5, West Point Foundry Records; "Cannon Manufacture at Pittsburgh," *Scientific American* 6 (March 15, 1862): 163; Ripley, *Artillery and Ammunition*, 109–16, 122, 370–71. The other sizable firms that made naval cannons were Hinckley, Williams, and Company in Boston; the Portland Company in Portland, Maine; Builders' Iron Foundry in Providence, Rhode Island; and the Scott Foundry in Reading, Pennsylvania. Schneller, *Quest for Glory*, 223.

111. Raber et al., *Forge of Innovation*, 144–47; Davis, *Arming the Union*, 50–56.

112. Skaggs, "Lukens, 1850–1870," 130–34; J. W. King, "Iron Rolling Mills of the West," Report 19, Annual Report of the Secretary of the Navy, Exec. Doc., 38th Cong., 1st sess., vol. 4, serial set 1183 (Washington, DC: Government Printing Office, 1864), 1104–9.

113. [Chief engineer] Alban C. Stimers to [assistant secretary of the navy] Gustavus V. Fox, February 1, 1863, Gustavus Vasa Fox Papers, New-York Historical Society, box 7, Letters Received 1863 H–Z, courtesy of William H. Roberts, personal correspondence, October 18, 2009; Roberts, *Civil War Ironclads*, 112, 211–12; William H. Roberts, personal correspondence, October 23, 2009.

114. Skaggs, "Lukens, 1850–1870," 130–34; King, "Iron Rolling Mills," 1106–8.

115. Raber et al., *Forge of Innovation*, 147; Richard Colton and Peter G. Smithurst, personal communications, October 23, 2009; Allan Nevins, *Abram S. Hewitt, with Some Account of Peter Cooper* (New York: Harper, 1935; reprint New York: Octagon Books, 1967), 193–94, 206–12. Raber et al. say Hewitt failed to win a major contract to produce barrel plate for the Springfield Armory; Nevins says he captured the sole contract. I have been unable to resolve this discrepancy.

116. Gregory Galer, Robert Gordon, and Frances Kemmish, "Connecticut's Ames Iron Works: Family, Community, Nature, and Innovation in an Enterprise of the Early American Republic," *Transactions [of the] Connecticut Academy of Arts and Sciences* 54 (December 1998): 143–54. "Ironically," writes naval historian Robert J. Schneller Jr., "ironmasters and ordnance experts soon learned that coke-smelted hot-blast iron actually produced stronger cannon than charcoal-smelted cold-blast iron. But because his guns proved so reliable, Dahlgren never tested his ideas about smelting fuel and blast temperature." At the end of the war, a "reliable heavy all-steel breechloading rifled gun was still decades away." Schneller, *Quest for Glory*, 145, 346.

117. Schneller, *Quest for Glory*, 68–69.

118. Ibid., 69–94, 205.

119. Letters Received from Foundries, 1855–1861, RG 74, Records of the [Navy] Bureau of Ordnance (hereafter Letters from Foundries), box 2, vol. 4 (January 1, 1856–March 26, 1856); John Dahlgren to Commodore Morris, July 30, 1855, Letters from Foundries, box 1, vol. 1.

120. Knap and Wade to Comm. Morris, January 14, 1856, box 2, vol. 4, Letters from Foundries.

121. Andrew A. Harwood to [George] Magruder, December 7, 1860, box 5, vol. 12, Letters from Foundries.

122. William Rogers Taylor to Capt. D. N. Ingraham, Chief of Ordnance, February 24, 1859, box 5, vol. 12, Letters from Foundries.

123. Senate, Joint Committee on the Conduct of the War, *Heavy Ordnance*, 38th Cong., 2nd sess., 1865, Rep. Com. 121, serial 1211, doc. 445, 2–3, quoted in Galer, Gorden, and Kemmish, *Connecticut's Ames Iron Works*, 144.

124. Weber, *Northern Railroads*, 15, 63, and passim.

125. Ibid., 63–65; Kenneth Warren, *Bethlehem Steel: Builder and Arsenal of America* (Pittsburgh: University of Pittsburgh Press, 2008), 21.

126. Nevins, *Abram S. Hewitt*, 195; Warren, *Bethlehem Steel*, 21–22; Fritz, *Autobiography*, 153.

127. Gordon, *American Iron*, 150.

128. John William Bennett, "Iron Workers in Woods Run and Johnstown: The Union Era" (PhD diss., University of Pittsburgh, 1977), 21–22; Ingham, *Making Iron and Steel*, 35. The mills were the Cambria Iron Works in Johnstown and the Pennsylvania Iron Works in Danville, Pennsylvania, respectively. Lesley HGIS; Daddow and Bannan, *Coal, Iron, and Oil*, 694.

129. Chris Evans, "Work and Workloads during Industrialization: The Experience of Forgemen in the British Iron Industry, 1750–1850," *International Review of Social History* 44 (1999): 209–11, quotation on 209.

130. Alan Birch, *The Economic History of the British Iron and Steel Industry, 1784–1879* (London: Frank Cass, 1967; reprint, New York: Augustus M. Kelley, 1968), 315, 319–25.

131. Birch, *British Iron and Steel*, 324–25; Fritz, *Autobiography*, 150–53; Lesley HGIS.

132. Grace Palladino, *Another Civil War: Labor, Capital, and the State in the Anthracite Regions of Pennsylvania, 1840–68* (Urbana: University of Illinois Press, 1990), 121; Francis G. Couvares, *Remaking of Pittsburgh: Class and Culture in an Industrializing City, 1877–1919* (Albany: State University of New York Press, 1984), 24.

133. David Montgomery, *Citizen Worker: The Experience of Workers in the United States with Democracy and the Free Market during the Nineteenth Century* (New York: Cambridge University Press, 1993), 92–93; Frederick L. Hitchcock, *War from the Inside: The Story of the 132nd Regiment Pennsylvania Volunteer Infantry in the War for the Suppression of the Rebellion, 1862–1863* (Philadelphia: J. B. Lippincott, 1904), accessed online via Project Gutenberg (eBook no. 29313) at www.gutenberg.org on September 26, 2009; Jerry Hunter, *Llwch Cenhedloedd: Y Cymry a Rhyfel Cartref America* (The Dust of Nations: The Welsh and America's Civil War) (Llanrwst, Wales: S4C, 2003), 38; Anne E. Mosher, *Capital's Utopia: Vandergrift, Pennsylvania, 1855–1916* (Baltimore: Johns Hopkins University Press, 2004), 23.

134. David Montgomery, *Beyond Equality: Labor and the Radical Republicans, 1862–1872* (New York: Alfred A. Knopf, 1967), 93–94, 97–98, 224–25, quotation on 103; Daniel J. Walkowitz, *Worker City, Company Town: Iron and Cotton-Worker Protest in Troy and Cohoes, New York, 1855–84* (Urbana: University of Illinois Press, 1978), 86–89; Bennett, "Iron Workers," 42.

135. Couvares, *Remaking of Pittsburgh*, 24; Walkowitz, *Worker City, Company Town*, 86–91, 104.

136. James C. Sylvis, *The Life, Speeches, Labors and Essays of William H. Sylvis* (1872; reprint, New York: Augustus M. Kelley, 1968), 137–38, quotation on 137.

137. Palladino, *Another Civil War*, 121–36, quotation on 123.

138. Evans, "Work and Work Loads."

139. Bennett, "Iron Workers," 180–81; Geary, *We Need Men*, 195n30, 97.

140. The works was the Caledonia Iron Works in Franklin County, Pennsylvania, owned by Thaddeus Stevens. Edwin D. Coddington, *The Gettysburg Campaign: A Study in Command* (New York: Charles Scribner's Sons, 1968), 166.

141. Walter Licht, *Industrializing America: The Nineteenth Century* (Baltimore: Johns Hopkins University Press, 1995), 97.

Conclusion

1. Samuel Harries Daddow and Benjamin Bannan, *Coal, Iron, and Oil, or The Practical American Miner* (Pottsville, PA: Benjamin Bannan, 1866), 684–93; Craig L. Bartholomew and Lance E. Metz, *The Anthracite Iron Industry of the Lehigh Valley*, ed. Ann Bartholomew (Easton, PA: Center for Canal History and Technology, 1988), 178–80.

2. Kenneth Warren, *Wealth, Waste, and Alienation: Growth and Decline in the Connellsville Coke Industry* (Pittsburgh: University of Pittsburgh Press, 2001), 22, 25–26.

3. Geoffrey Tweedale, *Sheffield Steel and America: A Century of Commercial and Technological Interdependence, 1830–1930* (Cambridge: Cambridge University Press, 1987), 16–17; Bartholomew and Metz, *Anthracite Iron Industry*, 182–83; John Fritz, *The Autobiography of John Fritz* (New York: John Wiley, 1912), 151; J. Peter Lesley, *The Iron Manufacturer's Guide to the Furnaces, Forges, and Rolling Mills of the United States* (New York: John Wiley, 1859), 225.

4. Anne Kelly Knowles, *Calvinists Incorporated: Welsh Immigrants on Ohio's Industrial Frontier* (Chicago: University of Chicago Press, 1997), 208–24; Vernon David Keeler, "An Economic History of the Jackson County Iron Industry," *Ohio Archaeological and Historical Quarterly* 42 (April 1933): 179–89.

5. Nancy L. Smith, "Iron Production between the River and the Front: A Revision of the Plantation Model in an Era of Proto-industrialization" (MSc thesis, Pennsylvania State University, 1992), 32–38.

6. Robert B. Gordon, *A Landscape Transformed: The Ironmaking District of Salisbury, Connecticut* (New York: Oxford University Press, 2001), 68–94.

7. John N. Ingham, *Making Iron and Steel: Independent Mills in Pittsburgh, 1820–1920* (Columbus: Ohio State University Press, 1991).

8. Anne Kelly Knowles and Richard G. Healey, "Geography, Timing, and Technology: A GIS-Based Analysis of Pennsylvania's Iron Industry, 1825–1875," *Journal of Economic History* 66 (September 2006): 626–30.

9. W. David Lewis, *Sloss Furnaces and the Rise of the Birmingham District: An Industrial Epic* (Tuscaloosa: University of Alabama Press, 1994), 40–41.

10. JL to Will [Dr. W. W. Lesley], July 1, 1857, J. Peter Lesley Papers, American Philosophical Society, Philadelphia.

11. Allan Nevins, *Abram S. Hewitt, with Some Account of Peter Cooper* (New York: Harper, 1935; reprint, New York: Octagon Books, 1967), 252–53; R. Bruce Council, Nicholas Honerkamp, and M. Elizabeth Will, *Industry and Technology in Antebellum Tennessee: The Archaeology of Bluff Furnace* (Knoxville: University of Tennessee Press, 1992), 170. Bluff Furnace, which was built to burn charcoal, was refitted to burn coke in 1860. For various reasons, it did not remain in production long. Council, Honerkamp, and Will, *Bluff Furnace*, 77–81.

12. Council, Honerkamp, and Will, *Bluff Furnace*, 170.

13. Henry M. McKiven Jr., *Iron and Steel: Class, Race, and Community in Birmingham, Alabama, 1875–1920* (Chapel Hill: University of North Carolina Press, 1995), 10–21; Giles Edwards to John Fritz, February 25, 1866, Fritz Papers; Ethel Armes, *The Story of Coal and Iron in Alabama* (Birmingham, AL: Chamber of Commerce, 1910; reprint, New York: Arno Press, 1972), 229, 353–54; Marjorie Longenecker White, *The Birmingham District: An Industrial History and Guide* (Birmingham, AL: Birmingham Historical Society, 1981), 127–28; Lewis, *Sloss Furnaces*, 142–43.

14. Armes, *Story of Coal and Iron*, 186, 310–15, quotation on 312; Lewis, *Sloss Furnace*, 41–42. See also Grace Hooten Gates, *The Model City of the New South: Anniston, Alabama, 1872–1900* (Huntsville, AL: Strode, 1978).

15. James Parton, "Pittsburgh," *Atlantic Monthly* 21 (January 1868): 21.

16. Nevins, *Abram S. Hewitt*, 256, 243.

17. John Gurda, *Bay View, Wis.*, centennial ed. (Milwaukee: University of Wisconsin Board of Regents, 1979); James Moore Swank, *Classified List of Rail Mills and Blast Furnaces in the United States* (Philadelphia: American Iron and Steel Association, 1873), 4–5. Indiana and Illinois went from being minor producers of charcoal iron to major producers of iron and steel after 1867 thanks to their deposits of block coal, which could be used raw in blast furnaces, and their proximity to lake transport and inland canals. C. D. Wilber, *Coal and Iron: Two Lectures Delivered before the Board of Trade of LaFayette, Indiana* (LaFayette, IN: Rosser, Spring, 1872), 33–34; Michael P. Conzen, *The Illinois and Michigan Canal National Heritage Corridor: A Guide to Its History and Sources*, ed. Kay J. Carr (DeKalb: Northern Illinois University Press, 1988).

18. Jos. S. Wilson, *Sketch of the Public Surveys in the State of Minnesota* (map), scale 18 miles to the inch (Washington, DC: Department of the Interior, General Land Office, 1868), Minnesota State Historical Society; E. C. Harder, "Iron Ores, Pig Iron, and Steel," in *Mineral Resources of the United States*, part 1, *Metallic Products, Calendar Year 1908*, Department of the Interior, US Geological Survey (Washington, DC: Government Printing Office, 1909), 66, 74, 94; Nevins, *Abram S. Hewitt*, 242. On the role of Marquette ores in the Pittsburgh iron industry, see Knowles and Healey, "Geography, Timing, and Technology."

19. Nevins, *Abram S. Hewitt*, 264–66, 291ff.

20. J. Peter Lesley journal, 1867, and "Notes at the Expos[ition] of 1867 as commissioner at Paris," J. P. Lesley Journals, Ames Family Historical Collection, AmesINK, Boulder, CO.

21. Abram S. Hewitt, "The Production of Iron and Steel in Its Economic and Social Relations," in *Selected Writings of Abram S. Hewitt*, ed. Allan Nevins (New York: Columbia University Press, 1937), 19–85, quotation on 29; Nevins, *Abram S. Hewitt*, 239.

22. Hewitt, "Production of Iron and Steel," 23–33, 57–58; Lesley, "Notes at the Expos[ition]," 18b.

23. Hewitt, "Production of Iron and Steel," 25–34, quotation on 33; Nevins, *Abram S. Hewitt*, 244–49, quotation on 241.

24. David J. Jeremy, *Transatlantic Industrial Revolution: The Diffusion of Textile Technologies between Britain and America, 1790–1830s* (Cambridge, MA: MIT Press, 1981).

25. Even the British and Swedish iron industries were seeded by foreign skilled labor, in both cases Walloon forgemen who emigrated from Belgium to Britain about 1600 and to Sweden in 1620 to 1655. Chris Evans, "The Industrial Revolution in Iron in the British Isles," in *Industrial Revolution in Iron: The Impact of British Coal Technology in Nineteenth-Century Europe*, ed. Chris Evans and Göran Rydén, 15–28 (Altershot, UK: Ashgate, 2005), 16; Chris Evans and Göran Rydén, *Baltic Iron in the Atlantic World in the Eighteenth Century* (Leiden: Brill, 2007), 75.

26. James E. Vance Jr., *The North American Railroad: Its Origin, Evolution, and Geography* (Baltimore: Johns Hopkins University Press, 1995); Jeremy, *Transatlantic Industrial Revolution.*

27. Louis C. Hunter, "Influence of the Market upon Technique in the Iron Industry in Western Pennsylvania up to 1860," *Journal of Economic and Business History* 1 (1928–29): 241–81.

28. *Memorial from Pennsylvania on the Manufacture of Iron* (Philadelphia: Convention of Iron Masters, 1850); Wilber, *Coal and Iron*, 34.

29. David Brody, "The Psychology and Method of Steelmaking," in his *Steelworkers in America: The Nonunion Era* (Cambridge, MA: Harvard University Press, 1960; reprinted Urbana: University of Illinois Press, 1998), 1–26.

30. John Bezís-Selfa, *Forging America: Ironworkers, Adventurers, and the Industrious Revolution* (Ithaca, NY: Cornell University Press, 2004); Edward Neal Hartley, *Ironworks on the Saugus: The Lynn and Braintree Ventures of the Company of Undertakers of the Ironworks in New England* (Norman: University of Oklahoma Press, 1957).

31. Kenneth Warren, *The American Steel Industry, 1850–1970: A Geographical Interpretation* (Pittsburgh: University of Pittsburgh Press, 1973), and Warren, *Bethlehem Steel: Builder and Arse-*

nal of America (Pittsburgh: University of Pittsburgh Press, 2008).

32. Lesley caught sight of the egg-shaped device in the works yard at Cooper Furnace in 1856 and told Susan he was eager to see one in operation, but he may never have done so, since his diary and letters make no mention of it. JPL to SIL, November 12, 1856; JPL to JL, November 17, 1856, Ames Family Historical Collection.

33. A. L. Holley's description, quoted in Fritz, *Autobiography*, 160.

34. Rebecca Harding Davis, *Life in the Iron Mills*, ed. Cecelia Tichi (Boston: Bedford Books, 1998), 45.

35. John William Bennett, "Iron Workers in Woods Run and Johnstown: The Union Era" (PhD diss., University of Pittsburgh, 1977), 31; Bartholomew and Metz, *Anthracite Iron Industry*, 184.

36. Bennett, "Iron Workers," 25, 31–32, quotation on 25; John A. Fitch, *The Steel Workers* (Pittsburgh: University of Pittsburgh Press, 1989), 33.

37. Bennett, "Iron Workers," 15, 48–49, 149–215, quotation on 49.

38. Michael L. Carlebach, *Working Stiffs: Occupational Portraits in the Age of Tintypes* (Washington, DC: Smithsonian Institution Press, 2002), 16.

39. Fitch, *Steel Workers*, 77–79; Francis G. Couvares, *The Remaking of Pittsburgh: Class and Culture in an Industrializing City, 1877–1919* (Albany: State University of New York Press, 1984), 24.

40. David Montgomery, *The Fall of the House of Labor: The Workplace, the State, and American Labor Activism, 1865–1925* (Cambridge: Cambridge University Press, 1989), 22; Couvares, *Remaking of Pittsburgh*, 25.

41. Davis, *Life in the Iron Mills*, 132–33; song originally published in the *National Leader Tribune*, March 30, 1875.

42. David Harvey, *The Limits to Capital* (Chicago: University of Chicago Press, 1982; Midway reprint, 1989), 383–84, quotation on 383.

43. Ewa T. Morawska, *For Bread with Butter: The Life-Worlds of East Central Europeans in Johnstown, Pennsylvania, 1890–1940* (Cambridge: Cambridge University Press, 1985); Warren, *Wealth, Waste, and Alienation.*

Appendix

1. Anne Kelly Knowles, ed., *Placing History: How Maps, Spatial Data, and GIS Are Changing Historical Scholarship*, digital supplement ed. Amy Hillier (Redlands, CA: ESRI Press, 2008); Knowles, ed., "Emerging Trends in Historical GIS," *Historical Geography* 33 (2005); Knowles, *Past Time, Past Place: GIS for History* (Redlands, CA: ESRI Press, 2002); and Knowles, "Historical GIS: The Spatial Turn in Social Science History," *Social Science History* 24, no. 3 (2000). For a summary of the emergence of historical GIS, see Knowles, "GIS and History," in *Placing History*, 1–26, and Ian N. Gregory and Paul S. Ell, *Historical GIS: Technologies, Methodologies, and Scholarship* (Cambridge: Cambridge University Press, 2008).

2. The *Guide* includes maps that locate all blast furnaces and rolling mills, but they are at too small a scale, and the point symbols overlap too much, to have been useful for constructing the HGIS.

3. *DeLorme Atlas and Gazetteer* volumes cover Mid-Atlantic states at approximately 1:150,000 scale; scale varies somewhat by state. We used editions from the late 1990s to 2003. The DeLorme atlases and the USGS topographic quadrangles locate many historical features, including abandoned ironworks, as well as small roads and streams, which were essential for identifying the probable location of vanished industrial works in rural areas. The Geonames site has been revised since I completed the spatial database in 2005. Digital raster graphics (scans of USGS topographic maps) are now available through the National Map; see http://geonames.usgs.gov/domestic/. This service is a boon for some research, but for my purposes the previous interface and map server were more useful, in part because they allowed one to select which scale of map to view. Carter

and Stamp also triangulated locations with GIS coverages of populated places, railroad stations, hydrography, and other geographic features that they had developed at University of Portsmouth. In all cases, we located individual furnaces and rolling mills as points specified by latitude and longitude values.

4. See http://memory.loc.gov/ammem/index.html.

5. Anne Kelly Knowles and Richard G. Healey, "Geography, Timing, and Technology: A GIS-Based Analysis of Pennsylvania's Iron Industry, 1825–1875," *Journal of Economic History* 66. no. 3 (2006): 608–34.

6. Pennsylvania Spatial Data Access, accessed online at www.pasda.psu.edu.

GLOSSARY

The following definitions of technical terms are adapted from Robert B. Gordon, *American Iron, 1607–1900* (Baltimore: Johns Hopkins University Press, 1996), by permission of the author.

anchony. Small square knob on the end of a billet or bloom made at a finery.

anthracite. Kind of coal noted for its hardness and lack of chemical impurities. Also called stone coal.

bituminous coal. Relatively soft coal with more chemical impurities than anthracite but purer and harder than lignite coal.

blast furnace. Pyramidal or columnar shaft furnace used to smelt iron from ore, in which air is forced under pressure from a bellows or blowing engine (hence "blast") into the burden of ore, fuel, and flux. The furnace produces liquid iron and slag.

bloom. Ball of iron and slag formed by direct reduction of ore or the fining of pig iron; finished bar of iron made at a finery.

bloomery. Furnace in which iron ore is reduced directly to solid iron and liquid slag with charcoal fuel. The works where iron is made by this process.

blowing engine. Large, power-driven machine to force air into a blast furnace.

blowing in. The process of initiating smelting at a blast furnace.

boiler. Water container of a steam engine, or the power-generating system as a whole.

boilerplate. Flat plate of metal produced at a rolling mill for use in building steam engines and other heavy machinery.

bosh (or boshes). Section of a blast furnace shaft where the walls taper outward; the widest part of the furnace.

burden. The weights of ore, fuel, and flux charged into a blast furnace.

coal measures. Deposits of coal, usually referring to an extensive area of deposits.

coke. The product of expelling volatile constituents from bituminous coal, used as fuel in blast furnaces.

cupola furnace. Shaft furnace used to melt pig iron.

cut nails. Nails produced by machines that stamped ("cut") standard shapes from sheets of rolled iron.

ductility. Capacity of metal to undergo plastic deformation without fracture.

finery (or finery forge). A hearth where charcoal is used to convert pig to wrought iron, producing a bloom or loop (loup). The works where iron is made by this process.

founder. Artisan in charge of smelting at a blast furnace.

foundry. Ironworks where pig iron, blooms, or other semifinished iron is cast or reheated and shaped into finished products.

helve hammer. Massive, power-driven hammer used to consolidate heated iron, typically by rapid, repeated blows.

hematite. Ore mineral, Fe_2O_3, containing 70 percent iron.

hot blast. A technology that preheated the air forced by a blowing engine into the blast furnace.

ironmaster. Term applied to someone who understood and governed operations at an ironworks; often the owner, sometimes the superintendent.

keeper. Artisan in charge of work done around the bottom part of a blast furnace; second in command to the founder. The keeper usually tapped the furnace.

magnetite. Magnetic mineral, Fe_3O_4, containing 72 percent iron.

merchant bar. Bar-shaped semifinished iron, used at rolling mills, foundries, and forges as well as by blacksmiths.

molder. Artisan who cast iron shapes at an ironworks.

Nasmyth hammer. Massive steam-driven hammer used to shape large machine parts, such as axles and shafts, at heavy foundries.

patternmaker. Artisan who designed and made molds for casting machine parts and iron wares.

pig iron. Long, narrow pieces of crude iron, the traditional shape into which liquid iron was cast as it flowed from the blast furnace. Name refers to the image of piglets (long row of impressions in the casting bed of sand) feeding from the sow (central channel running from the crucible at the bottom of the furnace out into the casting bed).

puddling. Process of stirring heated iron as it softened in a puddling furnace to expose the mass to oxygen and burn off impurities. Product was a spongy mass of white-hot iron.

rabble. Puddler's instrument, a long metal rod or paddle used to stir the iron in a puddling furnace.

rolling. Process of forcing heated iron between mechanically rotating rollers to squeeze out slag and to consolidate and shape the iron.

shingling. The work of operating the hammer (helve or Nasmyth) to consolidate iron from a puddling furnace.

steel. Iron that has been carburized (has absorbed carbon), to any of a variety of proportions, to increase hardness among other properties.

strap rails. Railroad rails made of wood with a strip of rolled iron attached on the top surface to lessen wear from train wheels.

T rail. Iron or steel railroad rail rolled in the shape of a capital T.

tuyere. Nozzle through which air is injected into a furnace.

wrought iron. Iron composed of ferrite and slag inclusions.

BIBLIOGRAPHY

Manuscript Collections

American Steel and Wire Collection. Baker Library Historical Collections, Harvard Business School.

Ames Family Historical Collection. AmesINK, Boulder, CO.

Benjamin Smith Lyman Collection. University of Massachusetts at Amherst, W. E. B. DuBois Library, Special Collections and Archives, Amherst, MA.

Chancery Court Records. Shelby County Museum and Archives, Columbiana, AL.

Christopher Fallon Papers. MS 1673. Historical Society of Pennsylvania, Philadelphia.

Confederate Record Books in the War Department Collection. Record Group 109, National Archives and Research Administration, Washington, DC.

Cooper, Hewitt and Company Papers. Manuscript Division, Library of Congress.

David Thomas Papers. National Canal Museum Archives, Easton, PA.

Deed books. Register and Recorder's Office. Lycoming County Courthouse, Lock Haven, PA.

Edward Thomas Letters. Heinz History Center, Pittsburgh.

Farrandsville/Letters of Edward and Bess Thomas. Clinton County Historical Society, Lock Haven, PA.

Forbes Family Business Records. Baker Library Historical Collections, Harvard Business School.

General Registry Office. Cardiff, Wales.
Historic Corporate Reports Collection. Baker Library Historical Collections, Harvard Business School.
J. Peter Lesley Papers. American Philosophical Society, Philadelphia.
John Fritz Papers. National Canal Museum Archives, Easton, PA.
John Henry Alexander Papers, 1800–1883. MS 10. Maryland Historical Society, Baltimore.
Lehigh Crane Iron Company Papers. Acc. no. 1198. Hagley Museum and Library, Wilmington, DE.
"Lonaconing Iron Works Records, 1837–1840." Typescript of George's Creek Coal and Iron Company Journal, MS 2292. Maryland Historical Society, Baltimore.
Lycoming Coal Company Records. Baker Library Historical Collections, Harvard Business School.
Monkton Iron Works Records. Bixby Memorial Free Library, Vergennes, VT.
Records of the Bureau of Ordnance [Navy]. Record Group 74, National Archives and Research Administration, Washington, DC.
Richmond City Court Hustings. Library of Virginia, Richmond.
Shelby Iron Company records. W. S. Hoole Special Collections Library, University of Alabama, Tuscaloosa.
Tredegar Company Papers. Library of Virginia, Richmond.
United States Census of Manufactures, 1850, 1860.
United States Census of Population, Population (1840–70) and Slave (1840–60) Schedules.
West Point Foundry Records, Putnam County Historical Society, Cold Spring, NY.
Wilbur Stout Collection. Vol. 408. Ohio Historical Society, Columbus.

Secondary Sources

Advantages of Ralston, Lycoming Co., Pennsylvania, for the Manufacture of Iron. New York: Tribune Job Printing Establishment, 1843.
Aiken, Charles S. *The Cotton Plantation South since the Civil War*. Baltimore: Johns Hopkins University Press, 1998.
Ainsworth, Fred C., and Joseph W. Kirkley, eds. *The War of the Rebellion: A Compilation of Official Records of the Union and Confederate Armies*. Series 4. Washington, DC: Government Printing Office, 1900.
Alger, Arthur M. *A Genealogical History of . . . Thomas Alger of Tauton and Bridgewater, in Massachusetts, 1665–1875*. Boston: David Clapp, 1876.
Allen, Jay D. "The Mount Savage Iron Works, Mount Savage, Maryland: A Case Study in Pre–Civil War Industrial Development." MA thesis, University of Maryland, 1970.
Allen, Robert C. "The Peculiar Productivity History of American Blast Furnaces, 1840–1913." *Journal of Economic History* 37 (September 1977): 605–33.
American National Biography online, accessed online at www.anb.org on March 27, 2006.
Ames, Mary Lesley, ed. *Life and Letters of Peter and Susan Lesley*. Vol. 1. New York: Putnam's, 1909.
Anderson, Ann. "Horace Ware." Calhoun-Shelby County, Alabama Archives Biographies. USGenWeb Archives. Accessed online at http://www.rootsweb.com/~usgenweb/al/alfiles.htm on March 25, 2006.
Anonymous [A Merchant of Philadelphia]. *Memoirs and Auto-biography of Some of the Wealthy Citizens of Philadelphia*. Philadelphia: Booksellers, 1846.
Armes, Ethel. *The Story of Coal and Iron in Alabama*. Birmingham, AL: Chamber of Commerce, 1910. Reprint, New York: Arno Press, 1972.

Ashton, Thomas Southcliffe. *Iron and Steel in the Industrial Revolution*. Manchester: Manchester University Press, 1963. Reprint, New York: Augustus M. Kelley, 1968.

Ashworth, John. *Slavery, Capitalism, and Politics in the Antebellum Republic*. Vol. 1, *Commerce and Compromise, 1820–1850*. Cambridge: Cambridge University Press, 1995.

Aycrigg, Benjamin. *Communication from B. Aycrigg, Civil Engineer, relative to the West Branch and Allegheny Canal. . . .* Harrisburg, PA, 1839.

Baer, Christopher T. *Canals and Railroads of the Mid-Atlantic States, 1800–1860*. Edited by Glenn Porter and William H. Mulligan Jr., cartographers Marley E. Amstutz and Anne E. Webster. Wilmington, DE: Regional Economic History Research Center, Eleutherian Mills–Hagley Foundation, 1981.

———. "PRR [Pennsylvania Railroad] Chronology: A General Chronology of the Pennsylvania Railroad Company, Predecessors and Successors and Its Historical Context." N.p., 1856. March 2005 edition, 2. Accessed online at http://www.prrths.com/Hagley/PRR1856%20Mar%2005.pdf on January 22, 2011.

Bartholomew, Craig L., and Lance E. Metz. *The Anthracite Iron Industry of the Lehigh Valley*. Edited by Ann Bartholomew. Easton, PA: Center for Canal History and Technology, 1988.

Bateman, Fred, James Foust, and Thomas Weiss. "The Participation of Planters in Manufacturing in the Antebellum South." *Agricultural History* 48, no. 2 (1974): 277–97.

Bateman, Fred, and Thomas Weiss. *A Deplorable Scarcity: The Failure of Industrialization in the Slave Economy*. Chapel Hill: University of North Carolina Press, 1981.

Beach, Ursula Smith. *Along the Warioto, or A History of Montgomery County, Tennessee*. Nashville, TN: McQuiddy Press, 1964.

Beers, Frederick W. *Atlas of Clinton County, New York*. New York: Beers, Ellis, and Soule, 1869.

Belhoste, Jean-François, and Denis Woronoff. "The French Iron and Steel Industry during the Industrial Revolution." Translated by Paul Smith. In *The Industrial Revolution in Iron: The Impact of British Coal Technology in Nineteenth-Century Europe*, edited by Chris Evans and Göran Rydén, 75–94. Aldershot, UK: Ashgate, 2005.

Bennett, John William. "Iron Workers in Woods Run and Johnstown: The Union Era." PhD diss., University of Pittsburgh, 1977.

Berlin, Ira, and Herbert G. Gutman. "Natives and Immigrants, Free Men and Slaves: Urban Workingmen in the Antebellum American South." *American Historical Review* 88 (1983): 1175–1200.

Berthoff, Rowland Tappan. *British Immigrants in Industrial America, 1790–1950*. Cambridge, MA: Harvard University Press, 1953.

Bezís-Selfa, John. *Forging America: Ironworkers, Adventurers, and the Industrious Revolution*. Ithaca, NY: Cornell University Press, 2004.

———. "A Tale of Two Ironworks: Slavery, Free Labor, Work, and Resistance in the Early Republic." *William and Mary Quarterly*, 3rd ser., 56, no. 4 (1999): 677–700.

Bining, Arthur Cecil. *Pennsylvania Iron Manufacture in the Eighteenth Century*. 1938. Reprint, New York: Augustus M. Kelley, 1970.

Birch, Alan. *The Economic History of the British Iron and Steel Industry, 1784–1879*. London: Frank Cass, 1967. Reprint, New York: Augustus M. Kelley, 1968.

Bishop, John Leander, *A History of American Manufactures from 1608 to 1860. . . .* Philadelphia: Edward Young, 1864.

Black, Robert C., III. *The Railroads of the Confederacy*. Chapel Hill: University of North Carolina Press, 1952.

Blake, William J. *The History of Putnam County, N.Y.* New York: Baker and Scribner, 1849.

Bodenhamer, David J., John Corrigan, and Trevor M. Harris, eds. *The Spatial Humanities: GIS and the Future of Humanities Scholarship*. Indianapolis: Indiana University Press, 2010.

Borrow, George. *Wild Wales: The People, Language, and Scenery*. London: J. M. Dent, 1906.

Brill, Ralph, Associates. "West Point Foundry Historic Landscape Report." Typescript. Putnam County Historical Society, 1979.

Brody, David. *In Labor's Cause: Main Themes on the History of the American Worker*. New York: Oxford University Press, 1993.

———. *Steelworkers in America: The Nonunion Era*. Cambridge, MA: Harvard University Press, 1960. Reprint, Urbana: University of Illinois Press, 1998.

Brown, John K. *The Baldwin Locomotive Works, 1831–1915*. Baltimore: Johns Hopkins University Press, 1995.

Brown, Sharon A. *Historic Resource Study: Cambria Iron Company*. Washington, DC: US Department of the Interior, National Park Service, 1989.

Bruce, Kathleen. *Virginia Iron Manufacture in the Slave Era*, New York: Century, 1930. Reprint, New York: Augustus M. Kelley, 1968.

Bundy, Carol. *The Nature of Sacrifice: A Biography of Charles Russell Lowell, Jr., 1835–64*. New York: Farrar, Straus and Giroux, 2005.

Bureau of Topographic and Geologic Survey. "Distribution of Pennsylvania Coals." Map, original scale unknown. Middletown, PA: Department of Conservation and Natural Resources, 2000. Accessed online at www.dcnr.state.pa.us/topogeo on June 8, 2008.

Bush-Brown, Margaret White Lesley. Untitled obituary. *New York Times*, November 18, 1944, 13.

Campbell, H. R. *Carron Company*. Edinburgh: Oliver and Boyd, 1961.

Campbell, John H. *History of the Friendly Sons of St. Patrick and of the Hibernian Society for the Relief of Emigrants from Ireland*. Philadelphia: Hibernian Society, 1892.

Cappon, Lester J., Barbara Bartz Petchenik, and John Hamilton Long, eds. *Atlas of Early American History: The Revolutionary Era, 1760–1790*. Princeton, NJ: Princeton University Press, 1976.

Carlebach, Michael L. *Working Stiffs: Occupational Portraits in the Age of Tintypes*. Washington, DC: Smithsonian Institution Press, 2002.

Carter, Edward C., II, John C. Van Horne, and Charles E. Brownell, eds. *Latrobe's View of America, 1795–1820: Selections from the Watercolors and Sketches*. New Haven, CT: Yale University Press, 1985.

Carter, Harold, and Sandra Wheatley. *Merthyr Tydfil in 1851: A Study of the Spatial Structure of a Welsh Industrial Town*. Social Science Monograph 7. Aberystwyth: University of Wales, Board of Celtic Studies, 1982.

Chandler, Alfred D., Jr. "Anthracite Coal and the Beginnings of the Industrial Revolution in the United States." *Business History Review* 46 (Summer 1972): 141–81.

Clark, Christopher. *The Roots of Rural Capitalism: Western Massachusetts, 1780–1860*. Ithaca, NY: Cornell University Press, 1990.

Coddington, Edwin D. *The Gettysburg Campaign: A Study in Command*. New York: Charles Scribner's Sons, 1968.

Cohn, Arthur B. *Lake Champlain's Sailing Canal Boats: An Illustrated Journey from Burlington Bay to the Hudson River*. Basin Harbor, VT: Lake Champlain Maritime Museum, 2003.

Collins, Steven G. "System in the South: John W. Mallet, Josiah Gorgas, and Uniform Production at the Confederate Ordnance Department." *Technology and Culture* 40 (July 1999): 517–44.

Colton, J. H. *Topographical Map of the City and County of New-York. . . .* Map, scale approx. 1:16,000. New York: J. H. Colton, 1836.

Conzen, Michael P. *The Illinois and Michigan Canal National Heritage Corridor: A Guide to Its History and Sources.* Edited by Kay J. Carr. DeKalb: Northern Illinois University Press, 1988.

———. "Spatial Data from Nineteenth-Century Manuscript Censuses: A Technique for Rural Settlement and Land Use Analysis." *Professional Geographer* 25 (1969): 337–43.

Cook, George H., and John C. Smock. *Northern New Jersey, Showing the Iron-Ore and Limestone Districts.* Map, horizontal scale 2 miles to the inch, vertical scale 2,000 feet to the inch. N.p., 1874.

Coste, Léon, and Auguste Perdonnet. *Mémoires métallurgiques sur le traitement minérais de fer, d'étain et de plomb en Angleterre.* Paris: Bachelier, 1830.

Cothren, William. "Farrand Family." In *History of Ancient Woodbury, Connecticut, from the First Indian Deed in 1659 to 1854. . . .* Waterbury, CT: Bronson Brothers, 1854.

Council, R. Bruce, Nicholas Honercamp, and M. Elizabeth Will. *Industry and Technology in Antebellum Tennessee: The Archaeology of Bluff Furnace.* Knoxville: University of Tennessee Press, 1992.

Couvares, Francis G. *The Remaking of Pittsburgh: Class and Culture in an Industrializing City, 1877–1919.* Albany: State University of New York Press, 1984.

Cowing, Craig. "The Katahdin Iron Works: The Early Years, 1846–1863." In *Entrepreneurship in Maine: Essays in Business Enterprise,* edited by Stuart Bruchey, 3–26. New York: Garland, 1995.

Coxe, Tench. "Digest of Manufactures, Communicated to the Senate, on the 5th of January, 1814." *American State Papers, Finance,* vol. 2, 13th Cong., 2nd sess. (Doc. 407).

"Crane's Patent for Smelting Iron with Anthracite Coal." *Journal of the Franklin Institute,* June 1838, 405–6.

Crockett, Walter H. *How Vermont Maple Sugar Is Made.* Bulletin 21. [Montpelier?]: Vermont Department of Agriculture, 1915.

Cronon, William. *Nature's Metropolis: Chicago and the Great West.* New York: Norton, 1992.

Cullum, George W. *Notices of the Biographical Register of Officers and Graduates of the U.S. Military Academy at West Point,* vol. 1. New York: J. Miller, 1879.

The Cyclopaedia of American Biography. Boston: Federal Book Company, 1903.

Daddow, Samuel Harries, and Benjamin Bannan. *Coal, Iron, and Oil, or The Practical American Miner.* Pottsville, PA: Benjamin Bannan, 1866.

Dahlstrom, Neil, and Jeremy Dahlstrom. *The John Deere Story: A Biography of Plowmakers John and Charles Deere.* DeKalb: Northern Illinois University Press, 2005.

Dalzell, Robert F., Jr. *Enterprising Elite: The Boston Associates and the World They Made.* Cambridge, MA: Harvard University Press, 1987.

Daniel Tyler: A Memorial Volume. . . . New Haven, CT: privately published, 1883.

Davies, John. *A History of Wales.* London: Penguin Books, 1990.

Davis, Aaron J. *History of Clarion County.* Syracuse, NY: D. Mason, 1887.

Davis, Carl L. *Arming the Union: Small Arms in the Civil War.* Port Washington, NY: Kennikat Press, 1973.

Davis, James J. *The Iron Puddler: My Life in the Rolling Mills and What Came of It.* Indianapolis: Bobbs-Merrill, 1922.

Davis, Rebecca Harding. *Life in the Iron Mills.* Edited by Cecelia Tichi. Boston: Bedford Books, 1998.

Davis, William C. *Battle at Bull Run: A History of the First Major Campaign of the Civil War.* Garden City, NY: Doubleday, 1977.

Degree, Kenneth A. *Vergennes in the Age of Jackson.* Vergennes, VT: author, 1996.

Dew, Charles B. "Black Ironworkers and the Slave Insurrection Panic of 1856." *Journal of Southern History* 41, no. 3 (1975): 321–38.

———. *Bond of Iron: Master and Slave at Buffalo Forge*. New York: Norton, 1994.

———. "David Ross and the Oxford Iron Works: A Study of Industrial Slavery in the Early Nineteenth-Century South." *William and Mary Quarterly* 31, no. 2 (1974): 189–224.

———. *Ironmaker to the Confederacy: Joseph R. Anderson and the Tredegar Iron Works*. New Haven, CT: Yale University Press, 1966.

———. "Slavery and Technology in the Antebellum Southern Iron Industry: The Case of Buffalo Forge." In *Science and Medicine in the Old South*, edited by Ronald L. Numbers and Todd L. Savitt, 107–26. Baton Rouge: Louisiana State University Press, 1989.

Dewick, Sarah A. *The Ancestry of John S. Gustin and His Wife Sarah McComb*. Boston: David Clapp, 1900.

Dexter, Franklin Bowditch. *Biographical Sketches of the Graduates of Yale College with Annals of the College History*. Vol. 5, *June 1792–September 1805*. New York: Henry Holt, 1911.

Deyrup, Felicia Johnson. *Arms Makers of the Connecticut Valley: A Regional Study of the Economic Development of the Small Arms Industry, 1789–1870*. Studies in History 33. Northampton, MA: Smith College, 1948.

Dictionary of American Biography. New York: Charles Scribner's Sons, 1928–58.

Dieffenbach, Susan, and Craig A. Benner. *Cornwall Iron Furnace: Pennsylvania Trail of History Guide*. Mechanicsburg, PA: Stackpole Books for the Pennsylvania Historical and Museum Commission, 2003.

Dilts, James D. *The Great Road: The Building of the Baltimore and Ohio, the Nation's First Railroad, 1828–1853*. Stanford, CA: Stanford University Press, 1993.

Dodge, Clifford H. "The Second Geological Survey of Pennsylvania: The Golden Years." *Pennsylvania Geology* 18 (February 1987): 9–15.

Donovan, Frank R. *Ironclads of the Civil War*. New York: American Heritage, 1964.

Dooley, Pat. "Traces of Queen's Mansion Remain at Farrandsville." *Grit*, November 15, 1981, news section, 53.

Doster, James F. "The Chattanooga Rolling Mill: An Industrial By-Product of the Civil War." *East Tennessee Historical Society's Publications* 36 (1964): 45–55.

Dublin, Thomas. *Women at Work: The Transformation of Work and Community in Lowell, Massachusetts, 1826–1860*. New York: Columbia University Press, 1979.

Dufrénoy, M. [Pierre Armand]. *On the Use of Hot Air in the Iron Works of England and Scotland*. Translated from a report made to the director general of mines in France, . . . in 1834. London: J. Murray, 1836.

Dufrénoy, M., Élie de Beaumont, Léon Coste, and Auguste Perdonnet. *Voyage métallurgique en Angleterre* [A metallurgical journey in England]. 2d ed. 2 vols. Paris: Bachelier, 1837.

Dunwell, Frances F. *The Hudson River Highlands*. New York: Columbia University Press, 1991.

Easson, Kay Parkhurst, and Roger R. Easson, eds. William Blake, *Milton*. Boulder, CO: Shambhala, 1978.

Edgar, William R. "History of Iron Mountain." Undated article accessed online at www.rootsweb.com/~mostfran/mine_history/iron_mountain.htm on March 26, 2006.

Edwards, Hywel Teifi. *Arwr Glew Erwau'r Glo: Delwedd y Glowr yn Llenyddiaeth y Gymraeg, 1850–1950* (Brave Hero of the CoalFields: The Image of the Miner in Welsh Literature, 1850–1950]. Llandysul: Gwasg Gomer, 1994.

Eggert, Gerald G. *Making Iron on the Bald Eagle: Roland Curtin's Ironworks and Workers' Community*. University Park: Pennsylvania State University Press, 2000.

———. *The Iron Industry in Pennsylvania.* Pennsylvania Historical Studies 25. University Park: Pennsylvania Historical Association, 1994.

Ellis, Stephanie. "The Piano Industry in America: 1775–1940." In *Atlas of Industrial America,* edited by Anne Kelly Knowles, 46–47. Middlebury, VT: privately published at Middlebury College, 2007.

Elsas, Madeleine, ed. *Iron in the Making: Dowlais Iron Company Letters, 1782–1860.* Cardiff: County Records Committee of the Glamorgan Quarter Sessions and County Council and Guest Keen Iron and Steel Company, 1960.

Emmons, Ebenezer. "Survey of the Second Geological District." In *Geology of New York,* part 2, 231–63, 291–308. Albany, NY: Carroll and Cook, 1842.

Evans, Chris. "The Industrial Revolution in Iron in the British Isles." In *The Industrial Revolution in Iron: The Impact of British Coal Technology in Nineteenth-Century Europe,* edited by Chris Evans and Göran Rydén, 15–28. Altershot, UK: Ashgate, 2005.

———. *The Labyrinth of Flames: Work and Social Conflict in Early Industrial Merthyr Tydfil.* Cardiff: University of Wales Press, 1993.

———. "Work and Workloads during Industrialization: The Experience of Forgemen in the British Iron Industry, 1750–1850." *International Review of Social History* 44 (1999): 197–215.

Evans, Chris, and Göran Rydén. *Baltic Iron in the Atlantic World in the Eighteenth Century.* Leiden: Brill, 2007.

———, eds. *The Industrial Revolution in Iron: The Impact of British Coal Technology in Nineteenth-Century Europe.* Aldershot, UK: Ashgate, 2005.

An Exposition of the Property of the Etowah Manufacturing and Mining Co. New York: George F. Nesbitt, 1860.

Fahlman, Betsy. *John Ferguson Weir: The Labor of Art.* Newark: University of Delaware Press, 1997.

"Farrandsville, Rich in History, Observes 150th Anniversary." *Express* (Lock Haven, PA), July 22, 1982.

Fell, James E., Jr. "Iron from 'The Bend': The Great Western and Brady's Bend Iron Companies." *Western Pennsylvania Historical Magazine* 67 (October 1984): 323–45.

The First Railroad in America: A History of the Origin and Development of the Granite Railway at Quincy, Massachusetts. Quincy, MA: privately published for the Granite Railway Company, 1926.

Fishlow, Albert. *American Railroads and the Transformation of the Ante-bellum Economy.* Cambridge, MA: Harvard University Press, 1965.

Fitch, John A. *The Steel Workers.* Pittsburgh: University of Pittsburgh Press, 1989.

Fleming, Brewster and Alley. "Birds-Eye Map of the Western and Atlantic R.R., the Great Kennesaw Route; Army Operations, Atlanta Campaign, 1864." Map, no scale. New York, 1887.

Fogel, Robert William. *Railroads and American Economic Growth: Essays in Econometric History.* Baltimore: Johns Hopkins University Press, 1964.

French, Benjamin F. *History of the Rise and Progess of the Iron Trade of the United States from 1621 to 1857.* New York: Wiley and Halsted, 1858.

Fritz, John. *The Autobiography of John Fritz.* New York: John Wiley, 1912.

Galer, Gregory, Robert Gordon, and Frances Kemmish. "Connecticut's Ames Iron Works: Family, Community, Nature, and Innovation in an Enterprise of the Early American Republic." *Transactions [of the] Connecticut Academy of Arts and Sciences* 54 (December 1998): 83–194.

Gandy, Matthew. *Concrete and Clay: Reworking Nature in New York City*. Cambridge, MA: MIT Press, 2002.

Garrett, Richard. *Weathersongs*. Vol. 1, *Days in Wales*. N.p.: Sunday Dance Music, 2006.

Gates, Grace Hooten. *The Model City of the New South: Anniston, Alabama, 1872–1900*. Huntsville, AL: Strode, 1978.

Geary, James W. *We Need Men: The Union Draft in the Civil War*. Dekalb: Northern Illinois University Press, 1991.

George's Creek Coal and Iron Company (1836). Rare PAM 10047. Maryland Historical Society.

Gibb, George Sweet. *The Saco-Lowell Shops: Textile Machinery in New England, 1813–1949*. Cambridge, MA: Harvard University Press, 1950.

Google Earth. Accessed at earth.google.com.

Gordon, Robert B. *American Iron, 1607–1900*. Baltimore: Johns Hopkins University Press, 1996.

———. *A Landscape Transformed: The Ironmaking District of Salisbury, Connecticut*. New York: Oxford University Press, 2001.

Gordon, Robert B., and Patrick M. Malone. *The Texture of Industry: An Archaeological View of the Industrialization of North America*. New York: Oxford University Press, 1994.

Gorgas, Josiah. "Notes on the Ordnance Department of the Confederate Government." *Southern Historical Society Papers* 12 (1884): 66–94.

"A Great Cannon Foundry." *Scientific American* 11 (September 10, 1864): 165.

Gregory, Ian N., and Paul S. Ell. *Historical GIS: Technologies, Methodologies, and Scholarship*. Cambridge: Cambridge University Press, 2008.

Gruenwald, Kim M. *River of Enterprise: The Commercial Origins of Regional Identity in the Ohio Valley, 1790–1850*. Bloomington: Indiana University Press, 2002.

Gurda, John. *Bay View, Wis.*, centennial edition. Milwaukee: Milwaukee Humanities Program, 1979.

Harder, E. C. "Distribution of Iron Ore in the United States." Plate 1 in *Mineral Resources of the United States*, part 1, *Metallic Products, Calendar Year 1908*. Department of the Interior, US Geological Survey. Washington, DC: Government Printing Office, 1909.

———. "Iron Ores, Pig Iron, and Steel." In *Mineral Resources of the United States*, part 1, *Metallic Products, Calendar Year 1908*, 61–126. Department of the Interior, US Geological Survey. Washington, DC: Government Printing Office, 1909.

Hardy, Eugene L. "A Topographical Map of the Etowah Property, Cass County, Georgia." Map, scale 2 inches to the mile, National Archives and Records Administration, Washington, DC, 1856.

Hartley, Edward Neal. *Ironworks on the Saugus: The Lynn and Braintree Ventures of the Company of Undertakers of the Ironworks in New England*. Norman: University of Oklahoma Press, 1957.

Harvey, David. *The Limits to Capital*. Chicago: University of Chicago Press, 1982; Midway reprint, 1989.

Harvey, Katherine A. *The Best-Dressed Miners: Life and Labor in the Maryland Coal Region, 1835–1910*. Ithaca, NY: Cornell University Press, 1969.

———, ed. *The Lonaconing Journals: The Founding of a Coal and Iron Community, 1837–1840*. Transactions of the American Philosophical Society, n.s., 67, part 2. Philadelphia: American Philosophical Society, 1977.

Healey, Richard G. *The Pennsylvania Anthracite Coal Industry, 1860–1902: Economic Cycles, Business Decision-Making, and Regional Dynamics*. Scranton, PA: University of Scranton Press, 2007.

Heberling, Paul M. "Status Indicators: Another Strategy for Interpretation of Settlement Pattern in a Nineteenth-Century Industrial Village." In *Consumer Choice in Historical Archaeology*, edited by Suzanne M. Spencer-Wood, 199–216. New York: Plenum Press, 1987.

Henning, L. F. *West View of Catasauqua*. N.p., [ca. 1852].

Henry, Matthew S. *History of the Lehigh Valley*. Easton, PA: Bixler and Corwin, 1860.

Herbert, Eugenia W. *Iron, Gender, and Power: Rituals of Transformation in African Societies*. Bloomington: Indiana University Press, 1993.

Hewitt, Abram S. "The Production of Iron and Steel in Its Economic and Social Relations." In *Selected Writings of Abram S. Hewitt*, edited by Allan Nevins, 19–85. New York: Columbia University Press, 1937.

A History of the Lehigh Coal and Navigation Company, published by order of the Board of Managers. Philadelphia: William S. Young, 1840.

Hitchcock, Frederick L. *War from the Inside: The Story of the 132nd Regiment Pennsylvania Volunteer Infantry in the War for the Suppression of the Rebellion, 1862–1863*. Philadelphia: J. B. Lippincott, 1904.

Hobsbawm, Eric J. "Artisan or Labour Aristocrat?" *Economic History Review*, 2nd ser., 37, no. 3 (1984): 355–72.

———. "The Tramping Artisan." *Economic History Review*, 2nd ser., 3 (1951): 299–320.

Hogg, Ian V. *Weapons of the Civil War*. New York: Military Press; Crown Publishers, 1987.

Hopkins, G. M. *Clark's Map of Litchfield County, Connecticut*. Map, scale 1:50,688. N.p., 1859, downloaded from Library of Congress, American Memory, http://memory.loc.gov/ammem, on July 9, 2007.

Höppe, Goran, and John Langton. *Flows of Labour in the Early Phase of Capitalist Development: The Time-Geography of Longitudinal Migration Paths in Nineteenth-Century Sweden*. Historical Geography Research Series 29. N.p.: Institute of British Geographers, Historical Geography Research Group, 1992.

Hudson, John C. *Plains Country Towns*. Minneapolis: University of Minnesota Press, 1985.

Hughes, R. Elwyn. "The Welsh Language in Technology and Science, 1800–1914." In *The Welsh Language and Its Social Domains, 1801–1911*, edited by Geraint H. Jenkins, 405–30. Cardiff: University of Wales Press, 2000.

Hughes, Thomas Ll. Untitled article. *Y Cyfaill o'r Hen Wlad* [The Friend from the Old Country], April 1854, 153.

Hume, J. R., and J. Butt. "Muirkirk, 1786–1802: The Creation of a Scottish Industrial Community." *Scottish Historical Review* 45 (1966): 160–83.

Hunt, Robert. "Map of Great Britain and Ireland, Shewing the Districts in Which Coal and Iron Are Respectively Found." Map, scale approx. 27 miles to the inch. In *Plans Referred to in the Report of the Commissioners Appointed to Inquire into the Application of Iron to Railway Structures*. London: William Clowes for Her Majesty's Stationery Office, 1849.

Hunter, Jerry. *Llwch Cenhedloedd: Y Cymry a Rhyfel Cartref America* (The Dust of Nations: The Welsh and America's Civil War). Llanrwst, Wales: S4C, 2003.

Hunter, Louis C. "Influence of the Market upon Technique in the Iron Industry in Western Pennsylvania up to 1860." *Journal of Economic and Business History* 1 (1928–29): 241–81.

———. *Steamboats on the Western Rivers: An Economic and Technological History*, 1949. Reprint, New York: Dover, 1993.

———. "A Study of the Iron Industry at Pittsburgh before 1860." PhD diss., Harvard University, 1928.

Hutslar, Donald A. *Log Construction in the Ohio Country, 1750–1850.* Athens: Ohio University Press, 1992.

Hyde, Charles K. *Technological Change and the British Iron Industry, 1700–1870.* Princeton, NJ: Princeton University Press, 1977.

Ince, Laurence. "The Neath Abbey Ironworks." *Industrial Archaeology* 11–12 (1977): 25–31.

Ingham, John N. *Making Iron and Steel: Independent Mills in Pittsburgh, 1820–1920.* Columbus: Ohio State University Press, 1991.

Isard, Walter, and John H. Cumberland. "New England as a Possible Location for an Integrated Iron and Steel Works." *Economic Geography* 26, no. 4 (1950): 245–59.

Isleib, Charles R., and Jack Chard. *The West Point Foundry and the Parrott Gun: A Short History.* Fleischmanns, NY: Purple Mountain Press, 2000.

Jackson, Joyce. *History of the Shelby Iron Company, 1862–1868.* Shelby, AL: Historic Shelby Association, 1990.

"James Ralston—the Origin of the Hot Blast." *United States Railroad and Mining Register,* June 18, 1864.

Jenkins, Jerry, with Andy Keal. *The Adirondack Atlas: A Geographic Portrait of the Adirondack Park.* Syracuse, NY: Syracuse University Press and the Adirondack Museum, 2004.

Jeremy, David J. *Transatlantic Industrial Revolution: The Diffusion of Textile Technologies between Britain and America, 1790–1830s.* Cambridge, MA: MIT Press, 1981.

Johnson, Carol Siri. "Prediscursive Technical Communication in the Early American Iron Industry." *Technical Communications Quarterly* 15, no. 2 (2006): 171–89.

Johnson, Kirk. "In 200 Years of Family Letters, a Nation's Story." *New York Times,* January 29, 2006, section A, 1, 20.

Jones, Bill, and Chris Williams. *B. L. Coombes.* Cardiff: University of Wales Press, 1999.

Jones, Edgar. *A History of GKN* [Guest, Keen and Nettlefolds]. Vol. 1, *Innovation and Enterprise, 1759–1918.* Houndsmills, UK: Macmillan, 1987.

Jones, William D. *Wales in America: Scranton and the Welsh, 1860–1920.* Cardiff: University of Wales Press, 1993.

"Joseph Parry (1841–1903), Composer and Musician." *Casglu'r Tlysau/Gathering the Jewels, the Website for Welsh Cultural History.* National Library of Wales. Accessed online at http://www.gtj.org.uk on July 27, 2007.

Journal of the House of Representatives of the Commonwealth of Pennsylvania. Harrisburg, 1826–40.

Journal of the Senate of the Commonwealth of Pennsylvania. Harrisburg, 1826–27, 1842.

Keeler, Vernon David. "An Economic History of the Jackson County Iron Industry." *Ohio Archaeological and Historical Quarterly* 42 (April 1933): 133–244.

Kennedy, T. J. "Map of Chester Co., Pennsylvania." Map, no scale. N.p., 1860.

Keyssar, Alexander. *Out of Work: The First Century of Unemployment in Massachusetts.* Cambridge: Cambridge University Press, 1986.

———. *The Right to Vote: The Contested History of Democracy in the United States.* New York: Basic Books, 2000.

Kimball, Gregg D. *American City, Southern Place: A Cultural History of Antebellum Richmond.* Athens: University of Georgia Press, 2000.

King, J. W. "Iron Rolling Mills of the West." Report 19, Annual Report of the Secretary of the Navy, Exec. Doc., 38th Cong., 1st sess., vol. 4, serial set 1183. Washington, DC: Government Printing Office, 1864.

Knowles, Anne Kelly. *Calvinists Incorporated: Welsh Immigrants on Ohio's Industrial Frontier.* Chicago: University of Chicago Press, 1997.

———. "GIS and History." In *Placing History: How Maps, Spatial Data, and GIS Are Changing Historical Scholarship*, edited by Anne Kelly Knowles, 1–25. Digital supplement edited by Amy Hillier. Redlands, CA: ESRI Press, 2008.

———. "Immigrant Trajectories through the Rural-Industrial Transition in Wales and the United States, 1795–1850." *Annals of the Association of American Geographers* 85 (1995): 246–66.

———. "Labor, Race, and Technology in the Confederate Iron Industry." *Technology and Culture* 42 (January 2001): 1–26.

———. "Wheeling Iron and the Welsh: A Geographical Reading of 'Life in the Iron Mills.'" In *Transnational West Virginia: Ethnic Work Communities during the Industrial Era*, edited by Ronald Lewis and Kenneth Fones-Wolf, 216–41. Morgantown: West Virginia University Press, 2002.

Knowles, Anne Kelly, and Richard G. Healey. "Geography, Timing, and Technology: A GIS-Based Analysis of Pennsylvania's Iron Industry, 1825–1875." *Journal of Economic History* 66 (September 2006): 608–34.

Kober, George M. "Iron, Steel, and Allied Industries." In *Industrial Health*, edited by George M. Kober and Emery R. Hayhurst, 176–84. Philadelphia: P. Blakiston's Son, 1924.

Krygier, John. "Sound and Geographic Visualization." In *Visualization in Modern Cartography*, edited by Alan MacEachren and D. R. F. Taylor, 149–66. New York: Pergamon, 1994.

Kury, Theodore W. "Labor and the Charcoal Iron Industry: The New Jersey–New York Experience." *Material Culture* 25, no. 3 (1993): 19–33.

Lahr, F. V., after James Queen. *Montour Iron Works, Danville, Pennsylvania.* Color lithograph. Philadelphia: P. S. Duval, [ca. 1855].

Lakwete, Angela. *Inventing the Cotton Gin: Machine and Myth in Antebellum America.* Baltimore: Johns Hopkins University Press, 2003.

Landes, David S. *The Unbound Prometheus: Technological Change and Industrial Development in Western Europe from 1750 to the Present.* Cambridge: Cambridge University Press, 1969.

Larkin, Jack. *Where We Lived: Discovering the Places We Once Called Home.* Taunton, MA: Taunton Press, 2006.

Larson, John Lauritz. *Bonds of Enterprise: John Murray Forbes and Western Development in America's Railway Age.* Rev. ed. Iowa City: University of Iowa Press, 2001.

Latour, Bruno. *Science in Action: How to Follow Scientists and Engineers through Society.* Cambridge, MA: Harvard University Press, 1987.

Laurant, Annie. *Des fers de Loire à l'acier Martin: Maîtres de forges en Berry et Nivernais.* Paris: Royer, 1995.

Leach, Frank Willing. "Old Philadelphia Families." In *The North American*, transcribed by Vince Summers as "Ralston Family History: Philadelphia and Chester Counties, Pennsylvania, (Philadelphia, 1912)." Accessed online at http://ftp.rootsweb.com/pub/usgenweb/pa/philadelphia/history/family/ralston.txt on September 22, 2007.

Leigh's Guide to Wales and Monmouthshire. London: M. A. Leigh, 1833.

Lemon, James T. *The Best Poor Man's Country: A Geographical Study of Early Southeastern Pennsylvania.* Baltimore: Johns Hopkins University Press, 1972.

Lesley, J. Peter. *The Iron Manufacturer's Guide to the Furnaces, Forges, and Rolling Mills of the United States.* New York: John Wiley, 1859.

———. *Manual of Coal and Its Topography.* Philadelphia: J. B. Lippincott, 1856.

[Lesley, J. Peter, ed.]. *Bulletin of the American Iron Association.* Philadelphia: American Iron Association, 1856–59.

Lesley, Joseph. Untitled obituary. *Philadelphia Enquirer*, February 12, 1889.

Lewis, W. David. "The Early History of the Lackawanna Iron and Coal Company: A Study in Technological Adaptation." *Pennsylvania Magazine of History and Biography* 96 (1972): 424–68.

———. *Sloss Furnaces and the Rise of the Birmingham District: An Industrial Epic.* Tuscaloosa: University of Alabama Press, 1994.

Lewis, Ronald L. *Coal, Iron and Slaves: Industrial Slavery in Maryland and Virginia, 1715–1865.* Westport, CT: Greenwood Press, 1979.

Licht, Walter. *Industrializing America: The Nineteenth Century.* Baltimore: Johns Hopkins University Press, 1995.

Linaberger, James. "The Rolling Mill Riots of 1850." *Western Pennsylvania Historical Magazine* 47 (January 1964): 1–18.

Livesay, Harold C. "Marketing Patterns in the Antebellum American Iron Industry." *Business History Review* 45, no. 3 (1971): 269–95.

Lloyd, Thomas W. *History of Lycoming County, Pennsylvania.* Vol. 1. Topeka, KS: Historical Publishing, 1929.

Lonn, Ella. *Foreigners in the Confederacy.* Chapel Hill: University of North Carolina Press, 1940.

Lord, Peter. *Arlunwyr Gwlad/Artisan Painters.* Aberystwyth: National Library of Wales, 1993.

———. *Hugh Hughes: Arlunydd Gwlad* [Artisan Painter], *1790–1863.* Llandysul: Gwasg Gomer, 1995.

———. *The Visual Culture of Wales: Industrial Wales.* Cardiff: University of Wales Press, 1998.

Luraghi, Raimondo. *A History of the Confederate Navy.* Translated by Paolo E. Coletta. Annapolis, MD: Naval Institute Press, 1996.

MacDougal, E. B. "Exploratory Analysis, Dynamic Statistical Visualization, and Geographic Information Systems." *Cartography and Geographic Information Systems* 19 (1992): 237–46.

MacEachren, Alan M., and Menno-Jan Kraak. "Exploratory Cartographic Visualization: Advancing the Agenda." *Computers and Geosciences* 23 (1997): 335–43.

Maddex, Lee R. "La Belle Iron Works." Historic American Engineering Record WV-47. Typescript. August 1990.

Mallet, J. W. "Work of the Ordnance Bureau of the War Department of the Confederate States, 1861–1865." *Alumni Bulletin of the University of Virginia* 3 (1910): 162–70.

Malloy, Mary. *"Boston Men" on the Northwest Coast: The American Maritime Fur Trade, 1788–1844.* Anchorage: University of Alaska Press, 1998.

Mathews, Alfred, and Austin N. Hungerford. *History of the Counties of Lehigh and Carbon, in the Commonwealth of Pennsylvania.* Philadelphia: Everts and Richards, 1884.

May, Earl Chapin. *Principio to Wheeling, 1715–1945: A Pageant of Iron and Steel.* New York: Harper, 1945.

Maynard, D. S. *Historical View of Clinton County.* Lock Haven, PA: Enterprise Printing, 1875.

McCurdy, Linda. "The Potts Family Iron Industry in the Schuylkill Valley." PhD diss., Pennsylvania State University, 1974.

McGrain, John W. *From Pig Iron to Cotton Duck: A History of Manufacturing Villages in Baltimore County.* Towson, MD: Baltimore County Public Library, 1985.

McKenzie, Robert Hamlet. "A History of the Shelby Iron Company, 1865–1881." PhD diss., University of Alabama, 1971.

McKiven, Henry M., Jr. *Iron and Steel: Class, Race, and Community in Birmingham, Alabama, 1875–1920*. Chapel Hill: University of North Carolina Press, 1995.

[McLane, Louis], Secretary of the Treasury. *Documents relative to the Manufactures in the United States* [HR Doc. 308, *McLane Report*]. 2 vols. Washington, DC: Duff Green, 1833.

McPherson, James M. *Battle Cry of Freedom: The Civil War Era*. New York: Oxford University Press, 1988.

Memorial from Pennsylvania on the Manufacture of Iron. Philadelphia: Convention of Iron Masters, 1850.

Meyer, David R. *Networked Machinists: High-Technology Industries in Antebellum America*. Baltimore: Johns Hopkins University Press, 2006.

———. *The Roots of American Industrialization*. Baltimore: Johns Hopkins University Press, 2003.

Montgomery, David. *Beyond Equality: Labor and the Radical Republicans, 1862–1872*. New York: Alfred A. Knopf, 1967.

———. *Citizen Worker: The Experience of Workers in the United States with Democracy and the Free Market during the Nineteenth Century*. New York: Cambridge University Press, 1993.

———. *The Fall of the House of Labor: The Workplace, the State, and American Labor Activism, 1865–1925*. Cambridge: Cambridge University Press, 1989.

Morawska, Ewa T. *For Bread with Butter: The Life-Worlds of East Central Europeans in Johnstown, Pennsylvania, 1890–1940*. Cambridge: Cambridge University Press, 1985.

Moravek, John Richard. "The Iron Industry as a Geographic Force in the Adirondack-Champlain Region of New York State, 1800–1971." PhD diss., University of Tennessee, Knoxville, 1976.

Mosher, Anne E. *Capital's Utopia: Vandergrift, Pennsylvania, 1855–1916*. Baltimore: Johns Hopkins University Press, 2004.

Mould, David H. *Dividing Lines: Canals, Railroads, and Urban Rivalry in Ohio's Hocking Valley, 1825–1875*. Dayton, OH: Wright State University Press, 1994.

National Cyclopedia of American Biography. New York: J. T. White, 1898.

Nevins, Allan. *Abram S. Hewitt, with Some Account of Peter Cooper*. New York: Harper, 1935. Reprint, New York: Octagon Books, 1967.

———, ed. *Selected Writings of Abram S. Hewitt*. New York: Columbia University Press, 1937.

Norris, James D. *Frontier Iron: The Story of the Maramec Iron Works, 1826–1876*. Madison: State Historical Society of Wisconsin, 1964.

Nye, David E. *America as Second Creation: Technology and Narratives of New Beginnings*. Cambridge, MA: MIT Press, 2003.

———. *American Technological Sublime*. Cambridge, MA: MIT Press, 1994.

Oberholtzer, Ellis Paxson. *The Literary History of Philadelphia*. Philadelphia: George W. Jacobs, 1906.

O'Connor, Thomas. *Civil War Boston: Home Front and Battlefield*. Boston: Northeastern University Press, 1997.

Ordnance Survey. "Merthyr Tydfil." Map, 42 sheets, scale 1:528 (1851). Glamorgan Archives, Cardiff.

Ostergren, Robert C. *Patterns of Seasonal Industrial Labor Recruitment in a Nineteenth Century Swedish Parish: The Case of Matfors and Tuna, 1846–1873*. Report 5 from the Demographic Data Base. Umeå, Sweden: Umeå University, 1990.

Palladino, Grace. *Another Civil War: Labor, Capital, and the State in the Anthracite Regions of Pennsylvania, 1840–68*. Urbana: University of Illinois Press, 1990.

Pamplin, William. "Cyfarthfa Works and Water Wheel" (1791–1800). Drawing, pencil on paper. Item 823.992, Cyfarthfa Castle Museum and Gallery, Merthyr Tydfil, Wales.

Parker, David B. "Mark Anthony Cooper (1800–1885)." *The New Georgia Encyclopedia.* Article dated September 3, 2002. Accessed online at www.georgiaencyclopedia.org on July 13, 2007.

Parton, James. "Pittsburgh." *Atlantic Monthly* 21 (January 1868): 17–36.

Paskoff, Paul. *Industrial Evolution: Organization, Structure, and Growth of the Pennsylvania Iron Industry, 1750–1860.* Baltimore: Johns Hopkins University Press, 1983.

Paullin, Charles O. *Atlas of the Historical Geography of the United States,* edited by John K. Wright. Publication 401. Washington, DC: Carnegie Institution of Washington and the American Geographical Society of New York, 1932.

Paxton, John A. *The Philadelphia Directory and Register for 1819.* Philadelphia, [ca. 1819].

Pelletreau, William S. *History of Putnam County, New York.* Philadelphia: W. W. Preston, 1886. Commemorative edition, Brewster, NY: Landmarks Preservation Committee, 1975.

Peterson, Fred W. *Homes in the Heartland: Balloon Frame Farmhouses of the Upper Midwest, 1850–1920.* Lawrence: University Press of Kansas, 1992.

Plan of Merthyr Tydfil, from Actual Survey. Attributed to John Wood, no scale (1836). Glamorgan County Record Office.

Population of the United States in 1860. Washington, DC: Government Printing Office, 1864.

Porter, Glenn, and Harold Livesay. *Merchants and Manufacturers.* Baltimore: Johns Hopkins University Press, 1971.

Pryce, W. T. R. "The Welsh Language, 1750–1961." In *National Atlas of Wales,* edited by Harold Carter, plate 3.1. Cardiff: University of Wales Press, 1980.

Raber, Michael S., et al. *Forge of Innovation: An Industrial History of the Springfield Armory, 1794–1968,* edited by Richard Colton. 1989. Reprint, Fort Washington, PA: Eastern National [Division, National Park Service], 2008.

Raymond, Robert. *Out of the Fiery Furnace: The Impact of Metals on the History of Mankind.* University Park: Pennsylvania State University Press, 1986.

Reader's Digest Complete Atlas of the British Isles. London: Reader's Digest Association, 1965.

Reeves, James E. *The Physical and Medical Topography . . . of the City of Wheeling.* Wheeling, WV: Daily Register, 1870.

Reid, Donald. *The Miners of Decazeville: A Genealogy of Deindustrialization.* Cambridge, MA: Harvard University Press, 1985.

Reiser, Catherine Elizabeth. *Pittsburgh's Commercial Development, 1800–1850.* Harrisburg: Pennsylvania Historical and Museum Commission, 1951.

Remington, W. Craig, and Thomas J. Kallsen. *Historical Atlas of Alabama.* Vol. 1, *Historical Locations by County.* Tuscaloosa: Department of Geography, University of Alabama, 1997.

Reps, John W. *Cities of the American West: A History of Frontier Urban Planning.* Princeton, NJ: Princeton University Press, 1979.

Rhys Davies obituary. *Richmond Enquirer,* September 14, 1838, 3.

Richards, John Adair. *A History of Bath County, Kentucky.* Yuma, AZ: Southwest Printers, 1961.

Rider, S. W., and A. E. Trueman. *South Wales: A Physical and Economic Geography.* London: Methuen, 1929.

Ripley, Warren. *Artillery and Ammunition of the Civil War.* New York: Promontory Press, 1970.

Robert and Sarah Ralston Family. Rootsweb. Accessed online at http://ftp.rootsweb.com/pub/usgenweb/pa/ philadelphia/history/family/ralston.txt on May 6, 2004.

[Roberts, Edward]. "David Thomas: The Father of the Anthracite Iron Trade." *Red Dragon,* October 1883, 288–99.
Roberts, Solomon W[hite]. "Gwneyd Haiarn â Maenlo" (Made Iron with Stone Coal). Obituary of George Crane, translated into Welsh. *Y Cyfaill o'r Hen Wlad* (The Friend from the Old Country), June 1846, 170–71.
Roberts, William H. *Civil War Ironclads: The U.S. Navy and Industrial Mobilization.* Baltimore: Johns Hopkins University Press, 2002.
Rodner, William S. *J. M. W. Turner: Romantic Painter of the Industrial Revolution.* Berkeley: University of California Press, 1997.
Rousey, Dennis C. "Aliens in the WASP Nest: Ethnocultural Diversity in the Antebellum Urban South." *Journal of American History* 79, no. 1 (1992): 152–64.
Schechter, Patricia. "Free and Slave Labor in the Old South: The Tredegar Ironworkers' Strike of 1847." *Labor History* 35 (Spring 1994): 165–86.
Schneller, Robert J., Jr. *A Quest for Glory: A Biography of Rear Admiral John H. Dahlgren.* Annapolis, MD: Naval Institute Press, 1996.
Scranton, Philip. *Proprietary Capitalism.* Cambridge: Cambridge University Press, 1983.
Scrivenor, Harry. *History of the Iron Trade.* London: Longman, Brown, Green, and Longmans, 1854.
Seaburg, Alan. "Thomas Handasyd Perkins." *Dictionary of Unitarian and Universalist Biography.* Accessed online at http://www25.uua.org/uuhs/duub on August 21, 2007.
Seaburg, Carl, and Stanley Paterson. *Merchant Prince of Boston: Colonel T. H. Perkins, 1764–1854.* Cambridge, MA: Harvard University Press, 1971.
Sears, John F. "Tourism and the Industrial Age: Niagara Falls and Mauch Chunk." In *Sacred Places: American Tourist Attractions in the Nineteenth Century,* by John F. Sears, 182–208. New York: Oxford University Press, 1989.
Shaw, William L. "The Confederate Conscription and Exemption Acts." *American Journal of Legal History* 6 (1962): 368–405.
Skaggs, Julian C. "Lukens, 1850–1870: A Case Study in the Mid-Nineteenth Century American Iron Industry." PhD diss., University of Delaware, 1975.
"Sketch Illustrating the Positions of the Commercial Cities and Towns of the Eastern, Middle, and Western States." Map, scale 58.5 miles to the inch. N.p., [ca. 1850].
Smith, Greg. "A History of the McIntyre Mine near Newcomb, N.Y." Accessed online at http://www.adirondack-park.net/history/mcintyre.mine.html on July 16, 2007.
Smith, James Larry. "Historical Geography of the Southern Charcoal Iron Industry, 1800–1860." PhD diss., University of Tennessee, Knoxville, 1982.
Smith, Merritt Roe. *Harpers Ferry Armory and the New Technology: The Challenge of Change.* Ithaca, NY: Cornell University Press, 1977.
Smith, Merritt Roe, and Leo Marx, eds. *Does Technology Drive History? The Dilemma of Technological Determinism.* Cambridge, MA: MIT Press, 1994.
Smith, Nancy L. "Iron Production between the River and the Front: A Revision of the Plantation Model in an Era of Proto-industrialization." MSc thesis, Pennsylvania State University, 1992.
Smucker, Anna Egan. *No Star Nights.* New York: Alfred A. Knopf, 1989.
Solocow, Arthur A., ed. "Pennsylvania." In *The State Geological Surveys: A History.* N.p.: Association of American State Geologists, 1988.
Southall, Humphrey R. "Towards a Geography of Unionization: The Spatial Organization and Distribution of Early British Trade Unions." *Transactions of the Institute of British Geographers* 13, no. 4 (1988): 466–83.
———. "The Tramping Artisan Revisits: Labour Mobility and Economic Distress in Early Victorian England." *Economic History Review* 44, no. 2 (1991): 272–96.

Stanley, Amy Dru. *From Bondage to Contract: Wage Labor, Marriage, and the Market in the Age of Slave Emancipation.* Cambridge: Cambridge University Press, 1998.

Stapleton, Darwin H. *The Transfer of Early Industrial Technologies to America.* Philadelphia: American Philosophical Society, 1987.

Starobin, Robert S. *Industrial Slavery in the Old South.* New York: Oxford University Press, 1970.

Steinfeld, Robert J. *Coercion, Contract, and Free Labor in the Nineteenth Century.* Cambridge: Cambridge University Press, 2001.

———. *The Invention of Free Labor: The Employment Relation in English and American Law and Culture, 1350–1870.* Chapel Hill: University of North Carolina Press, 1991.

Stilgoe, John R. *Common Landscape of America, 1580–1845.* New Haven, CT: Yale University Press, 1982.

———. *Metropolitan Corridor: Railroads and the American Scene.* New Haven, CT: Yale University Press, 1985.

Stout, Wilbur. "The Charcoal Iron Industry of the Hanging Rock Iron District: Its Influence on the Early Development of the Ohio Valley." *Ohio Archaeological and Historical Quarterly* 42 (January 1933): 72–104.

Stuart, Charles B. *The Naval Drydocks of the United States.* New York: Charles B. Norton, 1852.

Sullivan, William A. *The Industrial Worker in Pennsylvania, 1800–1840.* Harrisburg: Pennsylvania Historical and Museum Commission, 1955.

Swank, James Moore. *Classified List of Rail Mills and Blast Furnaces in the United States.* Philadelphia: American Iron and Steel Association, 1873.

———. *History of the Manufacture of Iron in All Ages.* American Classics of History and Social Science 6. Philadelphia: American Iron and Steel Association, 1892. Reprint, New York: Burt Franklin, 1965.

Sylvis, James C. *The Life, Speeches, Labors and Essays of William H. Sylvis.* 1872. Reprint, New York: Augustus M. Kelley, 1968.

Tarr, Joel A. "Searching for a 'Sink' for an Industrial Waste: Iron-Making Fuels and the Environment." *Environmental History Review* 18 (Spring 1994): 9–34.

Taussig, Frederick W. *The Tariff History of the United States.* 8th ed. New York: G. P. Putnam's Sons, 1931. Reprint, New York: Augustus M. Kelley, 1967.

Temin, Peter. *Iron and Steel in Nineteenth-Century America: An Economic Inquiry.* Cambridge, MA: MIT Press, 1964.

Templeton, Peter H. "'Energy, Enterprise, and the Sinews of Success': The Monkton Iron Company, 1807–1816." Student research paper, March 10, 1980. Typescript. Bixby Memorial Free Library, Vergennes, Vermont.

Thiesen, William H. *Industrializing American Shipbuilding: The Transformation of Ship Design and Construction, 1820–1920.* Gainesville: University of Florida Press, 2006.

Thomas, Brinley. *Migration and Economic Growth: A Study of Great Britain and the Atlantic Economy.* Cambridge: Cambridge University Press, 1954.

———. "Migration of Labour into the Glamorganshire Coalfield, 1861–1911." *Economica* 30 (1930): 275–94.

———. "Wales and the Atlantic Economy." *Scottish Journal of Political Economy* 6 (1959): 169–92.

Thomas, Samuel. "Reminiscences of the Early Anthracite-Iron Industry." *Transactions of the American Institute of Mining, Metallurgical, and Petroleum Engineers* 29 (1899): 901–28.

Thompson, Michael D. "The Iron Industry in Western Maryland." Unpublished paper, National Canal Museum, Morgantown, WV, 1976.

Toomey, John J., and Edward P. B. Rankin. *History of South Boston*. Boston: authors, 1901.
Tukey, John Wilder. *Exploratory Data Analysis*. Reading, PA: Addison-Wesley, 1997.
Tweedale, Geoffrey. *Sheffield Steel and America: A Century of Commercial and Technological Interdependence, 1830–1930*. Cambridge: Cambridge University Press, 1987.
United Nations, Economic Commission for Europe. *Problems of Air and Water Pollution Arising in the Iron and Steel Industry*. New York: United Nations, 1970.
United States Geological Survey. "Cumberland, Md.–Pa.–W.Va." Map, scale 1:24,000. Washington, DC: Government Printing Office, 1993.
———. "Farrandsville." Map, scale 1:24,000. Washington, DC: Government Printing Office, 1979.
———. "Frostburg, Md.–Pa." Map, scale 1:24,000. Washington, DC: Government Printing Office, 1998.
———. "Lonaconing, Md.–W.Va." Map, scale 1:24,000. Washington, DC: Government Printing Office, 1998.
United States Geological Survey Geographic Names Information System. Accessed online at http://geonames.usgs.gov/gnispublic, November 2005–March 2006.
Vance, James E., Jr. *The North American Railroad: Its Origin, Evolution, and Geography*. Baltimore: Johns Hopkins University Press, 1995.
Vandiver, Frank E. *Ploughshares into Swords: Josiah Gorgas and Confederate Ordnance*. Austin: University of Texas Press, 1952.
———, ed., *The Civil War Diary of General Josiah Gorgas*. Tuscaloosa: University of Alabama Press, 1947.
Van Slyck, J. D. *Representatives of New England*. Vol. 2, *Manufactures*. Boston: Van Slyck, 1879.
Wagner, Dean R., ed. *Historic Lock Haven: An Architectural Survey*. Lock Haven, PA: Clinton County Historical Society, 1979.
Walker, Joseph E. *Hopewell Village: The Dynamics of a Nineteenth Century Iron-Making Community*. Philadelphia: University of Pennsylvania Press, 1966.
Walkowitz, Daniel J. *Worker City, Company Town: Iron and Cotton-Worker Protest in Troy and Cohoes, New York, 1855–84*. Urbana: University of Illinois Press, 1978.
Wallace, Anthony F. C. *Rockdale: The Growth of an American Village in the Early Industrial Revolution*. New York: W. W. Norton, 1972.
Warren, Kenneth. *The American Steel Industry, 1850–1970: A Geographical Interpretation*. Pittsburgh: University of Pittsburgh Press, 1973.
———. *Bethlehem Steel: Builder and Arsenal of America*. Pittsburgh: University of Pittsburgh Press, 2008.
———. *Wealth, Waste, and Alienation: Growth and Decline in the Connellsville Coke Industry*. Pittsburgh: University of Pittsburgh Press, 2001.
Way, Peter. *Common Labour: Workers and the Digging of North American Canals, 1780–1860*. Cambridge: Cambridge University Press, 1993.
Wayne, William J. *Native Indiana Iron Ores and 19th Century Ironworks*. Department of Natural Resources Geological Survey Bulletin 42-E. Bloomington: State of Indiana, 1970.
Weber, Thomas. *The Northern Railroads in the Civil War, 1861–1865*. New York: King's Crown Press, 1952.
Webley, Derrick Pritchard. *East to the Winds: The Life and Work of Penry Williams (1802–1885)*. Aberystywyth: National Library of Wales, [ca. 1995].
White, Josiah. *Josiah White's History, Given by Himself*. Philadelphia: Lehigh Coal and Navigation Company, ca. 1910.

White, Marjorie Longenecker. *The Birmingham District: An Industrial History and Guide.* Birmingham, AL: Birmingham Historical Society, 1981.

Whitman, Walt. *Leaves of Grass.* New York: Signet Classic, 1958.

Wiggins, Sarah Woolfolk, ed. *The Journals of Josiah Gorgas, 1857–1878.* Tuscaloosa: University of Alabama Press, 1995.

Wilber, C. D. *Coal and Iron: Two Lectures Delivered before the Board of Trade of LaFayette, Indiana.* LaFayette, IN: Rosser, Spring, 1872.

Wilks, Ivor. *South Wales and the Rising of 1839: Class Struggle as Armed Struggle.* Urbana: University of Illinois Press, 1984.

Williams, Gwyn Alf. *The Merthyr Rising.* London: Croom Helm, 1978.

Williams, Michael. *Americans and Their Forests: A Historical Geography.* Cambridge: Cambridge University Press, 1989.

Wilson, Jos. S. *Sketch of the Public Surveys in the State of Minnesota.* Map, scale 18 miles to the inch. Washington, DC: Department of the Interior, General Land Office, 1868.

Witte, Stephen S., and Marsha V. Gallagher, eds. *The North American Journals of Prince Maximilian of Wied.* Vol. 1, *May 1832–April 1833,* translated by William J. Orr, Paul Schach, and Dieter Karch. Norman: University of Oklahoma Press, 2008.

Wood, J. G. *The Principal Rivers of Wales Illustrated.* N.p., 1813.

Woodall, Guy R. "The American Review of History and Politics." *Pennsylvania Magazine of History and Biography* 93, no. 3 (1969): 392–409.

Woodworth, Steven E., and Kenneth J. Winkle. *Oxford Atlas of the Civil War.* Oxford: Oxford University Press, 2004.

[Wright, Thomas] A Journeyman Engineer. *The Great Unwashed.* 1868. Reprint, London: Frank Cass, 1970.

———. *Some Habits and Customs of the Working Classes.* London: Tinsley Brothers, 1867.

Yates, W. Ross. "Discovery of the Process for Making Anthracite Iron." *Pennsylvania Magazine of History and Biography* 98 (1974): 206–23.

MAP SOURCES

The thematic maps in this book were constructed in two basic ways. Those that display data from the Lesley historical GIS were initially generated from queries of the database that were then compiled and designed in collaboration with the book's cartographer, Chester Harvey, who rendered all the final maps. The other maps were compiled from various cartographic and textual sources before going through the same design and rendering process. For a bibliography of all the sources used to construct the Lesley HGIS and to map the results of GIS analysis, see the appendix. The list below notes which maps include data from the Lesley HGIS in addition to providing all other sources used in each map.

Figures 2, 3, Lesley HGIS.

Figures 6–11, Lesley HGIS.

Figures 12–15, Lesley HGIS; E. C. Harder, "Distribution of Iron Ore in the United States," plate 1 in *Mineral Resources of the United States*, part 1, *Metallic Products, Calendar Year 1908*, Department of the Interior, US Geological Survey (Washington, DC: Government Printing Office, 1909), and Harder, "Iron Ores, Pig Iron, and Steel," in *Mineral Resources*, part 1, 61–126; Charles O. Paullin, *Atlas of the Historical Geography of the United States*, ed. John K. Wright, publication 401 (Washington, DC: Carnegie Institution of Washington and the American Geographical Society of New York, 1932), plate 6A.

Figures 16–19, Lesley HGIS.

Figure 27, Lesley HGIS; R. E. McGowins, "Pittsburgh 1858" (map), no scale, in Geo[rge] H. Thurston, *Allegheny County's Hundred Years* (Pittsburgh: A. A. Anderson, 1888). Research assistance by Richard Price, using Allegheny County Commissioners' Survey Maps, surveyed and computed under the direction of R. E. McGowin by A. Heddaeus and Charles Gramer, 1861, Heinz History Center, Pittsburgh.

Figure 31, Lesley HGIS; US Geological Survey topographic quadrangles, 7.5 minute series, scale 1:24,000, various dates, accessed online via the Geographic Names Information System (GNIS) at http://geonames.usgs.gov.

Figure 33, (*a*) Charles B. Dew, *Bond of Iron: Master and Slave at Buffalo Forge* (New York: Norton, 1994), 245; (*b*) T. J. Kennedy, "Map of Chester Co., Pennsylvania," no scale (n.p., 1860), downloaded from Library of Congress, American Memory, http://memory.loc.gov/ammem, on July 1, 2007; (*c*) G. M. Hopkins, *Clark's Map of Litchfield County, Connecticut* (map), scale 1:50,688 (n.p., 1859), downloaded from Library of Congress, American Memory at http://memory.loc.gov/ammem, on July 9, 2007.

Figure 35, (*a*) G. M. Hopkins, "Ashland, Md." (map), no scale, in *Atlas of Baltimore County, Maryland* (Philadelphia, 1877), 50; (*b*) A. Pallas, *Map of the Lands of the Farrandsville Co., Situated in Clinton County* (map), scale approximately 200 perches (3,300 feet) to the inch, in "Report to the Stockholders of the Farrandsville Company," January 1, 1856 (Philadelphia: Merrihew and Thompson's Steam-Power Book and Job Printing Office, 1856), 21 pp., Historic Corporate Reports Collection, Baker Library Historical Collections, Harvard Business School.

Figure 40, Robert Hunt, "Map of Great Britain and Ireland, Shewing the Districts in Which Coal and Iron, Are Respectively Found" (map), scale approx. 27 miles to the inch, in *Plans Referred to in the Report of the Commissioners Appointed to Inquire into the Application of Iron to Railway Structures* (London: William Clowes for Her Majesty's Stationery Office, 1849), unnumbered plate after p. iv; Alan Birch, *The Economic History of the British Iron and Steel Industry, 1784–1879* (London: Frank Cass, 1967; reprint, New York: Augustus M. Kelley, 1968); shaded relief map derived from NASA SRTM elevation data, courtesy of C. Scott Walker, Harvard Map Collection, Harvard University.

Figure 41, Ordnance Survey of Great Britain, *Merthyr Tydfil* (manuscript map), 42 sheets, scale 1:528 (ca. 1852), Glamorgan Record Office, Cardiff, Wales.

Figure 43, *Map of the Several Canals and Rail Roads by Which the Lycoming Coal Can Be Sent to Market* (map), no scale, in Lycoming Coal Company, *A Brief Description of the Property Belonging to the Lycoming Coal Company . . .* (Poughkeepsie, NY: P. Potter, 1828), 32 pp., digital scan courtesy of Rutgers University Special Collections; Bureau of Topographic and Geologic Survey, *Distribution of Pennsylvania Coals*, 3rd ed. (map), original scale unknown (Pennsylvania Department of Conservation and Natural Resources, 2000), accessed online at www.dcnr.state.pa.us/topogeo on May 6, 2007.

Figure 45, *Map of the Town of Morrisenae . . .* (map), various scales (Philadelphia: P. S. Duval, Lithographer, ca. 1839), item no. 2004.10.10, Clinton County Historical Society; Pallas, *Map of the Lands of the Farrandsville Co.* (see notes for fig. 35); United States Department of the Interior, Geological Survey, "Howard, Pennsylvania," and "Lock Haven, Pennsylvania," fifteen-minute series (1:62,500), US Geological Survey, 1923; Christopher T. Baer, "PRR [Pennsylvania Railroad] Chronology: A General Chronology of the Pennsylvania Railroad Company, Predecessors and Successors and Its Historical Context" (n.p., 1856; March 2005 edition); Brown and Runk, *History of Lycoming County, Pennsylvania . . .*, ed. John F. Meginness (1892; reissued Evansville,

IL: Unigraphic, 1979); Henry Schenck Tanner, *A Description of the Canals and Rail Roads of the United States . . .* (New York, NY: Tanner and J. Disturnell, 1840).

Figure 47, Lesley HGIS; *Distribution of Pennsylvania Coals* (see fig. 43, above); Christopher T. Baer, *Canals and Railroads of the Mid-Atlantic States, 1800–1860*, ed. Glenn Porter and William H. Mulligan Jr., cartographers Marley E. Amstutz and Anne E. Webster (Wilmington, DE: Regional Economic History Research Center, Eleutherian Mills–Hagley Foundation, 1981); Erwin Raisz, *Landforms of the United States*, 6th rev. ed. (map), scale approx. 75 miles to the inch (privately published, 1957); Tom Patterson, *Natural Earth I*, 500-meter world land cover map data for the coterminous United States, shaded relief, accessed online at http://www.shadedrelief.com/natural/pages/download.html.

Figure 50, Lesley HGIS; P. Desobry, *Line of the Morris Canal, New Jersey 1827* (map), scale approx. 3 inches to the mile (New York, NY: Imbert's Lithography, ca. 1827); Erwin Raisz, *Landforms of the United States* (see fig. 47, above); Tom Patterson, *Natural Earth I* (see fig. 47, above). Distances from Trenton Iron Works calculated in ArcGIS.

Figure 53, G. A. Aschbach, *Map of Lehigh County, Pennsylvania* (map), scale 6 miles to one inch (Allentown, PA: M. H. Traubel, Lithographer, 1862).

Figure 56, Lesley HGIS; various correspondence in Cooper, Hewitt and Company Papers, Manuscript Division, Library of Congress; Geological Survey of New Jersey, *Northern New Jersey, Showing the Iron-Ore and Limestone Districts* (map), scale 2 miles to the inch (New York: J. Bien Lithography, 1874), accessed online at http://mapmaker.rutgers.edu/HISTORICALMAPS on August 24, 2006; Tom Patterson, *Natural Earth I* (see fig. 47, above).

Figure 58, Warren Ripley, *Artillery and Ammunition of the Civil War* (New York: Promontory Press, 1970), 358–61; Ian V. Hogg, *Weapons of the Civil War* (New York: Military Press; Crown Publishers, 1987), 11–27, 34–46; Carl L. Davis, *Arming the Union: Small Arms in the Civil War* (Port Washington, NY: Kennikat Press, 1973), plates 15 and 16; Michael S. Raber et al., *Forge of Innovation: An Industrial History of the Springfield Armory, 1794–1968*, ed. Richard Colton (1989; reprint, Fort Washington, PA: Eastern National [Division, National Park Service], 2008), 80–81, 149; Josiah Gorgas, "Notes on the Ordnance Department of the Confederate Government," *Southern Historical Society Papers* 12 (1884): 70; J. W. Mallet, "Work of the Ordnance Bureau of the War Department of the Confederate States, 1861–1865," *Alumni Bulletin of the University of Virginia* 3 (1910): 162–70; telephone interview with US Park Service ranger Creighton Waters, Harpers Ferry National Historical Park, December 18, 2008; Raimondo Luraghi, *A History of the Confederate Navy*, trans. Paolo E. Coletta (Annapolis, MD: Naval Institute Press, 1996), 34–50; Tom Patterson, *Natural Earth I* (see fig. 47, above).

Figure 61, Manuscript Census, Merthyr Tydfil, Glamorganshire, 1841, microfilm ref. no. HO 107/1415, enumeration district 23, Public Record Office, Kew Gardens; Ethel Armes, *The Story of Iron and Coal in Alabama* (Birmingham, AL: Chamber of Commerce, 1910; reprint, New York: Arno Press, 1972), 172–75; Manuscript Population Census, Hanover Township, Lehigh County, Pennsylvania, 1850; Giles Edwards to John Fritz, April 20, 21, 25, and 28, 1855, John Fritz Papers, National Canal Museum Archives; R. Bruce Council, Nicholas Honerkamp, and M. Elizabeth Will, *Industry and Technology in Antebellum Tennessee: The Archaeology of Bluff Furnace* (Knoxville: University of Tennessee Press, 1992), 62–65, 67–74; Erwin Raisz, *Landforms of the United States* (see fig. 47, above); Tom Patterson, *Natural Earth I* (see fig. 47, above).

Figure 62, Letters of J[ohn] M. Tillman, 1862–65, Shelby Iron Company records, W. S. Hoole Special Collections Library, University of Alabama, box 885, folders 108, 114; box 969, unnumbered folder "Statements 1862–64"; William K. Hubbell, "The Railroads of the Confederate States as of June 1, 1861" (map), no scale, in Robert C. Black III, *The Railroads of the Confederacy* (Chapel Hill: University of North Carolina Press, 1952), sheet map in back pocket; Erwin Raisz, *Landforms of the United States* (see fig. 47, above); Tom Patterson, *Natural Earth I* (see fig. 47, above).

Figure 63, Lesley HGIS; Josiah Gorgas, "Notes on the Ordnance Department of the Confederate Government," *Southern Historical Society Papers* 12 (1884): 67–94; Raimondo Luraghi, *A History of the Confederate Navy* (see fig. 58, above), 34–50; James A. Henretta et al., *America's History* (Chicago: Dorsey Press, 1987), 477; "Conquest of the South, 1861–1865," in Charles O. Paullin, *Atlas of the Historical Geography of the United States* (see figs. 12–15, above), plate 163; Tom Patterson, *Natural Earth I* (see fig. 47, above).

Figure 65, Samuel Harries Daddow and Benjamin Bannan, *Coal, Iron, and Oil, or The Practical American Miner* (Pottsville, PA: Benjamin Bannan, 1866), 684–95; Tom Patterson, *Natural Earth I* (see fig. 47, above).

Figure 69, Jos. S. Wilson, *Sketch of the Public Surveys in the State of Minnesota* (map), scale 18 miles to the inch (Washington, DC: Department of the Interior, General Land Office, October 2, 1866), Minnesota Historical Society, St. Paul.

INDEX

Page numbers followed by an *f* or a *t* indicate a figure or a table, respectively.